An Introduction to Relativistic Gravitation

This is an introductory textbook on applications of general
relativity to astrophysics and cosmology. The aim is to provide
advanced students with a tool-kit for understanding
astronomical phenomena that involve velocities close to that of
light or intense gravitational fields. The approach taken is first
to give the reader a thorough grounding in special relativity,
with space–time the central concept, following which general
relativity presents few conceptual difficulties. Examples of
relativistic gravitation in action are drawn from the
astrophysical domain. The book can be read on two levels: first
as an introductory fast-track course, and then as a detailed
course reinforced by problems which illuminate technical
examples. The book has extensive links to the literature of
relativistic astrophysics and cosmology.

An Introduction to Relativistic Gravitation

Rémi Hakim
Département d'Astrophysique Relativiste et de Cosmologie
Observatoire de Meudon (France)

CAMBRIDGE
UNIVERSITY PRESS

PUBLISHED BY THE PRESS SYNDICATE OF THE UNIVERSITY OF CAMBRIDGE
The Pitt Building, Trumpington Street, Cambridge, United Kingdom

CAMBRIDGE UNIVERSITY PRESS
The Edinburgh Building, Cambridge CB2 2RU, UK   www.cup.cam.ac.uk
40 West 20th Street, New York, NY 10011-4211, USA   www.cup.org
10 Stamford Road, Oakleigh, Melbourne 3166, Australia
Ruiz de Alarcón 13, 28014 Madrid, Spain

First published 1999

Printed in the United Kingdom at the University Press, Cambridge

Typeface Times 10/13pt   *System* LaTeX   [UPH]

A catalogue record for this book is available from the British Library

Library of Congress Cataloguing in Publication data
Hakim, Rémi, 1936–
    An introduction to relativistic gravitation / Rémi Hakim
        p.   cm.
    Includes bibliographical references and index.
    ISBN 0 521 45312 7 – ISBN 0 521 45930 3 (pbl.)
    1. General relativity (Physics) 2. Astrophysics. 3. Cosmology.
    I. Title.
QC173.6.H3 1998
530.11-dc21   97-17392 CIP

ISBN 0 521 45312 7 hardback
ISBN 0 521 45930 3 paperback

In memory of Claude René Woog, assassinated in July 1944 in Auschwitz

Contents

Preface   *page* xi

1 NEWTONIAN GRAVITATION   **1**

1.1 Newtonian space–time   1
1.2 Simultaneity and distance measures   4
1.3 Newton's absolutes and the notion of the ether   7
1.4 The principle of inertia   9
1.5 The laws of dynamics and Galilean relativity   12
1.6 Inertia and relativity principles as seen by Galileo   16
1.7 Newtonian gravitation   17
1.8 Measuring the gravitational constant   22
1.9 Limits of the Newtonian theory of gravity   28
1.10 The finiteness of the velocity of light   30
1.11 Michelson's experiment   34

2 MINKOWSKI SPACE–TIME   **41**

2.1 The space–time of special relativity   43
2.2 The Lorentz transformation   46
2.3 Remarks   48
2.4 Causality and simultaneity   52
2.5 Times and distances measured by inertial observers   58
2.6 Global properties of space–time   60
2.7 Experimental verification of Special Relativity   61

3 THE RELATIVISTIC FORM OF PHYSICAL LAWS   **68**

3.1 Tensor formalism   68
3.2 The Doppler effect and aberration   75
3.3 The kinematic description of particle motion   77
3.4 Relativistic dynamics: $E=mc^2$ 82
3.5 Minkowski space in curvilinear coordinates   85

4 GRAVITATION AND SPECIAL RELATIVITY   **95**

4.1 The gravitational redshift   97
4.2 Light bending   98

4.3 The advance of the perihelion of Mercury 99
4.4 The need for nonlinear equations for gravitation 101

5 ELECTROMAGNETISM AND RELATIVISTIC HYDRODYNAMICS 103

5.1 Densities and currents 103
5.2 The equations of electromagnetism 106
5.3 The energy–momentum tensor 109
5.4 Relativistic hydrodynamics 110

6 WHAT IS CURVED SPACE? 116

6.1 Some manifestations of curvature 116
6.2 Curvature of two-dimensional surfaces 119
6.3 The meaning of intrinsic curvature 125
6.4 Surfaces in $\mathbf{R}^n$ – Riemann spaces 126
6.5 Intrinsic curvature of a manifold 129
6.6 Properties of the curvature tensor 130
6.7 Space–time as a Riemannian manifold 132
6.8 Some properties of tensors in curved space 133
6.9 Three arguments for curved space–time 134

7 THE PRINCIPLE OF EQUIVALENCE 139

7.1 The weak equivalence principle and the Eötvös–Dicke experiments 140
7.2 The equivalence principle and minimal coupling 149
7.3 The gravitational redshift 157
7.4 Geodesic motion 159
7.5 Geodesic deviation 162
7.6 The metric tensor in spherical symmetry 163
7.7 Overview of the PPN formalism 165
7.8 The classical tests 167
7.9 Gravitational lenses 174

8 EINSTEIN'S RELATIVISTIC GRAVITATION (GENERAL RELATIVITY) 187

8.1 Einstein's equations 188
8.2 Other derivations of the Einstein equations 192
8.3 The Schwarzschild solution 195
8.4 The local geometry of Friedman spaces 200
8.5 Other metrics of astrophysical interest 206
8.6 The linearised Einstein equations 207
8.7 Gravitational radiation 210

Appendix A   Tensors   227
    1 Dual of a vector space   227
    2 Tensor products of vector spaces   228
    3 Criteria for being a tensor   229
Appendix B   Exterior Differential Forms   232
    1 Exterior calculus   234
    2 Differential forms   236
    3 Volume element: dual forms   238
    4 Differentiation of forms   241
    5 Maxwell's equations in differential forms   242
    6 Integration of differential forms – Stokes' theorem   243
Appendix C   Variational Form of the Field Equations   245
Appendix D   The Concept of a Manifold   249
    1 Differentiable manifolds   252

References   255
Physical Constants   270

Preface

This book is devoted to general relativity, i.e. to the synthesis of special relativity and gravitation. This Relativistic Gravitation, as it is sometimes called, appears to be of uppermost importance in all those astronomical phenomena that involve velocities close to that of light or intense gravitational fields. The study of the latter constitutes a new subject, Relativistic Astrophysics, an expression due to Alfred Schild (1967).

The content of this book is the *minimum minimorum* needed to approach this relatively recent domain.

It may be interesting at this point to recall how and why this new subject started. Once the classical tests of general relativity were performed (bending of light rays by the Sun (1919), gravitational redshift (in white dwarfs) [W.S. Adams (1925)]; perihelion advance of Mercury), the subject became very formal, as current technology did not provide contact with experiment or astronomical observation. Although much research had great conceptual interest (unified theories of gravity and electromagnetism, for example), general relativity became rather arid [see J. Eisenstaedt (1986)] because of the lack of laboratory experiments or observations of relativistic objects, which were in any case unknown to theory before the 1930s, and even then ignored in the 1940s and 1950s. Thus, cosmology was regarded more as a "free area for thinking about relativity" [J. Eisenstaedt (1989)] than a field for astronomical verifications of general relativity, or even the "Science of the Universe" [E.R. Harrison (1981)].

The discovery of the Mössbauer effect (1958) allowed laboratory measurement of the gravitational redshift, while experimental progress led to a repetition of the Eötvös experiment with unrivalled precision. At the same time, discussions of nucleosynthesis of elements after the second world war led G. Gamow and his collaborators to the idea that the Universe itself was hot in the distant past, giving rise to blackbody background radiation. This radiation was discovered[1] in 1965. This aroused great interest in cosmology, which thereafter began to develop markedly. A few years earlier (1963) quasars – the most distant objects in the Universe – had been discovered. A little later (1967) the first pulsars were found, and very soon identified by T. Gold, (1969) with the neutron stars envisaged in 1932 by Landau. Thus the 1960s revived the experimental and observational basis of relativity[2] by providing the astrophysical

---

[1] The history of this discovery is related by S. Weinberg (1978); see also E.R. Harrison (1981).
[2] We should add that simultaneous progress in space technology allowed experiments impossible on Earth.

objects of study that had hitherto been lacking. At the same time the notions of black hole and gravitational radiation were being clarified. These were specifically relativistic concepts, which might produce all kinds of astrophysical effects which were potentially observable. Clearly, these observational discoveries encouraged theoretical research and new observations (for example, deep surveys of the Universe to look for the large-scale structure envisaged by theory).

Among the various relativistic theories of gravitation, Einstein's general relativity is without doubt the first to give correct theoretical results, and the best verified experimentally. Accordingly this book uses this theory without studying dense stars, gravitational collapse or black holes: the interested reader is invited to look at the book by S.L. Shapiro and S.A. Teukolsky (1983) on these subjects. Cosmology is not considered either: there exist a large number of books like those of P.J.E. Peebles (1993), N. Straumann (1984), Ya.B. Zeldovich and I.D. Novikov (1975) or S. Weinberg (1972), etc. It is not a monograph on general relativity – there are many excellent books at various levels, but rather an account of the concepts and basic techniques, allowing easy access to Relativistic Astrophysics. It should also serve as an introduction to the current literature. Apart from the elements of general relativity, two astrophysical applications are briefly studied: *gravitational lensing*, as observed in recent years, and *gravitational waves*, which have not yet been directly detected[3], but whose existence is demonstrated by observations of the binary pulsar 1913+16, and which in the future may give rise to a new branch of astronomy.

For the beginner, relativity, whether the special or general theory, presents (like other physical theories) some conceptual difficulties, and some purely technical ones.

On the conceptual level the main problem is "thinking relativistically"[4], i.e. in terms of *space–time*, which is a central idea. One might hope that this view of space–time, in which its structures are determined by the light cone, would be acquired from the beginning of university physics courses. This is however not the case, as for many students (and lecturers) special relativity is confined to the Lorentz transformation, length contraction and time dilatation. However, general relativity presents few conceptual difficulties initially for someone who has mastered the special theory. For this reason this book gives a treatment of this theory, based only on the notion of space–time, itself introduced at the classical level. It is then easy to show qualitatively that by bending the light cone of special relativity, gravity requires the introduction of curved space–time. The other central concept of general relativity, the Principle of Equivalence, causes less difficulty, at least in the simpler versions which are adequate for a first study. However, this principle also involves a number of subtle points, of which we mention a few in order to illustrate their connection with geophysical, astrophysical and other phenomena. It is important also in showing that the theory is not closed, but other possibilities remain.

On the technical level, tensor calculus in Minkowski space is fairly easy to master,

---

[3] A French–Italian cooperation the VIRGO project may perhaps show their existence by the year 2000.
[4] E.F. Taylor, J.A. Wheeler (1966).

and we have provided exercises for this purpose. However this is more difficult in curved spaces. The combination of these technical difficulties with conceptual ones constitutes the main problem in learning relativistic gravitation. For this reason we have introduced curvilinear coordinates in special relativity[5]. Extending this to curved space–time then becomes very simple: apart from a few details the formulae are essentially the same.

We should finally add that the contents of this book are based on courses taught for several years at very different levels, ranging from the second year of a French university course to the fourth or fifth, naturally somewhat modified according to the audience. This diversity has led to the possiblility of using the book on two different levels.

The first level, which is elementary, aims at a general view of the subject and is based on intuitive reasoning and a heuristic rather than a technical approach.

The second level, although far below what is needed for advanced work, requires more effort. Concepts are outlined along with their experimental bases and logical structures, but require new techniques which are nonetheless at a realistic level of mathematical difficulty. Really reaching this second level demands use of the exercises at the ends of the chapters. The aim here is a minimum level of understanding of general relativity suitable for readers interested primarily in astrophysics, which will make the current literature accessible. Obviously, only more advanced texts (monographs or papers) can provide a fuller grasp of the subject.

This second level is open in the sense that the subject is still developing. Thus, the reader will find references both to the Founding Fathers and to treatments of various points such as proofs not given in the book, experimental or observational details, "iconoclastic" views, or simply to give credit where it is due. Thus, the references are not exhaustive, nor are they intended to be.

At *level 1*, the reader may limit himself to the whole of Chapters 1, 2, and 4; in Chapter 6 the sections on manifestations of curvature, curvature of two-dimensional surfaces, the significance of intrinsic curvature, three arguments for curved space–time; in Chapter 7 the weak equivalence principle and the Eötvös experiment, the equivalence principle and minimal coupling; and in Chapter 8, the Schwarzschild metric (first part).

At *level 2*, which naturally includes level 1 and represents about 75 teaching hours including the exercises, the reader should study the various technical aspects and their applications, and attempt the relevant problems. The exercises given at the ends of the chapters include simple proofs not given in the text, as well as purely technical points or illustrations of various points. Whatever the reason for their presence, they are all very useful for learning the material presented, and indispensable for the reader wishing to go further in relativistic astrophysics. It may be useful to consult the relativity texts cited in the bibliography.

As an example, confining attention to books published in the last twenty years, we mention the following ones (in approximate ascending order of difficulty and with no attempt at completeness) J. Schwartz, M. McGuinness (1979); J. Foster,

---

[5] Following V.I. Fock (1966).

J.D. Nightingale (1979); M.G. Bowler (1976); B. Schutz (1988); N. Straumann (1984); S. Weinberg (1972); R.M. Wald (1984).

In addition to these books the reader may find the problem book by Lightman *et al.* (1975) useful. The encyclopaedic book by C. Misner, K. Thorne and J.A. Wheeler (1973) will be used as a kind of dictionary for the limited purposes of the present book, and will be referred to very often.

Acknowledgments
It is a pleasure to thank Mesdames Sabine Collé and Dominique Lopes for the care they have given to the typing of the manuscript.

I hope that my colleagues M.M. Alvarez, L. Bel, L. Blanchet, B. Carter, T. Damour, N. Deruelle, E. Gourgoulhon, D. Gerbal, F. Hammer, J.P. Lasota, J.P. Luminet, J.A. Marck, P. Peter, D. Priou, D. Polarski, J. Schneider, C. Vanderriest and M. Zonabend will find in this book my thanks for their many remarks and suggestions, which greatly improved this introduction as well as removing some errors. I have also derived great benefit from the remarks and criticisms of my students in recent years.

Finally I would like to thank Dr. Andrew King for the difficult task of translating the French version of this book into English.

1 Newtonian Gravitation

For classical physics, space and time provide the arena in which the phenomena of nature unfold. These phenomena do not change the space–time frame, which is *inert* and absolutely fixed for all time. Moreover, space and time are regarded as completely distinct and having *no connection* with each other. Relativity theory links space and time, and reaches its culmination in General Relativity, which connects the space–time properties with the dynamical processes occurring there.

1.1 Newtonian space–time

Physical space possesses the usual properties of continuity, homogeneity and isotropy which we attribute to the space $\mathbf{R}^3$ when equipped with its affine structure (parallelism, existence of straight lines) and its usual metric structure (Pythagoras' "theorem"). However, we must understand the physical significance of the mathematical concepts connected with $\mathbf{R}^3$. Thus, the existence of physical phenomena which can be represented by straight lines (mathematics) leads to the (experimental) notion of alignment: three points are (physically) aligned if we can find a viewing point from which they appear to coincide. From this it follows that light constitutes our standard of straightness; it is only by a further step (which may prove to be incorrect) that we can identify the trajectory of a light ray with a straight line in $\mathbf{R}^3$. Similarly, the mathematical concept of parallelism in $\mathbf{R}^3$ is directly related to the (physical) notion of rigid transport and of distance. Finally, we must recognise that the (mathematical) properties of homogeneity and isotropy of physical space only express our experience of mechanical systems: that these remain unaltered when placed in any position or place. In every case our theoretical ideas arise from properties which are experienced on a *local* scale.

The usual idea of time is quantified by a simple real parameter, and so $\mathbf{R}$ represents time. $\mathbf{R}$ is the quantitative expression of the qualitative notion of *change*. Physically, this requires knowledge of a *standard* of change, whose "successive" states (in some intuitive sense) can be quantified so as to define a unit of time. The principal physical characteristic of time is that it flows *uniformly*. This assumes the existence of standards which remain recognisably similar to themselves during their successive changes of state. This hypothesis is based on the empirical observation that different clocks give compatible measurements of the flow of time. Finally, the fact that $\mathbf{R}$ is well-ordered is a mathematical expression of the causal nature of time.

The physical properties of space (isotropy, homogeneity) and time (uniformity) are expressed mathematically by invariance under various transformation groups (rotations, translations in space and time) which imply the existence of first integrals or constants of the motion (angular momentum, momentum and energy).

We have seen that in the framework of classical physics space and time can be represented mathematically by $\mathbf{R}^3$ and $\mathbf{R}$ respectivly. It is then possible to construct a four-dimensional classical (i.e. prerelativistic) space–time.

$$\mathscr{M} \equiv \mathbf{R}^{3+1} = \mathbf{R}^3 \times \mathbf{R}, \tag{1.1}$$

which is a purely mathematical construct of little use, there being no physical necessity for it.

In fact, even the prerelativistic notions of space and time are intimately connected. Minkowski already noted this in his famous article of 1908: *"The objects we perceive invariably imply space and time simultaneously. No-one has ever observed a position except at a given time, or measured a time other than at a given place."* Further, we can add that the presence of matter is indispensable if we want to observe anything at all. Although it is in relativity that the connection between space and time is made manifest, it also exists in the Newtonian context, as we shall briefly discuss.

A rather deeper discussion [see H. Reichenbach (1958); M. Bunge (1966)] of Newtonian time clearly shows that we cannot conceive of time without recourse to spatial concepts. Further, if we were to analyse the various methods of measuring time, the connection between space and time would also appear clearly.

The genesis of the main properties of physical space also implicitly involves temporal phenomena. Consider for example the *Euclidean* character of physical space. This property arises from the possibility of *rigid* motions (if necessary uniformly in a straight line) and also the process of *distance measurement*. The latter implies either the transport of rigid measuring rods, or the use of *light signals*. The Euclidean character of physical space is thus implicitly based on observations using the notion of time, either directly (motions of a rod) or indirectly (propagation of light signals). Other properties of physical space lead to similar conclusions.

Thus classical space–time $\mathbf{R}^{3+1}$ appears physically much more fundamental than at first sight: we can therefore ask if its mathematical structures correspond to particular aspects of reality.

This structure is shown in **Fig. 1.1**. Time, universal and absolute, is represented by directed straight lines (the direction is purely conventional and only has to obey a minimal principle of coherence we shall make precise later) which cut each of the Euclidean 3-planes representing space. *A priori*, the angle of the straight lines to the spatial 3-planes shown on the figure has *no* significance. In fact $\mathbf{R}^{3+1}$ has no Euclidean metric structure, unlike Special Relativity[1] – which could give a meaning to the notion of an angle in this space[2]. For this reason *all* timelike straight lines, i.e. those not

---

[1] Pseudo-Euclidean, see *Chapter 2*.
[2] *via* a scalar product of the type $X.Y = |X||Y|\cos\theta$.

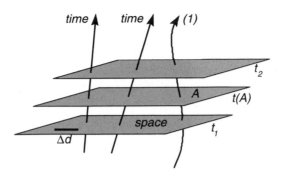

Fig. 1.1. Classical space–time. This has a structure of spacelike Euclidean strata (Δd is the distance between two points of the same stratum, of the same physical space). Time is represented by straight lines cutting the various strata. (1) shows the trajectory of a particle with an arbitrary motion in the Newtonian space–time.

contained in a spatial 3-plane, constitute possible time axes which are all completely equivalent. Similarly, the apparent angle between two of these time axes on the figure has no absolute meaning: it is a result of the inadequacy of our representation of a non-Euclidean space, classical space–time, by means of a Euclidean space, i.e. that of a sheet of paper representing $\mathbf{R}^2$. The structure of *strata of simultaneous events* in $\mathbf{R}^{3+1}$ can be interpreted as its *causal structure*. Thus the spatial hyperplane $t(\mathbf{A})$ passing through event $\mathbf{A}$ (see the figure) separates Newtonian spacetime into two regions: the future $t > t(\mathbf{A})$, which *can* be influenced by phenomena which occur at $\mathbf{A}$, and the past, $t < t(\mathbf{A})$ which *can* have influenced A; the frontier between the past and the future is the present, i.e. the spatial hyperplane $t(\mathbf{A})$, which can itself influence the event $\mathbf{A}$ because of the possibility in Newtonian physics of *instantaneous action at a distance*.

We now turn to the metric structures of $\mathbf{R}^{3+1}$: each spacelike stratum has a Euclidean structure, namely the usual distance measure. Similarly, each timelike fibre has a Euclidean metric, i.e. duration. On the other hand, $\mathbf{R}^{3+1}$ has *no* Euclidean structure, and in particular certainly not that of $\mathbf{R}^4$.

For a Euclidean distance in $\mathbf{R}^4$ of the type

$$\Delta \ell^2 = \Delta \mathbf{x}^2 + \Delta t^2 v_0^2$$

to make sense, we need an absolute universal parameter v_0, with the dimensions of a velocity, and obviously possessing some physical significance.

We still have to ask if the timelike lines which we used as time axes have physical meaning. The *principle of inertia* which we study below allows us to give them a physical significance which until now has been lacking.

In summary, Newtonian space–time has four dimensions, a causal structure determined by *geometric objects* belonging to it (spacelike 3-planes and timelike lines, all similarly oriented) and some quite complex metric and affine structure. A class of geometric objects , timelike straight lines (representing the motions of free particles) reinforces the connection between space and time. We shall see in *Chapter 2* which structures remain in Special Relativity, and which have to be modified.

1.2   **Simultaneity and distance measures**

In the foregoing analysis of classical space–time we treated simultaneity as something obvious. Does this correspond in reality to something *actually measurable*, at least in principle? Is there actually some implicit hypothesis going further than what one can draw directly from the analysis of real experiments, even idealised ones? We shall study this question more closely[3].

We begin by noting that the idea of simultaneity at a point of space has almost no concrete sense[4]. It can only be realised even approximately for collisions of elementary particles and is at best a logical identity, i.e. a definition of a space–time event. The really important case concerns the simultaneity of spatially distinct events.

Clearly the temporal comparison of distant events is possible only when a *signal* is sent from one place to another which allows the synchronisation of clocks situated at the two places. To make this comparison, we must know both the spatial distance of the two objects and the signal velocity.

How do we measure this velocity? If P_1 and P_2 denote the two spatial points, and t_1 and t_2 are the departure and arrival times of the signal, we have

$$v = \frac{P_1 P_2}{t_2 - t_1}.$$

However, for the latter expression to make sense the clocks at P_1 and P_2 must have been synchronised already. Evidently Fizeau's experiment to measure the velocity of light (see **Fig. 1.2**) only needs one clock: at time t_1 a light signal is sent to a mirror at distance ℓ, where it is reflected at times t_2 back to its point of departure, where it arrives at time t_3; we thus have

$$c = \frac{2\ell}{t_3 - t_1},$$

and the time t_2 never appears ...

In fact Fizeau's experiment only allows one to measure the harmonic mean of the velocities c_1 and c_2 of the light moving towards and away from the mirror:

$$\frac{2}{c} = \frac{1}{c_1} + \frac{1}{c_2}.$$

It follows that the use of a *single clock* is justified only by the supplementary *hypothesis* $c_1 = c_2$. To verify this experimentally we would have to have two clocks at two different points which had been synchronised beforehand **(Fig. 1.3)**.

We have thus arrived at the following vicious circle: simultaneity requires the knowledge of a velocity which in turn requires us to have established the simultaneity of two events!

---

[3] Moreover, the critique of the notion of classical simultaneity was an important part of Einstein's discovery of Special Relativity.

[4] We follow the treatment by H. Reichenbach (1927).

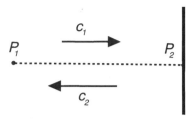

Fig. 1.2. Measuring the velocity of light (Fizeau). A light signal emitted at P_1 at time t_1 is reflected at P_2 and returns to P_1 at time t_3; in doing this we measure only the mean of the velocities in the directions $P_1 \, P_2$ and $P_2 \, P_1$. To measure t_1 and the arrival time t_2 at P_2 requires previous synchronisation of the two clocks at P_1 and P_2, and thus the sending of a light signal between the two points... whose propagation velocity must be known in advance! In fact Newtonian physics allows the synchronisation of clocks by slow transport, for example using signals with infinite propagation velocity.

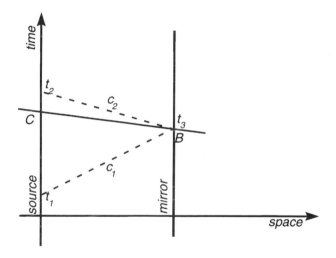

Fig. 1.3. Measuring the velocity of light (Fizeau) in space–time. C and B are simultaneous but C is arbitrary (with $t_1 < t_3 < t_2$), as t_3 depends on the way that simultaneity is defined.

This is not a contradiction: *simultaneity is not primarily an experimental question; but also one of definition*, which, like all definitions, contains an arbitrary element.

Thus, in Fizeau's experiment, the moment of arrival of the light ray at the mirror will *by definition* be

$$t_2 = t_1 + \frac{1}{2}(t_3 - t_1).$$

Of course we could also have adopted the definition

$$t_2 = t_1 + \varepsilon(t_3 - t_1),$$

with $0 < \varepsilon < 1$; this would not lead to any contradiction [H. Reichenbach (1927)].

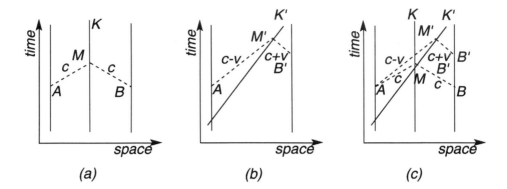

Fig. 1.4. The relativity of simultaneity (considered in Newtonian space–time). Figure (a) shows the simultaneity of distant events A and B: an observer K, at rest with respect to the measuring rod, only regards as simultaneous points from which light rays (space–time trajectories represented by vertical lines) from the events arrive at the same instant. Figure (b) shows what the simultaneous events are for observer K' (oblique space–time trajectories on the figure) who is moving in uniform straight-line motion with respect to K, using the definition shown in (a). Figure (c) shows the superposition of cases (a) and (b) and clearly demonstrates that events which are simultaneous for M (such as A and B) are not simultaneous for M' (e.g. A and B').

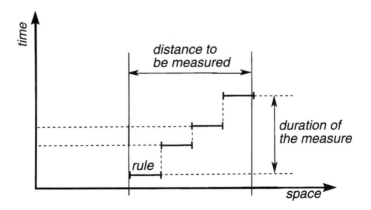

Fig. 1.5. Measurement of distance using a rigid rod. Each displacement of the rod along itself requires a certain time. However, in prerelativistic physics, the duration of each measurement must at least in principle be arbitrarily reducible.

In practice, light signals, (as the most rapid known) are used to *define* simultaneity. Thus, in Einstein's definition of *physical simultaneity*, two events at distant points A and B are simultaneous if light signals from A and B arrive at the same instant at the middle M of the segment AB.

Things get more complicated when we wish to establish the simultaneity of two events in a frame moving with respect to another frame. A pair of events which are simultaneous in one reference system are no longer so in the second one: see **Fig. 1.4.**

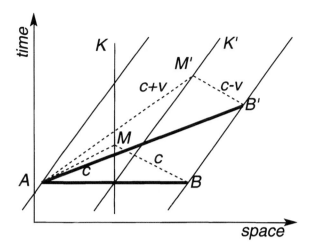

Fig. 1.6. Relativity of measures of distance. In his own system, the observer K' measures the distance AB'; on the other hand, observer K measures $AB \neq AB'$.

Figure 1.4 (a) shows that events A and B are simultaneous for M. **Figure 1.4 (b)** shows the trajectories of the points A and B respectively; an observer moving with respect to K follows the trajectory denoted by K'. At M' he sees the light signals from the two points A and B' which are obviously not simultaneous in K'. **Figure 1.4 (c)** shows the two situations together: two events which are simultaneous for M in K, (i.e. A and B), appear non-simultaneous in K'.

We come now to the meaurement of distances. To measure a length we have to displace a rigid rod a certain number of times: the measurement process therefore takes a certain time (see **Fig. 1.5**) which in the classical case can be arbitrarily reduced. However, if the object whose length is to be measured is moving with respect to a reference frame K, this procedure is not easy to put into practice. What is much more important is that we must measure spatial distances physically simultaneously. Since simultaneity is an essentially relative notion, so is the measurement of distance (see **Fig. 1.6**).

1.3 Newton's absolutes and the notion of the ether

The classical space and time whose properties we have discussed above make no use of the notion of the ether[5] and are purely *relative* to pairs of events. The motion of a mechanical system can only be referred to another material system. This "relative" viewpoint – due to Ernst Mach – corresponds very closely to the measurement and observation processes which form the basis of modern concepts of space and time.

Newton, on the other hand, had a very different view, in which space and time are *absolute*, existing by themselves independently of the material objects they contain. He

---

[5] The idea of the ether dates from the 19th century, and appears to be due to J.C. Maxwell.

thus distinguished relative motions – exactly those which we observe – and absolute motions, made relative to absolute space with an absolute scale of time. For him, relative space (as well as relative time) is only a "measure" of absolute space. Thus he wrote:

> *Absolute space in its own nature, without relation to anything external, remains always similar and immovable. Relative space is some movable dimension or measure of the absolute spaces; which our senses determine by its position to bodies; and which is commonly taken for immovable space ...*
>
> *Absolute time, true and mathematic, without any relation to anything external flows uniformly and is called duration.*
>
> *An absolute motion is the displacement of a body from an absolute place to another, and the relative motion the displacement from a relative place to another ...*
>
> *[quoted by M. Jammer (1954)]*

The existence of absolute space and time, even though undetectable by mechanical methods, is a logical and ontological necessity for the validity of the principle of inertia, at least for Newton. If a state of rest presupposes the existence of absolute space, the requirement that an inertial system should not be a pure fiction but might at least in principle be conceived and effectively used, necessarily implies the existence of absolute space and time.

We shall see later that all inertial systems can be derived from each other by Galilean transformations. For Newton, only one frame among all these was truly immobile in an absolute sense. But how were we to distinguish this from the others? *"That the centre of the system of the world is immovable. This is ackowledged by all, while some contend that the earth, others that the sun, is fixed in that centre."* In fact Newton placed the centre of the Universe at the centre of mass of the solar system[6] where it was incapable of any kind of verification.

Many philosophers (Leibniz, Huyghens, etc.) opposed Newton's absolute notions, because, as he himself recognised, absolute space and time, in effect, can never be experienced by our senses. Nevertheless, for theological reasons[7] [Newton attributed divine qualities to space and time: God is not eternity and infinity but rather eternal and infinite in Newton's philosophy. See Max Jammer (1954)], and despite the objections of Leibniz and Huyghens, Newton's absolute notions were saluted as a remarkable achievement of "natural philosophy", i.e. of science.

However, the development of mechanics in the 18th century (Lagrange, Laplace, Poisson) showed that Newton's absolute notions played at most a minor role in describing and predicting natural phenomena. It became progressively clear that in practice one could dispense with them. Here is J.C. Maxwell's opinion, for example:

---

[6] Although Halley announced in 1718 that certain stars such as Sirius, Aldebaran, and Betelgeuse, previously regarded as fixed, had changed their positions since Ptolemy's time, Newton took no account of the possible motion of the solar system (imagined earlier by Giordano Bruno) which would have disqualified it as the "centre of the Universe" ...

[7] The so-called ontological proof of the existence of God and other traditional proofs began to be regarded as logically deficient.

Absolute space is conceived as remaining always similar to itself and immovable. The arrangement of the parts of space can no more be altered than the order of the portions of time. To conceive them to move from their places is to conceive a place to move from itself. But as there is nothing to distinguish one portion of time from another except the different events which occur in them, so there is nothing to distinguish one part of space from another except by reference to some other event, or the place of a body except by reference to some other body. All our knowledge, both of time and space, is essentially relative.

[quoted by M. Jammer (1954)]

By the end of the 19th century it had become completely clear that the idea of absolute space was totally impervious to observation [see the discussion of the Galilean principle of relativity below]; in particular the idea that absolute space could act, but not suffer any action that would modify it (E. Mach).

At this stage there remained only one real possibility of demonstrating the existence of absolute space, the *ether*. After the discoveries of (i) the *finite* nature of the velocity of light and (ii) its wavelike nature, it appeared certain that a material medium, the *ether*, the last avatar of absolute space, was required in order to support the propagation of electromagnetic waves. But, as we shall see, even this possibility was to be excluded by experiment, including that of Michelson and many others.

1.4 The principle of inertia

In many books on mechanics, the treatment begins with a statement of Newton's three laws[8], and in particular with the principle of inertia[9] in the form *"a body subject to no force is either at rest or in straight-line uniform motion"*, or in some similar form. The principle of inertia thus appeared to be an immediate consequence of Newton's second law, in which the force had been set equal to zero! In fact, the principle of inertia has a much less trivial significance, which is independent of the second law, and which we shall now consider.

Far from being a banal triviality, the principle of inertia affirms *the existence of privileged motions*, the "free" or *inertial motions*. These motions correspond to *all* possible straight line motions in $\mathbf{R}^3$, traversed uniformly in time. It is worth emphasizing the revolutionary importance of this principle: for medieval thinkers [e.g. Buridan, in the 14th century] and their successors, inertial motion was a result of an intrinsic quality of the moving body, for Newton, (and for his predecessors who had gradually arrived at the principle[10]) [see M.A. Tonnelat (1971) or M. Jammer (1961)], motions

---

[8] i.e. (i) the principle of inertia; (ii) the equality of force and rate of change of momentum, and (iii) the equality of action and reaction.

[9] Newton's formulation is as follows: *"Every body persists in its state of rest or straight-line uniform motion so long as no force acts to change it."*

[10] The definition we give here is kinematic. It is also possible to give a dynamical definition of the principle. However, this has the inconvenience of making much more delicate the discussion of concepts linked to classical space–time, as the latter then contains a dynamical element which has no very obvious justification in Newtonian physics. With such a dynamical definition a Galilean reference frame is one in which the second law can be written in the usual form $F = m\gamma$.

of this type were on the one hand equivalent to a state of rest and on the other hand neither engendered nor stopped by exterior action except at their beginning or end.

In the framework of classical space–time, the principle of inertia has the consequence that the affine structure we discussed above actually represents a physical reality: timelike straight lines, thus not lying in the spacelike 3-planes $t = $ const., correspond to physically realisable motions, or rather to motions conceivable as idealisations of real motions, the motions of "free" particles. Of course, the idea of "free" motion is an abstraction which nevertheless reflects a certain reality, within limits depending on the scale of the systems considered.

This privileged class of motions is thus intrinsically defined in space–time: it is the set of timelike straight lines, thus *geometric objects* in $\mathbf{R}^{3+1}$, *independent of any coordinate system*, and also independent of the nature of the "clocks" used to measure time and the "measuring rods" allowing distance measurements. The introduction of classical space–time is necessary for the intrinsic geometrical characterisation of inertial motions. Conversely, it is the existence of this class of motions, or at least the recognition of their importance, which lies at the basis of the concept of space–time.

In practice, the existence of inertial motions is revealed by that of *inertial reference systems*, or *Galilean reference systems*, i.e. scales of time which flow uniformly, *and* of Cartesian coordinates in $\mathbf{R}^3$ (also moreover in $\mathbf{R}^{3+1}$, where "free" motions are straight lines; further, we can always associate a *physical* system in "free" motion with *these* Cartesian coordinates).

It is clear that any event in classical space–time $\mathbf{R}^{3+1}$ can be described using arbitrary coordinates. However, the possibility of using Cartesian coordinates on the one hand, and the fact that these coordinates have a physical meaning (suitable "free" motions), amounts to a postulate, in fact precisely the principle of inertia.

Inertial motions thus determine *canonical* coordinates in $\mathbf{R}^{3+1}$, and there exists a set of reference frames at each point (i.e. independent vectors) determined by this class of Galilean coordinates. These are *Galilean* or *inertial frames*. These frames will serve as the basis for determining analytically the vector fields we shall use, such as velocities, accelerations, forces, etc...

Given Galilean coordinates[11] $(\mathbf{x}, t)$, it is easy to see that we can deduce an infinity of others through the relations

$$\begin{cases} \mathbf{x}' = \mathscr{R}\mathbf{x} - \mathbf{v}t + \mathbf{x}_0 \\ t' = t + t_0 \end{cases} \tag{1.2}$$

where $\mathscr{R}$ is an arbitrary rotation, $\mathbf{v}$ an arbitrary velocity, $\mathbf{x}_0$ is any vector of $\mathbf{R}^3$ and t_0 any duration. These transformations, the *Galilean transformations*, are linear, and thus transform a straight line in $\mathbf{R}^{3+1}$ into another straight line of $\mathbf{R}^{3+1}$: inertial motions thus transform into inertial motions. They confirm the fact that space is isotropic:

---

[11] What follows mainly concerns Galilean coordinates. However, as we shall use Cartesian coordinates (x, t), we shall speak without distinction of Galilean coordinates (or inertial frames), which in reality is an abuse of language.

if $\mathbf{v} = \mathbf{0}$, $\mathbf{x}_0 = \mathbf{0}$ and $t_0 = 0$, rotation of a straight line in $\mathbf{R}^3$ produces another straight line. Spatial homogeneity is similarly implied. These transformations form a continuous group of ten parameters [three rotations, three spatial translations, three *pure* (or restricted) Galilean transformations

$$\begin{cases} \mathbf{x}' = \mathbf{x} - \mathbf{v}t \\ t' = t, \end{cases} \tag{1.3}$$

and one giving translations of the time origin] whose properties are clearly seen in the physical domain.

To these transformations we can add changes of units

$$\mathbf{x}' = \alpha\mathbf{x}, \quad t' = \beta t, \tag{1.4}$$

as well as *reflections (parity)* and *time reversal*

$$\begin{cases} \mathbf{x}' = -\mathbf{x}, \ t' = t \\ \mathbf{x}' = \mathbf{x}, \ t' = -t \end{cases} \tag{1.5}$$

none of which change the inertial character of free motions either.

At this point we can ask if true inertial systems exist in Nature. In the laboratory, "free" horizontal motions approximate inertial motions if friction is negligible. A better inertial system is linked to the motion of the Earth in its orbit about the Sun. But here too this is an approximation valid only for short timescales. The mean speed of the Earth is about 29 km/s, but this varies somewhat (a few km/s) so that this too is not a truly inertial system. In its turn, a system linked to the centre of gravity of the solar system (in practice, to the Sun) may be regarded as a good approximation to a Galilean reference frame.

> Incidentally, the adoption of geocentric (Ptolemaic) or heliocentric (Copernican) coordinates is equivalent from a mathematical point of view. However, a heliocentric system makes analytic calculations of the motion of the planets much easier. Moreover, the Copernican system is a better inertial system. However, the persecution of the Copernicans [see E. Namer (1975)] (of whom Galileo is the best known example) had its distant origins in Plato's cosmic ideas [see A. Koestler (1960)] and was philosophical and religious: there has never been a dispute about the choice of inertial system!

But the Sun, whose velocity is of order 220 km/s with respect to the fixed stars, participates in the general rotation of our Galaxy (the Milky Way, a typical spiral galaxy), and is also subject to the gravitational attraction of the other stars making it up: this too is not a true inertial system. Similarly, our Galaxy also does not constitute an inertial system: it is situated in a local cluster containing about twenty galaxies, which interact with it. Thus its motion cannot be (and is not) rectilinear or uniform. Of course, this hierarchy of ever more inertial systems does not stop here: the Local Group is part of the Virgo Group, which in turn moves within the Virgo supercluster...

We thus arrive at the idea that the consideration of ever more distant objects, whose proper motions cannot be observed, provides a better and better *approximation* to an inertial system, which is a theoretical and thus *ideal* concept. We can regard a "good" inertial system as being formed by the Solar System and three quasars, which are pointlike objects (as seen from Earth), and the most distant currently known.

It should be clear that a true inertial system cannot exist in Nature. This is a theoretical concept whose main virtue is that it allows us to go beyond purely kinematic considerations, from simply describing physical systems: that is, to study dynamics. It is precisely because there is no truly isolated object (nor could there be) that the concept of an inertial system was so difficult to free from *a priori* ideas. We should add that the kinematic presentation (i.e. in terms of space and time) of the principle of inertia that we have given here is also a simplification, as this principle only acquires its full meaning in a dynamical context, where the notions of force, mass, etc. are developed.

A last theoretical remark. It might be interesting from a mathematical point of view to give a space–time, i.e. a largely arbitrary manifold, and an invariance group acting transitively on it, which would define an inertial class of motions (cf. $\mathbf{R}^{3+1}$ and the Galilean group). Although this idea is suggestive of generalisations, both mathematical and theoretical, it does not correspond to the historical development of the subject.

1.5   The laws of dynamics and Galilean relativity

Given Galilean reference frames, (or Galilean reference systems and Galilean coordinates), defined physically, then conceptually and finally mathematically, we can study not only the properties of motion of a mechanical system (kinematics), but also write down *dynamical laws*[12] which hold in these inertial frames.

We can thus write down the fundamental law of dynamics (Newton's second law) in such a system in the form

$$\mathbf{F} = m\frac{d^2\mathbf{x}}{dt^2},\tag{1.6}$$

where $\mathbf{F}$ is the applied force on a particle of mass m situated at $\mathbf{x}$ at time t.

> In fact, Newton did not write his second law in this form, but actually in a form which in modern notation reads
>
> $$\mathbf{F}dt = d\mathbf{p}$$
> $$\mathbf{p} = m\mathbf{v},$$
>
> where p is the momentum of the particle. The latter form, which emerged from, among other things, earlier analyses by Huyghens, Marci, Wallis and Wren of collisions of particles (where the total momentum is conserved in the collision), carries over into special relativity (with the appropriate modification of the relation between momentum and velocity), whereas the earlier one does not.

Newton's second law is a *postulate* which, we repeat, required the invention of the principle of inertia before it could be formulated analytically. It unifies a great range of concepts which required much effort, time and discussion to emerge, and is

---

[12] i.e. general laws allowing us to go beyond a kinematic description by "summarizing" classes of motions having the same cause.

certainly not trivial despite the simplicity of its formulation[13]. It is therefore worth brief consideration.

First, the law relates two concepts of different natures; one is kinematic (the acceleration γ) and the other dynamical (the force **F**). This *already* constitutes a revolution, as since Aristotle and up to Newton, force had always been seen as generating velocity and not acceleration.

> If we tried to write down a dynamical law corresponding to pre-Newtonian mechanical ideas, we would produce something like
>
> $$\mathbf{F} \propto \mathbf{v}.$$
>
> This is not totally absurd: Aristotle inferred a law of this type by observing fishermen pulling a boat, as the frictional force was evidently proportional to the velocity (at low velocity). This type of concept, plus the lack of sufficiently developed mathematical techniques, prevented Kepler from arriving at the law of universal gravitation (see below) even though he had all the elements necessary (the three laws bearing his name).

In the dynamical equation, the *inertial mass m* appears as the coefficient of the response (acceleration) to outside excitation (force). It thus characterises the particle's *resistance to motion*, an intrinsic property characteristic of the particle's *matter content*: the more massive a body, the greater the force needed to bring it to a given velocity. Again, the concept of inertia required a long time (several centuries) to be clarified [see Max Jammer (1961)]: from the first analyses of Buridan (14th century) to Kepler's *qualitative* concept of inertia, to the particle collision experiments quoted above, to those of Huyghens on centrifugal force, was a long path to Newton's extraordinarily complex systematisation and development.

> The proportionality between force and acceleration of a particle could differ from the usual form $\mathbf{F} = m\gamma$. Specifically, confining ourselves to linear relations, we could write
>
> $$F^i = \sum_j m^i{}_j \gamma^j \quad (i, j = 1, 2, 3). \tag{1.7}$$
>
> Of course, the tensor $m^i{}_j$, if not proportional to $\delta^i{}_j$, would destroy the invariance under spatial rotations: the two or three eigenvectors of $m^i{}_j$ would introduce privileged directions in physical space, which would no longer be isotropic. This amounts to saying that the existence of a mass tensor $m^i{}_j \neq m \, \delta^i{}_j$ ($\delta^i{}_j$ is the Kronecker symbol, equal to 1 if $i = j$ and 0 otherwise) has to be settled by experiment. This has been done (see *Chapter 3*) with the result that $m^i{}_j = m\delta^i{}_j$ to a very good approximation: this is the Hughes–Drever experiment [V.W. Hughes *et al.* (1960), R.W.P. Drever (1961); see also R.H. Dicke (1965)]. Because of its theoretical importance, which goes much further than the anisotropy of mass [see C.M. Will (1980)] – we briefly indicate the basis of the experiment, quoting S. Weinberg (1972): Hughes *et al.* and Drever, independently *"observed the resonant absorption of photons by a Li[7] nucleus in a 4700 Gauss magnetic field. The ground state has spin 3/2, so it splits in a magnetic field into four energy levels, which should be equally spaced if the laws of nuclear physics are rotationally invariant. In this case the three transitions among neighbouring states should have the same energy and*

---

[13] The invention of the calculus by Leibniz and Newton was also required...

the photon absorption coefficient should show a single sharp peak at this energy. However, if inertia were anisotropic then the four magnetic substates would not be exactly equally spaced, and there would be not one but three closely spaced resonance lines." This is the principle of the experiment, subsequently repeated many times with different materials. The Hughes–Drever result was $\Delta m/m \lesssim 10^{-20}$, which has been greatly improved since then. This result can be interpreted in different ways [C.M. Will (1981); R.H. Dicke (1965)] and, in particular, as a test of the precision of local Lorentz invariance.

Finally the notion of force itself needs a few remarks.

First, if we wish only the initial position and velocity of the particle to completely specify its motion, we require that $\mathbf{F}$ should depend solely on $\mathbf{x}$ (the particle position at time t), on $\mathbf{v}$ (its velocity at time t) and on time. We thus get a second-order differential system. If $\mathbf{F}$ depended, for example, on $\gamma \equiv (d/dt)v$, we would also need to know the initial acceleration.

Thus $\mathbf{F}$ is a vector of $\mathbf{R}^3$, which seems clear, because we require it to be proportional to $m\gamma$. However, even this property is not obvious: before we can write down the usual dynamical law, we need to recognise and have tested its vectorial character. We shall return later to other important characteristics of the Newtonian force, notably in connection with the Galilean principle of relativity and gravitation.

We now come to the *Galilean principle of relativity*. For this purpose we consider a mechanical system consisting of N point masses m_i ($i = 1, 2, \ldots, N$), mutually interacting under two-body forces of the type

$$\mathbf{F}_{(ij)}\left(\mathbf{x}_i - \mathbf{x}_j\right),$$

so that in an inertial system the equations of motion of the ith particle can be written as

$$m_i \frac{d^2 \mathbf{x}_i}{dt^2} = \sum_{j \neq i} \mathbf{F}_{(ij)}\left(\mathbf{x}_i - \mathbf{x}_j\right) ; i = 1, 2, \ldots, N. \tag{1.8}$$

How do we write the equations of motion for this mechanical system in another inertial system? To see this, let us make a Galilean transformation: the ith particle is described by the coordinates $\left(\mathbf{x}'_i, t'\right)$ and the last equation becomes

$$m_i \frac{d^2 \mathbf{x}'_i}{dt'^2} = \sum \mathbf{F}_{(ij)}\left(\mathbf{x}'_i - \mathbf{x}'_j\right) ; \tag{1.9}$$

where we have limited ourselves to "pure" Galilean transformations, although this is not strictly necessary. The new equation is formally identical to the first! Put another way, the mechanical system considered is described in the same way, by analogous equations in *all* Galilean reference systems. More concretely, this means that uniform rectilinear motion has no detectable effect on the *mechanical* properties of the system: this motion does not give rise to supplementary fictitious forces (such as the centrifugal force in rotational motion), which would thus produce observable effects. This is what Galileo expressed in the form *"motion is like nothing."*

Note that we have implicitly assumed the invariance of the masses m_i of the particles under transformations of the Galilean group:

$$m_i = m'_i.$$

In fact, if we regard mass as an intrinsic property (a measure of the resistance to changes of motion), it is clear that this must be so.

We can now state the Galilean principle of relativity. As M.A. Tonnelat (1971) remarked quite correctly, this has two aspects: one physical and one descriptive. Its physical content is as follows: *inertial motion of a mechanical system has no effect on its mechanical properties.* The analytic consequence of this is given by the descriptive statement: *a mechanical system is described by laws having the same form in all Galilean reference systems.*

> Let us imagine that the laws of mechanics had different forms in two Galilean systems, i.e.:
>
> $$\mathbf{F} = m\gamma,$$
>
> in the first system, and
>
> $$\mathbf{F}' = m\gamma' + \mathbf{K},$$
>
> in the second, where $\mathbf{F}'$ is the transformed $\mathbf{F}$ under an arbitrary Galilean transformation. The additional term $\mathbf{K}$ would thus show the formal non-invariance of the equations of mechanics. As it could be regarded as a force, it would give rise to measurable effects (at least in principle) and would thus allow one to show that the two inertial systems were in relative motion. This would evidently contradict the principle of relativity.

At this point we can ask ourselves why we refer to the "principle of relativity" and not the "relativity theorem"? It indeed appears that we have actually proved a "relativity theorem" rather than appealed to a principle. In reality we have actually used the Galilean relativity principle, although only implicitly, particularly when we considered interaction forces of the form $\mathbf{F}_{(ij)}(\mathbf{x}_i - \mathbf{x}_j)$. We would not have been able to prove a relativity "theorem" for mechanical systems subject to arbitrary forces. Conversely, the principle allows us to *restrict* the class of physically admissible forces.

We show this briefly with two simple examples

(i) We consider first the motion of a particle of mass m under the action of an exterior force field

$\mathbf{F}(\mathbf{x}, \mathbf{v}; t)$:

$$m\frac{d^2\mathbf{x}}{dt^2} = \mathbf{F}(\mathbf{x}, \mathbf{v}; t).$$

Uniformity of the flow of time,

$$\mathbf{F}(\mathbf{x}, \mathbf{v}; t + t_0) = \mathbf{F}(\mathbf{x}, \mathbf{v}; t), \quad \text{(for all } t_0) \tag{1.10}$$

requires that $\mathbf{F}$ does not depend on time. Similarly, *homogeneity* of space requires that $\mathbf{F}$ should not depend on position. The same reasoning shows that $\mathbf{F}$ cannot depend on $\mathbf{v}$ if $\mathbf{F}$ is to be the same in all inertial frames. Finally, *isotropy* of space

$$\mathscr{R}\mathbf{F} = \mathbf{F} \tag{1.11}$$

(for all rotations $\mathscr{R}$)

requires that $\mathbf{F} \equiv \mathbf{0}$! The only motion compatible with the Galilean relativity principle is thus free motion... This does not of course mean that external forces cannot exist as approximate models for complex dynamical systems, where we are not interested in all degrees of freedom. For example, if we consider the motion of a charged particle in the magnetic field of a solenoid, the true interaction satisfying the principle of relativity is that of the charged particle with all the particles making up the solenoid: the magnetic field simply expresses this interation without requiring us to write down (and solve!) the equations for the $\approx 10^{23}$ particles of the solenoid.

(ii) Now consider a system consisting of two particles interacting *via* a force which is independent of their velocities. For simplicity (although the final result holds independently) we assume that the force is derivable from a potential: $V(\mathbf{x}_1, \mathbf{x}_2; t)$:

$$\mathbf{F}_{(ij)} = -\frac{\partial}{\partial \mathbf{x}_i} V.$$

The same reasoning as in the earlier example leads to the general form for V

$$V = V\left(|\mathbf{x}_j - \mathbf{x}_j|\right) ; \; i, j = 1, \tag{1.12}$$

which immediately gives

$$\mathbf{F}_{(ij)} = -\mathbf{F}_{(ji)} \propto \mathbf{x}_i - \mathbf{x}_j. \tag{1.13}$$

In this case, the principle of relativity requires that the force is central, and moreover obeys the principle of action and reaction:

$$\mathbf{F}_{(ij)} + \mathbf{F}_{(ji)} = 0.$$

1.6   Inertia and relativity principles as seen by Galileo

A classic citation from Galileo *"Dialogue on two principal systems of the World – the Ptolemaic and the Copernican"* (1632), shows the 17th century view of these principles.

> *With some friends, shut yourself in a large enclosed chamber in the depths of a large ship; furnish yourselves with flies, butterflies and other small flying animals; take a large bowl of water and put some fish in it; hang a small bucket from the ceiling from which drops of water leak into a jar with a narrow neck, placed on the floor. When the ship is stationary, observe carefully how the small animals fly with equal velocities in all directions in the chamber; the fish swim equally in all directions; all the drops fall into the jar; and if you throw something to a friend, you do not need to throw it harder in any one direction rather than another, for equal distances; if you jump with your feet together you travel equal distance in all directions ...*
>
> *Now set the ship in motion, as speedy as you choose. Now if the movement is uniform, and not oscillating, you will not notice the slightest change in the effects described; and none of you will be able to say if the ship is moving or not; when you jump you will reach the same distances as before, and even if the ship moves at high speed you will not be able to make bigger leaps towards the stern than towards the bow, even though while you are in the air, the floor of the chamber is moving in the opposite direction from your jump; and if you throw something to your friend you will not need to throw it harder, whether or not he is astern of you; the drops will continue to fall into the jar below, even though the ship*

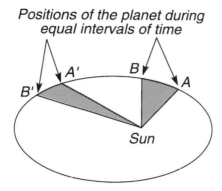

Positions of the planet during
equal intervals of time

Fig. 1.7. The law of areas. If the Earth (or any other planet) describes the arcs **AB** and **A′B′** in the same times, the curvilinear triangles **SAB** and **SA′B′** have the same areas (**S** is the position of the Sun at one of the foci of the ellipse which is the trajectory of the planet considered).

> *moves a fair distance while each drop is falling; the fish will swim into all parts of their bowl without tiring...; and finally the butterflies and the flies will continue to fly equally in all directions... The reason for all this is that the motion is common to the ship and all that it contains, including the air...*

1.7   Newtonian gravitation

Newton's theory of gravity arose from Kepler's study of the motion of the planets, itself based on the remarkably precise observations[14] of Tycho Brahé (1588). Of course, they also used the theoretical development of the notions of inertia, motion, mechanical systems force, etc...

> We recall Kepler's three laws: (i) the orbits of the planets are *ellipses* (until Kepler, they had been regarded as circular; Tycho's observations required great care, not to speak of excellent vision, to decide beyond question between circles and ellipses); (ii) in equal times the radius vector sweeps out equal areas [this is the *law of areas* (**Fig. 1.7**); it is equivalent to the conservation of angular momentum $|\mathbf{x}\wedge\mathbf{v}|$ =constant ($\mathbf{v}$: velocity, $\mathbf{x}$: radius vector)[15]; (iii) the square of the period of revolution is proportional to the cube of the semi-major axis of the ellipse:

$$T^2 = \frac{4\pi^2 a^3}{GM_\odot}. \tag{1.14}$$

> ($M_\odot$: mass of the Sun; a : semi-major axis of the ellipse described by the planet considered)

---

[14] Let us remember that they were made with the naked eye...

[15] Peripatetic ideas of motion and of force as generating velocity would have led to a gravitational force varying as $1/r$; as $\mathbf{F} \propto \mathbf{v}$ and $\mathbf{v} \propto 1/r$, Kepler would have been able to deduce $\mathbf{F} \approx 1/r$.

Newton's theory is based on the law of universal attraction, given for the first time by Newton[16] (1687)

$$\mathbf{F}_{(12)} = -G\frac{m_1 m_2}{r_{12}^2}\mathbf{n}_{12} \tag{1.15}$$

where $\mathbf{F}_{(12)}$ is the gravitational force between two bodies of masses m_1 and m_2, r_{12} is their distance, $\mathbf{n}_{12}$ is a unit vector pointing from one to the other, and G is the gravitational constant

$$G = 6.67 \times 10^{-8} \text{ CGS}.$$

Let us briefly examine the characteristic consequences of the gravitational force.

The first is that the action between the two bodies occurs *instantaneously and at distance*. This is completely contrary to physical intuition, which holds that the action of one body on another requires a certain time, a *finite* propagation velocity, before it can take effect. This objection, of which Newton was perfectly aware[17], considerably slowed the acceptance of the *Principia*: it took almost fifty years before Newtonian gravity was accepted on the continent.

The second characteristic is that, for the first time, the motion of the planets and of the Moon or of falling bodies could be attributed to the *same* cause, described by the *same* law. This was one of the great achievements of 17th century science.

Thus, gravitational forces appeared as *extremely weak, long range* forces.

For two electrons of mass $m = 0.91 \times 10^{-27}$ g, the ratio between gravitational and electrostatic forces is

$$F_{\text{grav}}/F_{\text{es}} \approx 2.4 \times 10^{-43} \text{ !}$$

The law of universal attraction presupposed once more the development of the concept of mass. As we discussed above, the notion of inertial mass – the mass appearing as the coefficient of the acceleration in Newton's second law – arises from two classes of experimental facts: (i) the elastic or inelastic collisions studied by Marci, Wallis, Wren and Huyghens; and (ii) uniform rotation of bodies (Huyghens noted that the ratio of centrifugal forces of two bodies with the same angular velocities was the same as the ratio of their weights). We should add to this a discovery by Jean Richer in 1671: the weight of a body varies with latitude (in fact he studied the period of a pendulum in Paris and Guyana). The inertial mass is defined experimentally by collision with a test particle whose mass is used to define the unit of mass. This uses the conservation of momentum and measurements of the velocity changes of the two particles; this type of law cannot be used to give an "operational" definition of mass, contrary to the claims of Mach [see M. Bunge (1966)]. Thus the mass of the Sun

---

[16] In fact many philosophers of the era were inclined towards a force law proportional to $1/r^2$ (Hooke, for example), but they were incapable of actually demonstrating it.

[17] *"That gravity should be innate, inherent and essential to matter, so that one body may act upon another at a distance through a vacuum, without the mediation of anything else, by and through which their action and force may be conveyed from one to another, is to me so great an absurdity, that I believe no man who has in philosophical matters a competent faculty of thinking, can ever fall into it."* [cited by B. Hoffman, H. Dukas (1972)].

is not estimated by collisions, but using Kepler's third law and the whole theoretical apparatus included in the theory (Newton's second law, universal attraction, equality of inertial and gravitational mass).

The two masses m_1 and m_2 play different roles in the law of universal attraction. Moreover, they play *a priori* a different role from that of inertial mass: they represent the *gravitational coupling* of particles 1 and 2. Let us consider a system of two particles interacting gravitationally. Distinguishing the inertial and gravitational masses, the equations of motion read

$$m_{I_1}\gamma_1 = -Gm_{P_1}m_{A_2}\frac{\mathbf{x}_1 - \mathbf{x}_2}{|\mathbf{x}_1 - \mathbf{x}_2|^3} \equiv \mathbf{F}_1 \tag{1.16}$$

$$m_{I_2}\gamma_2 = -Gm_{P_2}m_{A_1}\frac{\mathbf{x}_2 - \mathbf{x}_1}{|\mathbf{x}_1 - \mathbf{x}_2|^3} \equiv \mathbf{F}_2 \tag{1.17}$$

where $\mathbf{x}_1$ and $\mathbf{x}_2$ are the positions of particles 1 and 2. In these equations the m_I are the *inertial masses*, the m_P the *passive gravitational masses* (i.e. the coupling coefficients of the particle which *feels* the gravitational field of the other particle), and the m_A the *active gravitational masses* (which characterise the intensity of the gravitational field of the particle which *creates the field*). We can easily see that if Newton's principle of action and reaction (3rd law) holds

$$\mathbf{F}_{(12)} + \mathbf{F}_{(21)} = \mathbf{0}, \tag{1.18}$$

we necessarily have $m_A = m_P$. In fact, this principle could be experimentally wrong[18]; for this reason we must test the equality of the two masses (passive and active) experimentally. We find [L.B. Kreuzer (1968); see also the interpretation by C. Will (1976)]

$$\left|\frac{m_A - m_P}{m_P}\right| < 5 \times 10^{-5}.$$

We turn now to the possible equality of the inertial mass m_I and the passive gravitational mass m_P. In a uniform gravitational field $\mathbf{g}$, Newton's second law is

$$m_I\gamma = m_P\mathbf{g}; \tag{1.19}$$

this would mean that *freely falling bodies would not follow the same motions*: we would find

$$\mathbf{z}(t) = \frac{1}{2}\frac{m_p}{m_I}\mathbf{g}\,t^2\,, \tag{1.20}$$

where $\mathbf{z}(t)$ is the height at time t of a massive particle released from rest at $t = 0$. The experiment was performed by Galileo[19] – actually using inclined planes rather than the tower of Pisa – and it is known that we have $m_p \approx m_I$. Similarly, if $m_p \neq m_I$, *pendula of*

---

[18] For example, it is not true for electromagnetic forces.

[19] In the 5th century AD the grammarian and philosopher Philiponos had performed experiments on the fall of massive bodies; but unlike Galileo he did not understand how to neglect inessential phenomena, such as air resistance [see B. Cohen (1960)].

the same length would have different periods proportional to $(m_p/m_I)^{1/2}$: Newton carried out this experiment and found that

$$\left|\frac{m_I - m_p}{m_p}\right| < 10^{-3}.$$

More recent undisputed experiments (see *Chapter 7*)[20] give

$$\left|\frac{m_I - m_p}{m_p}\right| < 10^{-11},$$

and it is hoped in the near future to reach an accuracy of better than 10^{-17} or even 10^{-20}, using satellite experiments.

We may thus set

$$m_I = m_p, \tag{1.21}$$

to a precision almost unequalled in physics, and regard this equality as a fundamental principle, *the weak equivalence principle*. What are its main consequences?

To see these we consider a system of particles interacting via two-body forces $\mathbf{F}_{(ij)}(\mathbf{x}_i - \mathbf{x}_j)$, immersed in a *homogeneous and uniform* gravitational field $\mathbf{g}$. In a Galilean reference system we can then write

$$m_i \frac{d^2 \mathbf{x}_i}{dt^2} = \sum_{j \neq i} \mathbf{F}_{(ij)}\left(\mathbf{x}_i - \mathbf{x}_j\right) + m_i \mathbf{g}, \tag{1.22}$$

for the equation of motion satisfied by the *i*th particle. We now use a reference frame which is *freely falling* in the field $\mathbf{g}$, and hence in a non-Galilean system, as it is *accelerating* with respect to the inertial system in which we wrote the equations of motion, i.e.

$$\begin{cases} t' = t \\ \mathbf{x}' = \mathbf{x} - \frac{1}{2}\mathbf{g}t^2. \end{cases} \tag{1.23}$$

In this reference frame it is easy to see that the dynamical equations of the system are

$$m_i \frac{d^2 \mathbf{x}'_i}{dt'^2} = \sum_{j \neq i} \mathbf{F}_{(ij)}\left(\mathbf{x}'_i - \mathbf{x}'_j\right). \tag{1.24}$$

Put another way, in *this* freely falling reference frame the external gravitational field does not appear, and plays no role in the dynamics of the system.

> If we considered a massive particle in a freely falling reference frame, it would (like every other particle feeling no forces other than gravity) have described a straight-line trajectory with uniform motion. This reference system might thus be considered as inertial in some sense.

---

[20] V.B. Braginskii and V.I. Panov (1972), claimed $|\delta m/m| < 10^{-12}$, but this result is generally considered doubtful.

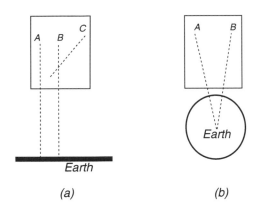

Fig. 1.8. Einstein's lift. In an effectively homogeneous gravitational field (a) (the dimensions of the lift are much smaller than the scale on which the gravitational field varies), a freely falling lift constitutes an inertial system: two particles at rest, **A** and **B**, remain so: and a free particle **C** performs uniform rectilinear motion in this frame. By contrast, when the field is not homogeneous (b), an observer in the lift can decide that he is situated in an external gravitational field by looking at the behaviour of particles **A** and **B**: as they fall towards the centre of the Earth **O** – following the lines of force of the field – they tend to approach one another.

Quite generally, we can always *locally* remove the effect of a Newtonian gravitational field, even if it is not homogenenous (such as the Earth's field), by choosing a freely falling reference system (and adapted coordinates of course) provided that the characteristic size of the system is much smaller than the characteristic length over which the gravitational field varies (**Fig. 1.8**).

The second consequence of the equality of gravitating and inertial masses is of great theoretical importance. It is no longer possible to find any inertial system as any material body necessarily interacts gravitationally with the rest of the Universe! In other words, contrary to the case of electromagnetic forces, where electrically neutral bodies exist, *there exists no matter which has mass but is gravitationally neutral*. This absence of a global inertial system implies also that accelerations no longer have any absolute meaning, at least locally. If, however, we accept the idea of absolute space, accelerations must still have at least an absolute meaning, even globally. For Newton and his successors this would have been another reason to postulate an ether.

Since the dynamical effects of sufficiently homogeneous gravitational fields do not depend on the intrinsic properties of individual bodies (such as their mass or their charge), they appear as *geometric characteristics* of space, which then ceases to be Euclidean[21]: this point of view can be made compatible with Newtonian theory [P. Havas, (1964), A. Trautman (1963)], and is one of the foundations of Einstein's theory. However this theory only takes on its full meaning (see *Chapter 7*) within a complete unification, not only qualitatively, of the concepts of space and time.

---

[21] See Gauss, Lobachevski, Clifford, Riemann in J.L. Coolidge (1963).

A last remark on the law of universal attraction is that the gravitational force exerted by a particle of mass m at point $\mathbf{x}$, at a point with coordinates $\mathbf{y}$, is derivable from the *gravitational potential*.

$$\varphi(|\mathbf{x} - \mathbf{y}|) = -G\frac{m}{|\mathbf{x} - \mathbf{y}|}, \tag{1.25}$$

so that the gravitational potential at $\mathbf{x}$ produced by a set of particles at points with coordinates $\mathbf{x}_i$ can be written

$$\varphi(\mathbf{x}) = -G\sum_i \frac{m_i}{|\mathbf{x} - \mathbf{x}_i|}, \tag{1.26}$$

where we have implicitly made the *additivity hypothesis* for the interactions of each particle. This hypothesis is not self-evident, and actually fails on the microphysical scale, at least in this form.

We note finally that by applying the Laplacian operator

$$\Delta \equiv \sum \frac{\partial^2}{\partial x_i^2} \equiv \nabla^2 \equiv \left|\frac{\partial}{\partial \mathbf{x}}\right|^2, \tag{1.27}$$

to the preceding expression for φ, we find easily that

$$\Delta\varphi(\mathbf{x}) = 4\pi G\rho(\mathbf{x}), \tag{1.28}$$

i.e. Poisson's equation. Here $\rho(\mathbf{x})$ is the mass density of particles at $\mathbf{x}$.

Setting

$$\rho(x) = \sum_i m_i \delta^{(3)}(\mathbf{x} - \mathbf{x_i}), \tag{1.29}$$

where $\delta^{(3)}(\mathbf{x} - \mathbf{x}_i)$ is the Dirac distribution (for a density of $+1$ situated at $\mathbf{x}_i$), and using the fact that

$$\Delta\frac{1}{|\mathbf{x}|} = -4\pi\delta^{(3)}(\mathbf{x}), \tag{1.30}$$

we get Poisson's equation. Using these definitions[22] one obtains

$$\varphi(\mathbf{x}) = -G\int d^3x' \frac{\rho(\mathbf{x}')}{|\mathbf{x} - \mathbf{x}'|}. \tag{1.31}$$

1.8   Measuring the gravitational constant

The value of the gravitational constant appearing in Newton's law of universal attraction plays an important role in all gravitational effects. We might try to measure G using the oscillations of a simple pendulum. However, this would not be very accurate, as what one actually measures is the local acceleration due to gravity, g, which is related to the gravitational constant by the relation

$$g = \frac{GM_\oplus}{R_\oplus{}^2}, \tag{1.32}$$

---

[22] Recall that $\int d^3x'\delta^{(3)}(\mathbf{x} - \mathbf{x}')f(\mathbf{x}') = f(\mathbf{x})$, where $f(\mathbf{x})$ is any continuous function.

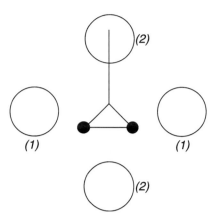

Fig. 1.9. Measurements of G**.** A torsional pendulum whose moving part consists of two "small" test masses, is subject to the gravitational attraction of two "large" masses, in the two positions (1) and (2). In the Cavendish (1798) experiment, the deviation of the moving part of the pendulum was measured as the positions of the "large" masses were varied. In the measurement by Heyl (1942) the variation of the oscillation period of the pendulum is measured: the proximity (or distance) of the "large" masses produces an extra restoring force resulting from the gravitational attraction on the "small" masses, which is thus proportional to G.

where $M_\oplus$ is the mass of the Earth and $R_\oplus$ is its radius[23], and where we have neglected the effect of centrifugal force caused by the Earth's rotation (the effect is only a few per cent: g = 9.83 at the equator and g = 9.78 at the poles). Unfortunately neither $M_\oplus$ nor $R_\oplus$ are known with sufficient accuracy.

Thus, determining G is impossible if we use only large gravitational forces, i.e. those of the planets and their satellites, as their masses are not known accurately enough. Thus, only laboratory experiments allow us to find G. In turn, the weakness of the interaction means that we currently only know G to three significant figures.

Historically[24], the first accurate measurement of G was by Cavendish (1798), using a torsional pendulum, independently invented by Coulomb and Michell in about 1750. The principle of the experiment is as follows: two masses are suspended from the moving part of a torsional pendulum (see **Fig. 1.9**). Two much larger masses are placed close to the apparatus: their gravitational attraction deviates the pendulum, and measuring this deviation allows us to find G.

> The Cavendish experiment (**Fig. 1.10**) was probably the first truly modern experiment in physics, particularly in its systematic study of the possible causes of error, and the means of reducing them. We can list some of the problems that Cavendish encountered and solved. First, there are *temperature gradients* inside the apparatus, which cause *convection currents*. The pendulum had therefore to be enclosed as tightly as possible. Moreover, the whole experiment was placed inside a larger container. To avoid disturbing the

---

[23] $M_\oplus = 5.977 \times 10^{27}$ g; $R_\oplus = 6.3718 \times 10^8$ cm.
[24] We follow here the treatment of F. Everitt (1975).

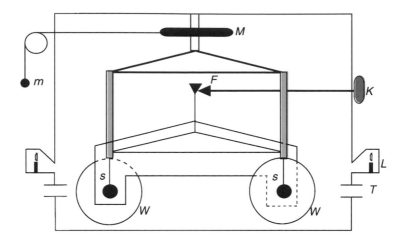

Fig. 1.10. Cavendish's experiment (1798). Two small masses **s** are suspended from the moving part of a torsional pendulum whose position is measured (i) in the absence of large masses **W** and (ii) in their presence, using a small telescope **T**. The distance between the two small masses is about two metres. **L** is a candle illuminating the apparatus. **FK** is a rigid shaft allowing the moving part of the pendulum to be reset to zero. The pulley **M** and wire **m** allow one to adjust the distance of the large masses **W**.

apparatus, it had to be *controlled from a distance*. Thus, the position of the large masses (**W**) was fixed by a pulley (**M**) connected by a wire to the outside of the container (**m**); similarly, the *resetting to zero* of the moving part of the pendulum, which involved the small masses (**s**), was performed using a rigid shaft (**FK**) and gearing; finally, *reading off the deviation* of the pendulum was performed by using a vernier illuminated from outside (**L**), the gradations being read by means of a telescope (**T**). Another significant cause of error was the effect of the *Earth's magnetic field*: it was necessary to use non-magnetic materials, although residual effects were possible [insufficiently pure lead for the large masses (**W**), for example]. It was also necessary to estimate *the effect of convection* caused by temperature gradients in the apparatus [the lead masses (**W**) did not cool in the same way as the other materials at night, for example]; the effect on the moving pendulum of *Brownian motion*[25], resulting from molecular motions [negligible, in view of the precision of the experiment]; the effects of *seismic disturbances* (also negligible), etc. . . .

Later, several important improvements to the precision of the measurements were made by Boys (1889). These included the use of quartz fibres (instead of silver, as in Cavendish's experiment), and the use of a Poggendorf mirror ("optical lever") for readouts. Finally, Boys realised that the accuracy of the experiment would be improved by making it smaller, contrary to immediate expectation. This is because if the "large" masses can be increased the angular precision remains the same. Further, a smaller size limits the convection currents, as the temperature gradients are smaller. Various

---

[25] Discovered in 1827 by the botanist R. Brown. This was a source of error that Cavendish could hardly know . . .

Table 1.1. *Values found for G*

Cavendish	1798	6.754
Reich	1838	6.70
von Jolly	1881	6.465
Wilsing	1889	6.596
Poynting	1894	6.698
Boys	1895	6.658
Braun	1896	6.658
von Eötvos	1896	6.65
Richard *et al.*	1898	6.685
Burgess	1901	6.64
Crémieu	1909	6.67
Heyl	1930	6.670
Zahradnicek	1933	6.659
Heyl *et al.*	1942	6.673
Rose *et al.*	1969	6.674
Facy *et al.*	1972	6.6714
Sagitov *et al.*	1977	6.6745
Luther *et al.*	1982	6.6726

(G in units of 10^{-8} CGS)

improvements were later made (see Table 1.1), until the most recent experiments, which are based on different principles.

The first modern experiment was that of Heyl (1942). Instead of measuring the deviation of a torsional pendulum, one here measures its period for two different positions of the "large" masses: the period change gives G. As measuring a period is more accurate than measuring a deviation, a better value of G results.

The second experiment was by R.D. Rose *et al.* (1969)[26]. Its principle is still that of the Cavendish experiment, but the whole apparatus can turn about a vertical axis. The two "large" masses tend to turn the moving part of the torsional pendulum (see **Fig. 1.11**) *but* a servo mechanism then turns the apparatus in the opposite sense, so as to reduce the deviation of the torsional pendulum to zero. The apparatus thus turns about a vertical axis (the fibre holding the moving part of the pendulum) with effectively uniform acceleration (of the order of 5×10^{-6} radians/s^2) which can be measured, thus giving G directly.

Table 1.1 gives the result of several experiments. The figures give 10^8 G in CGS units. According to Beams (1971), the relative error in G is probably of the order of 5×10^{-3} at least, except for the last measurement, where it is estimated as being of order 10^{-3}.

All of these experiments *assume* the validity of the law of universal attraction[27] in

---

[26] See also Beams (1971).

[27] A near–exhaustive list of the various measurements of G and their main sources of error can be found in G.T. Gillies (1983).

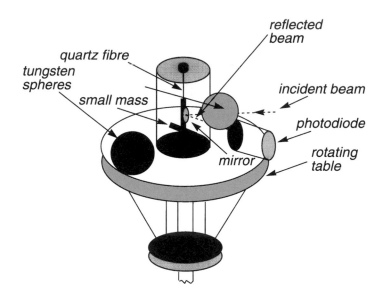

Fig. 1.11. The experiment of R.D. Rose *et al.* (1969).

the form mm'/r^2. In reality this law is only well verified on the scale of the solar system; we can ask if this is really true on the laboratory scale. [D.R. Long (1974)]. If the law of universal attraction is not $\propto 1/r^2$ at small distances, it can be written in the general form

$$F = G(r)\frac{mm'}{r^2}, \tag{1.33}$$

where the gravitational "constant" is a slowly varying function (unknown, in general, at least from theoretical assumptions) of distance. We thus have to measure G at various distances: a gravitational attraction varying as $1/r^2$ would give the same value of G for all r, within experimental error. To get a first idea, we consider earlier measurements of G, as was done by D.R. Long (1974). Table 1.2 shows the results. **Figure 1.12** [D.R. Long (1974)] shows a slight tendency for G to vary with distance, which experiment seems to confirm [D.R. Long (1976)]. However, taking account of the errors, and of other experiments [V.I. Panov and V.N. Frontov (1980); etc.], it appears that we can indeed regard G as constant, and that the $1/r^2$ law holds at the laboratory scale.

Various theoretical ideas put forward in recent years [see references 1 to 15 of E.G. Adelberger *et al.* (1990)], suggest the possibility of a "fifth force", in addition to the known ones (the gravitational, electromagnetic, strong and weak nuclear forces). This fifth force would add to Newton's universal attraction and modify it. Many experiments have attempted to find it, without any definite conclusions. Among the ideas most often considered is the possibility that the fifth force could be derived from

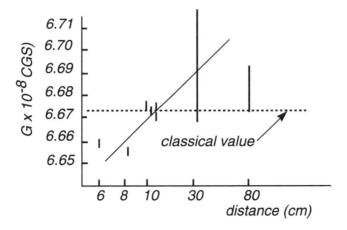

Fig. 1.12. Values of G as a function of distance [from D.R. Long (1974)]. If G does not vary
with distance, we would expect the measurements to be distributed about a horizontal straight
line. The dotted line shows $G = const.$, with the accepted value. The continuous line shows the
least-squarcs fit; this suggests a tendency for G to increase with distance. However, the error
bars arc too large for any firm conclusion at present. The coordinates are semi-logarithmic.

Table 1.2. *Measurements of G as functions of the distance of the masses*

Boys (1894)	6.6576 ± 0.002	6.3 cm
Braun (1896)	6.655 ± 0.002	8.6 cm
Poynting (1891)	6.6984 ± 0.029	32 cm
Richard *et al.* (1898)	6.685 ± 0.011	80 cm
Heyl (1930)	6.670 ± 0.005	13 cm
Heyl *et al.* (1942)	6.673 ± 0.003	13 cm
Rose *et al.* (1969)	6.674 ± 0.004	12 cm

a potential of the form

$$v(r_{12}) = \pm \frac{g_5^2}{4\pi}(q_5)_1(q_5)_2 \frac{rme^{-r_{12}/r_0}}{r_{12}},$$

or a linear combination of such expressions: g_5^2 is a universal constant while q_5 would
depend on the body considered. Models envisage q_5 depending on baryon number,
lepton number, or both, etc... It follows that this new force would not obey the weak
equivalence principle, so that Eötvös-type experiments [see *Chapter 7*] would constrain
its intensity. Moreover, the existence of a preferred length r_0 shows that Newton's law
would be modified at this scale. This is the reason for the many verifications of this
law recently. More details are contained in the article by E.G. Adelberger *et al.* (1990)
cited above.

Another possibility is that the law of universal attraction might hold, but that the
gravitational constant G might have varied over time. However, such a variation, which

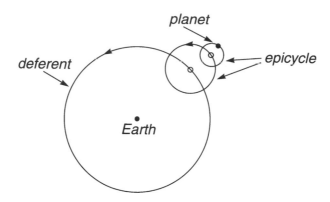

Fig. 1.13. Planetary motion according to Ptolemy. The planets (only one is shown) rotate about the Earth in circles ("perfect figures"). The largest circle is the "deferent"; the smaller ones are "epicycles". To achieve an accurate representation of the observational data, the deferent is not centered exactly at the Earth and there are several epicycles.

would necessarily be very small, would have many consequences[28], such as a variation of the Earth's radius in the past, geophysical effects, or changes of the motions of satellites or of the Moon. [T.C. Van Flandern (1981)]. It appears that one of the most precise upper limits to this variation is provided by the motion of the *binary pulsar* PSR 1913+16 [T. Damour *et al.* (1988)]. In all cases it is found that the variation is consistent with a zero value, and in any case

$$\frac{\delta \log G}{\delta t} < 10^{-11} \text{ year}^{-1}.$$

1.9   Limits of the Newtonian theory of gravity

Newton's laws of motion and of universal gravitation provided a unique and rational theory for describing the motions of the planets and their satellites, a problem that had been unsolved for two thousand years. There was no need to use complicated and completely *ad hoc* epicycles to describe their motion and make predictions; only the $1/r^2$ law was needed, constituting a complete revolution.

Nevertheless, towards the middle of the nineteenth century a few discrepancies began to surface (such as the residual perihelion[29] advance of Mercury by 38″ per century noted by Leverrier in 1850) which were measured more precisely by Newcomb at the end of the century (42″9 for Mercury, 8″ for Mars, 10″ for Venus, etc.).

We make this rather more precise. A single planet moves about the Sun in an ellipse, whose perihelion is an *a priori* fixed point. Because of the weak perturbations resulting from various effects (attraction of the other planets, possible solar oblateness, precession

---

[28] See *Chapter 7*.
[29] The perihelion is the point of the planet's orbit closest to the Sun.

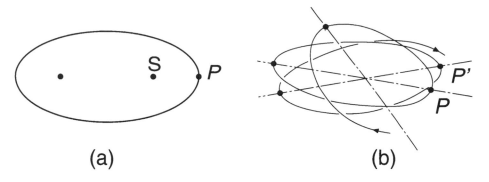

Fig. 1.14. Perihelion advance of a planet. (a) P is the perihelion and S the Sun; (b) advance of the perihelion (exaggerated) to successive positions $P, P', \ldots$

of the Earth, etc.), the real trajectory is a kind of rosette (**Fig. 1.14**) corresponding to an ellipse whose major axis slowly rotates in its plane. Naturally the perihelion rotates also.

> In classical mechanics one can show that the only central forces which give rise to *closed* test particle trajectories are of the forms $F \propto 1/r^2$ or $F \propto r$. Thus, any perturbation, however small, of the $1/r^2$ law produces a perihelion advance (or retardation). The ellipse "opens" and rotates slowly. (**Fig. 1.14**).

For example, we have

$$\text{Mercury}: \quad 42''56 \pm 0''94$$
$$\text{Venus}: \quad 8''4 \pm 4''8$$
$$\text{Earth}: \quad 4''6 \pm 2''7$$

for the residual perihelion advances, left *unexplained* in Newtonian theory. For example, for Mercury there is a perihelion advance of $532''$ per century caused by planetary perturbations and a possible solar oblateness, the remainder of the advance ($5067''$ per century) resulting from the precession of the equinoxes ($50''26$ per century along the ecliptic). The residual advance left unexplained by Newtonian theory is certainly significant even after taking account of observational error [G.M. Clemence (1947)].

Although similar discrepancies exist for the motion of the Moon or comet Encke, classical explanations can be found (slowing of the Earth's rotation caused by the tides affects the Moon's motion; a perturbing meteor stream influences the comet). The main discrepancies with Newtonian theory occur in the motion of the planets, particularly Mercury.

There were several attempts to remove these disagreements. These were either too arbitrary or in contradiction with other data: for example a belt of minor planets, Seeliger's proposal concerning the zodiacal light, modifications of the law of universal attraction, non-Euclidean geometry of space, etc.

Only relativistic theories (of which Einstein's is currently the most plausible as well

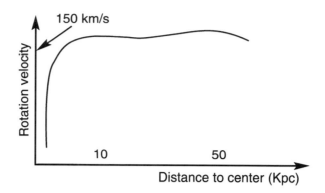

Fig. 1.15. Rotation curve of a spiral galaxy. Note the characteristic plateau at large distances.

as being historically the first) give a proper solution to this problem, i.e. one which is not simply *ad hoc* or in disagreement with other data.

Newtonian gravitation has recently been questioned again [Milgrom, Sanders, etc.; *see* J. Bekenstein (1988) for a review and further references] in connection with the *rotation curves* of spiral galaxies (see **Fig. 1.15**). At large distances these show a characteristic flattening, whereas one would expect a decrease of velocity with distance.

> Consider a rotating spiral galaxy, like ours. A corotating test star at distance r from the centre has
>
> $$mv^2 = \frac{GmM}{r},$$
>
> where m is the mass of the test star and M is the total mass of the galaxy. Newton's law thus cannot explain the characteristic plateau in all rotation curves, as $v^2 \approx 1/r$.

It is possible that the presence of "dark matter"[30] or "missing mass" in the form of exotic particles (unusual ions, strange matter, etc.) might explain this plateau without the need to modify the law of universal attraction. However, the dark matter has not so far been identified.

1.10   The finiteness of the velocity of light

The discovery that the velocity of light is finite was a very important part of the origins of relativity theory. It was not at all obvious in the 17th century that this velocity was finite. For example, Descartes had a *"more than moral certainty"* that light propagated instantaneously, contrary to Père Mersenne[31] or Galileo, who tried to measure its velocity.

The first measurement (see **Fig. 1.16**) of the velocity of light was by the Danish astronomer O. Römer (1675) who was observing the occultation period[32] of Io, one of

---

[30] Or "invisible matter" or "hidden matter", *etc.*
[31] …who was the first to measure the velocity of sound.
[32] He observed the date of the occultation. Römer first used Cassini's observations.

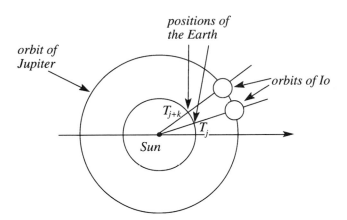

Fig. 1.16. Römer's measurement of the occultation period of Io (one of the satellites of Jupiter). T_j and T_{j+k} denote the positions of the Earth after j and $j+k$ orbits of Io.

the satellite of Jupiter (about 42 hours)[33]. He noticed that over six months this period, which should have been constant, lengthened by about 1300 seconds. Io should have reappeared at regular intervals, but was later and later as compared with predictions. Only one inference could explain this: Römer pointed out that over six months the Earth–Jupiter distance varied by roughly the diameter of the Earth's orbit about the Sun (Jupiter's period is about 12 years, and moves only about 12° in six months). The observed time delay corresponded to the extra time needed for light to travel this distance (the diameter of the Earth's orbit is of order 3×10^{13} cm). Later, many other methods were used.

The basis of Römer's measurement is as follows: an observer on Earth measures the time intervals δt_k between k consecutive reappearances of Io from the shadow of Jupiter (see **Fig. 1.16**). Let $\Delta_k \ell$ be the change of distance between the Earth and Jupiter in time δt_k. Assuming that the velocity of light c is finite, we have

$$\delta t_k = k T_0 + \frac{\Delta_k \ell}{c}, \tag{1.34}$$

where T_0 is the occultation period of Io. This equation has two unknowns, T_0 and c. T_0 can be found as follows. If after m consecutive occultation periods T returns to its initial value, then $\Delta_m \ell = 0$, so $\delta t_m = m T_0$. The quantities δt_m and m are directly observable. To find c, we have to observe the number n of reappearances of Io occurring between two diametrically opposed positions of the Earth in its orbit. If D is the diameter of this orbit, we then have $\Delta_n \ell = D$, and thus

$$c = \frac{D}{\delta t_n - n T_0} = \frac{D}{\delta t_n - \frac{n}{m} \delta t_m}; \tag{1.35}$$

we note that this equation contains only directly observable quantities. With the values known to him, Römer found $c \sim 200\,000$ km/s, which is of the correct order of magnitude.

---

[33] Galileo's invention of the astronomical telescope in 1610 allowed him to discover four of the satellites of Jupiter.

This lengthening of the occultation period [M. Jammer (1979)] is a cumulative effect that can be interpreted as a kind of Doppler effect. To see this, consider the Earth at times six months apart; during this time Jupiter has moved very little in its orbit. If T is the measured period and v is defined by

$$v \equiv \frac{D}{T},\tag{1.36}$$

then setting $v = T^{-1}$ and $v_0 = T_0^{-1}$, the equation

$$c = \frac{D}{t_n - nT_0}\tag{1.37}$$

can be rewritten as

$$v = v_0(1 - \beta),\tag{1.38}$$

with $\beta \equiv v/c$. This is the Doppler formula for a *fictitious* Earth travelling along the axis $T_1 T_2$ (see **Fig. 1.17**).

The corresponding data are: $D \approx 2 \times 10^{13}$ cm; $T_0 \approx 42$ hours; period of Jupiter ≈ 12 yrs; change of period in 1675 $\approx 22'$ (today $\approx 16'$).

With hindsight, the importance of the finite velocity of light is very clear. If this velocity were infinite, Maxwell's equations

$$\begin{cases} \nabla \cdot \mathbf{E} = 0, & \nabla \cdot \mathbf{B} = 0 \\[2mm] \nabla \wedge \mathbf{E} = -\dfrac{\partial \mathbf{B}}{\partial t}, & \nabla \wedge \mathbf{B} = \dfrac{1}{c^2}\dfrac{\partial \mathbf{E}}{\partial t}, \end{cases}\tag{1.39}$$

would not contain the "displacement current" $(-\partial \mathbf{E}/\partial t)$ and the electromagnetic equations would be invariant under the Galilei group. They would then also satisfy the Galilean relativity principle, keeping the same form in two arbitrary Galilean reference frames, and one could not detect their relative motion by electromagnetic means. One can show that a non-relativistic limit of Maxwell's equations is found when the velocity of light is increased without limit, which is equivalent to suppressing the displacement current [M. Le Bellac and J.M. Lévy–Leblond (1973)][34].

Thus, Maxwell's equations are not invarient under the Galilei group, so it must be possible, at least in principle, to demonstrate the relative motion of two inertial reference systems by electromagnetic means alone.

Maxwell's equations (1.39) easily give

$$\Box \mathbf{E} = 0, \quad \Box \mathbf{B} = 0,\tag{1.40}$$

where

$$\Box \equiv \frac{1}{c^2}\frac{\partial^2}{\partial t^2} - \Delta,\tag{1.41}$$

which reduce to

$$\Box \varphi = 0,\tag{1.42}$$

---

[34] See however this article for various subtleties involved in taking the limit.

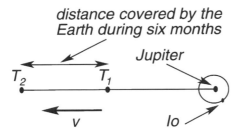

Fig. 1.17. The observed variation of the occultation period of Io can be interpreted as a kind of Doppler effect for a fictitious Earth moving along the straight line $T_1 T_2$ every six months.

in a given inertial system. In another inertial system, we can write

$$\varphi(\mathbf{x}, t) \equiv \varphi\left[g^{-1}(\mathbf{x}, t)\right]$$

where g denotes a transformation of the Galilei group, and we have used the same symbols for the space and time coordinates in the two inertial systems. Put another way, the function φ and its transform can be regarded as elements of the same functional space. Thus, limiting ourselves to pure Galilean transformations – which in fact is no restriction on the generality of the argument –

$$\mathbf{x}' = \mathbf{x} - \mathbf{v}t, \quad t' = t, \tag{1.43}$$

we have

$$\varphi'(\mathbf{x}', t') \equiv \varphi\left(\mathbf{x}' + \mathbf{v}t', t'\right), \tag{1.44}$$

or, using the same notation for the variables in the two inertial systems,

$$\varphi'(\mathbf{x}, t) \equiv \varphi\left(\mathbf{x} + \mathbf{v}t, t\right). \tag{1.45}$$

Noting that we can also write

$$\varphi(\mathbf{x}, t) \equiv \varphi'\left(\mathbf{x} - \mathbf{v}t, t\right), \tag{1.46}$$

the wave equation $\Box \varphi = 0$ becomes

$$\left\{ \frac{1}{c^2} \frac{\partial^2}{\partial t^2} + \left(\frac{\mathbf{v} . \nabla}{c} \right)^2 - 2 \frac{\mathbf{v} . \nabla}{c} \frac{1}{c} \frac{\partial}{\partial t} - \Delta \right\} \varphi'(\mathbf{x}, t) = 0 \tag{1.47}$$

where we have used the fact that

$$\frac{\partial}{\partial t} \varphi(\mathbf{x}, t) = -\mathbf{v} . \nabla \varphi'(\mathbf{x} - \mathbf{v}t, t) + \frac{\partial}{\partial t} \varphi'(\mathbf{x} - \mathbf{v}t, t). \tag{1.48}$$

The wave equation thus has different forms in two inertial systems moving with respect to each other with velocity $\mathbf{v}$.

Maxwell's identification of light waves with electromagnetic waves appeared to require the existence of a medium for their propagation: the ether, a concept due to Maxwell. At the end of his life (1879) he discussed Römer's measurement and noted that the same measurements made at intervals of six years (i.e. when Jupiter had travelled around one half of its orbit) could be used to demonstrate the motion of the solar system through the hypothetical ether.

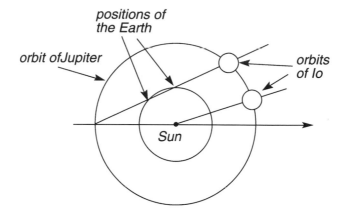

positions of
the Earth

orbit of Jupiter

orbits
of Io

Sun

Fig. 1.18. Römer's measurement at an interval of six years. This would in principle allow a demonstration of the global motion of the solar system with respect to the ether.

If v is the velocity of the solar system with respect to the ether and D is the diameter of the Earth's orbit, the delay in the reappearance of Io (after occultation by Jupiter) is

$$\delta t = \frac{D}{c+v}. \tag{1.49}$$

By contrast, six years later (see **Figs. 1.17** and **1.18**)

$$\delta t' = \frac{D}{c-v}, \tag{1.50}$$

so that the difference in the two delays, which is in principle observable, is

$$\begin{aligned}
\Delta t &\equiv \delta t' - \delta t \\
&= D\left[\frac{1}{c-v} - \frac{1}{c+v}\right] \\
&= \frac{2D}{v}\frac{\beta^2}{1-\beta^2}
\end{aligned} \tag{1.51}$$

where $\beta \equiv v/c$. This effect is second order in v/c and is thus very small, of order $(30/300\,000)^2 \approx 10^{-8}$! One consequence is that Römer's measurement gave a reasonably correct estimate of the velocity of light, given the then current accuracy of astronomical observations.

1.11   Michelson's experiment

At the end of the 19th century one of the major problems of physics was to demonstrate the existence of the ether. At the beginning of the century, physicists such as Arago (1818), Fresnel (1818), Fizeau (1852), etc. had tried to find *first order* effects in v/c[35] without success. More sensitive second-order experiments were therefore required: this

[35] See M.A. Tonnelat (1971).

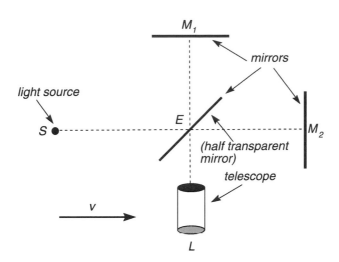

Fig. 1.19. The Michelson interferometer (1881). The beam **SE** is split in two (**EM₁** and **EM₂**) by the semitransparent mirror **E**. The reflected beams **M₁ E** and **M₂ E** are recombined at **EL** and interfere because of their different path lengths, which is observed at **L**.

was the aim of the famous experiment by Abraham Michelson (1881), repeated by Michelson and Morley (1887), and perfected by others.

We note in passing that Römer's measurement was incapable of demonstrating second-order effects, given the inaccuracy of astronomical data.

Michelson's interferometer consists essentially of a light source **S** (see **Fig. 1.19**) which projects a light beam on to a semitransparent mirror **E**, which splits the incident beam into two orthogonal beams, which are in turn reflected by two mirrors **M₁** and **M₂** onto a telescope **L**. The path difference between the two beams caused by the unequal lengths $\mathbf{EM_1} = \ell_1$ and $\mathbf{EM_2} = \ell_2$ makes the beams interfere in a way that also depends on the motion of the apparatus through the ether. If the apparatus is rotated through 90° so as to exchange the roles of ℓ_1 and ℓ_2, there should be a displacement of the interference fringes caused by the global motion of the apparatus.

We examine this more closely. Let us first orient the apparatus so that the arm **EM₂** is parallel to the motion. The time for light to travel along **EM₁E** is

$$t_1 = \frac{\ell_1}{c+v} + \frac{\ell_1}{c-v} = \frac{2\ell_1}{c}\frac{1}{1-\beta^2}. \tag{1.52}$$

By contrast, the time for the second beam to travel along **EM₁E** is

$$t_2 = \frac{1}{c}\left[\mathbf{EM_1} + \mathbf{M_1E'}\right], \tag{1.53}$$

for in time t_2 the mirror has moved to **E'** (see **Fig. 1.20**), through a distance

$$\mathbf{EE'} = vt_2, \tag{1.54}$$

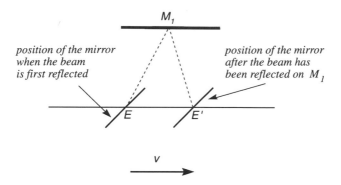

Fig. 1.20. Ray paths in the Michelson interferometer. One of the arms moves with velocity v with respect to the ether.

so that

$$\mathbf{EM_1 + M_1E'} = 2\left[\ell_1^2 + \frac{(vt_1)^2}{4}\right]^{1/2} = \left[4\ell_1^2 + v^2t_2^2\right]^{1/2}. \tag{1.55}$$

Finally we get

$$t_2 = \frac{2\ell_2}{c}\frac{1}{\sqrt{1-\beta^2}}. \tag{1.56}$$

The difference in path length of the two beams is thus

$$\Delta t\,(\ell_1,\ell_2;\beta) = t_2 - t_1 = \frac{2}{c}\left[\frac{\ell_2}{\sqrt{1-\beta^2}} - \frac{\ell_1}{1-\beta^2}\right]. \tag{1.57}$$

To remove inaccuracies caused by ℓ_1 and ℓ_2, we now rotate the apparatus through 90°: this is equivalent to exchanging ℓ_1 and ℓ_2. We thus get a displacement of the fringes corresponding to the difference

$$\delta t = \Delta t\,(\ell_1,\ell_2;\beta) - \Delta t\,(\ell_2,\ell_1;\beta) \tag{1.58}$$

$$= \frac{2\,(\ell_1 + \ell_2)}{c}\left[\frac{1}{1-\beta^2} - \frac{1}{\sqrt{1-\beta^2}}\right]$$

$$\approx \frac{(\ell_1+\ell_2)}{c}\beta^2. \tag{1.59}$$

If the source emits at wavelength λ we will observe a displacement of the fringes when

$$\delta t \approx \frac{\lambda}{c}, \tag{1.60}$$

and hence when

$$\ell_1 + \ell_2 \approx \lambda\beta^{-2}. \tag{1.61}$$

Thus, for radiation at 5μm and $v \approx 30$ km/s we find $\ell_1 + \ell_2 \approx 50$ m: such a distance can be obtained by using multiple reflections.

Michelson's experiment (1881), as noted by F. Everitt (1975), was very sensitive to thermal vibrations and distortions of the arms EM_1 and EM_2 of the apparatus: the problem is essentially a *"problem of mechanical stability. A measurement to a twentieth of a fringe corresponds to a 130 Å displacement of any mirror. A steel I-beam 100 cm long and 4 cm deep deflects under its own weight by 1 arc-second, which means that a 2 cm mirror resting on it is tilted back at its upper edge by 2000 Å through beam flexure. Low frequency seismic vibrations of amplitude 10^{-3} g then cause motions of 2 Å and vibrations are even more serious. A temperature difference of 0.02° across the beam warps it by 1 arc-second. Such were the effects Michelson and Morley had to contend with. When, as in the Stanford Gyro Relativity experiment, one aims for measurements of 0.001 arc-second accuracy, questions of mechanical stability shape the whole experimental approach."* [F. Everitt, W.W. Hansen 1975]. For this reason Michelson repeated his experiment with Morley (1887) so as to limit this source of error.

The Michelson–Morley experiment could have detected an "ether wind" of the order of 10 km/s, well below the 29 km/s due to the motion of the Earth in its orbit. However its result was *completely negative* at whatever time of year or altitude it was performed. Later experiments have improved the accuracy to 1.5 km/s and even beyond that in the latest versions [J.P. Cedarholm *et al.* (1958); A. Brillet and J.L. Hall (1979, 1982)].

This negative result was a major problem for the conceptual basis of Newtonian physics which various *ad hoc* hypotheses failed to solve. Only Einstein was able to resolve these difficulties, at the end of a penetrating analysis of the notions of space and time.

Exercises

1.1 Use a system of units in which $\hbar = c = 1$ and the MeV is the energy unit.

(i) What units are the following expressed in: time, length, mass, momentum, pressure, particle density, mass density, temperature, Boltzmann's constant, area, electric charge?

(ii) Same questions, if instead of adopting an energy unit, we adopt a length (the fermi; 1 fm $= 10^{-13}$ cm). In what units is an energy expressed?

(iii) Same questions as in (i) and (ii), this time adopting the second as a unit.

1.2 Verify that the Galilean transformations have six parameters and form a group.

1.3 If $g \equiv \{\mathscr{R}, \mathbf{x}_0, t_0, \mathbf{v}\}$ is a Galilean transformation, where $\mathscr{R}$ is a rotation, what are (i) the identity transformation, (ii) g^{-1}, (iii) $g_1 \cdot g_2$?

1.4 Why must the transformations between inertial reference frames form a group?

1.5 Verify explicitly that the Galilean transformations conserve the various structures (affine, metric, causal) of Newtonian space–time $\mathbf{R}^{3+1}$.

1.6 Consider two particles of masses m_1 and m_2, interacting via the potential $V(|\mathbf{x}_1 - \mathbf{x}_2|)$, where $\mathbf{x}_i$ is the position vector of particle i. Write down the Schrödinger equation for this system and show that it is invariant under the Galilei group.

1.7   Consider the sphere S^3 of unit radius, embedded in $\mathbf{R}^4$:

$$x^2 + y^2 + z^2 + w^2 = 1.$$

Given that $\mathbf{R}^4$ is Euclidean (i.e. $ds^2 = dx^2 + dy^2 + dz^2 + dw^2$), calculate the metric induced on the sphere S^3 by that of $\mathbf{R}^4$.

1.8   Consider a particle of mass m acted on by a central force derived from a potential $V(r)$.

(i) Recall why the motion is in a plane. Denote by $\mathbf{J}$ the angular momentum of the particle.

(ii) In the plane of the motion, use polar coordinates (r, φ). Show that the Lagrangian of the system is given by:

$$L = \frac{1}{2}m[\dot{r}^2 + r^2\dot{\varphi}^2] - V(r).$$

Write down the equations of motion and deduce Kepler's second law.

(iii) Show that the energy of a particle can be put in the form

$$E = \frac{1}{2}m\dot{r}^2 + \frac{\mathbf{J}^2}{2mr^2} + V(r).$$

(iv) Discuss the nature of the trajectories given by various values of E for the case $V(r) = -\text{const.}/r$.

(v) In which cases are the trajectories ellipses? Show that the parameter p and the eccentricity e of the ellipses are given by

$$p = \frac{\mathbf{J}^2}{m \times \text{const.}} \quad, \quad e = \sqrt{1 + \frac{2E\mathbf{J}^2}{m(\text{const.})^2}}.$$

1.9   Let $\rho(\mathbf{x})$ be the mass density of a body of total mass M.

(i) Using the formal solution of Poisson's equation for the gravitational potential $\varphi(\mathbf{x})$, show that the latter quantity can be written in the form

$$\varphi(\mathbf{x}) = -\frac{GM}{r} - \frac{G}{2}\sum_{i,j} Q_{ij}\frac{x^i x^j}{r^5} + \cdots$$

(multipole expansion)

where the coordinates $x^i (i = 1, 2, 3)$ have origin at the centre of gravity of the mass distribution. Q_{ij} is a 3×3 matrix, the quadrupole tensor.

(ii) How is Q_{ij} modified if the origin of coordinates is shifted from the centre of gravity to an arbitrary point? What happens to the previous expression for φ?

(iii) What is $\varphi(\mathbf{x})$ when $\rho(\mathbf{x})$ is spherically symmetric?

(iv) What is $\varphi(\mathbf{x})$ for a body, all of whose multipole moments vanish *except* the quadrupole term? Give the corresponding expression for the gravitational force.

(v) *Application*: The Sun has a rotation period of $T \approx 25$ days, its equatorial diameter is slightly larger than its polar diameter (it is therefore slightly *oblate*), and thus has a nonzero quadrupole moment. Explain why. Next, *assume that all the multipole moments of the Sun beyond the quadrupole moment vanish*

(a) What are the nonzero components of Q_{ij} if the Sun is axisymmetric?

(b) Consider a massive body moving in the Sun's equatorial plane, orthogonal to its rotation axis. Give an expression for the force produced by the Sun acting on this body. What is the associated potential energy? What effect does this produce on the body (qualitatively)?

(c) Setting

$$J_2 = -Q_{33} / \left(2 M_\odot R_\odot^2 \right),$$

and we are given that

$$J_2 = 2.5 \times 10^{-5},$$

$$M_\odot = 2 \times 10^{33} \text{g},$$

$$R_\odot = 7 \times 10^{10} \text{cm},$$

evaluate the order of magnitude of the perihelion advance of Mercury resulting from this effect (consider the energies involved). The data for Mercury are

perihelion distance: 4.6×10^{12} cm
period: 0.24 years,
eccentricity of the orbit: 0.21.
Compare with the result given by classical mechanics [cf. L. Landau, E. Lifschitz, *Mechanics* (1960)].

1.10   Consider a planet of mass m acted on by the potential $V(r) = -GM_\odot m/r$. It moves in an ellipse. This motion is now perturbed by a "small" potential $\varepsilon(r)$: "small" means

$$\left(\frac{\varepsilon(r)}{V(r)} \right)^2 \ll \left(\frac{\varepsilon(r)}{V(r)} \right).$$

The planet is no longer subject to a $1/r$ potential and the trajectory opens slightly: the planetary perihelion varies by a small amount $\delta\varphi$ on each revolution. This exercise aims to calculate this quantity.

(i) Integrating the conserved angular momentum and, using conservation of energy, show that

$$\Delta\varphi \equiv (\varphi - \varphi_0) = \int \frac{J \, dr}{r^2 \{ 2m[E - V(r) - \varepsilon(r)] - J^2/r^2 \}^{1/2}},$$

where r varies from r_{min} to r_{max}, the zeros of the denominator. Why is $\delta\varphi = 2\Delta\varphi$?

(ii) Show that we also have

$$\delta\varphi = -2\frac{\partial}{\partial J}\int_{r_{min}}^{r_{max}} dr\left\{2m[E - V(r) - \varepsilon(r)] - \frac{J^2}{r^2}\right\}^{1/2}.$$

(iii) Thus calculate $\delta\varphi$, the first-order variation of the perihelion in $\varepsilon(r)$.

(iv) *Application*: (a) $\varepsilon(r) = \beta/r^2$, (b) $\varepsilon(r) = \gamma/r^3$.

[Reference: L. Landau, E. Lifschitz, *Mechanics* (1960).]

1.11   Consider a planet in a circular orbit about the Sun, and *assume* that the gravitational constant G varies very slowly over time. Let R be the radius of the planet's orbit and Ω its angular velocity.

(i) What does "very slowly" mean? Make this precise.

(ii) Show that the radius of the orbit varies as

$$\frac{1}{R}\frac{dR}{dt} = -\frac{1}{G}\frac{dG}{dt},$$

while the angular velocity of the planet obeys

$$\frac{1}{\Omega}\frac{d\Omega}{dt} = \frac{2}{G}\frac{dG}{dt}.$$

(iii) We assume that $G^{-1}dG/dt \sim 10^{-11}$ per year. Find the variation of the radius of the Earth's orbit in 10^4 years, 10^6 years and 10^8 years, and the variation of its rotation period about the Sun.

(iv) After 25 years of measurements of the reflection of a laser beam from the surface of the Moon, the Earth–Moon distance is known to better than one centimetre. How much time would be needed until one could hope to demonstrate a variation $|\Delta G/\Delta T.G|$ of order 10^{-11} ?

1.12   Consider masses m_1, m_2 in circular orbits about each other with angular velocity ω. Calculate the quadrupole moment of this system.

1.13   Same question as the preceding one, for the case $m_1 \gg m_2$, with m_2 having an elliptical orbit resulting from the gravitational interaction.

1.14   Consider two equal masses m, interacting via a force $F = -k|x_1 - x_2|$, and moving along the x axis. Calculate the quadrupole moment of this system.

2 Minkowski Space–Time

We have already indicated the main experimental facts leading to the invention of the Special Theory of Relativity: measurement of a finite velocity of light, the inability to demonstrate the existence of the ether by first-order effects, and finally the inability to do this to the second order in v/c (experiments by Abraham Michelson). As well as these experimental facts, theoretical discussions concerning the ether, the non-invariance of Maxwell's equations [H. Poincaré (1904)] under the Galilei group, Lorentz's electron theory [see D. Bohm (1965)], and the critiques of E. Mach of the foundations of classical mechanics profoundly influenced the development of Einstein's ideas [see S. Feuer, (1978)].

We should however be clear that classical (i.e. prerelativistic) ideas were not abandoned before all the possibilities open at the end of the 19th century were applied to try and save them. For example, Fitzgerald and Lorentz independently proposed to explain the negative result of the Michelson–Morley experiment in terms of a contraction of the interferometer arm moving parallel to the velocity; this was to be an effect of the ether:

$$\ell = \ell_0 \sqrt{1 - v^2/c^2}. \tag{2.1}$$

However, although the Fitzgerald–Lorentz contraction is indeed a possible explanation, such a *physical* effect of the ether would imply that the refractive index of a solid, the resistance of a conducting wire, the vibration frequency of a bar of quartz, etc. would all be modified by motion with respect to the ether. Experiments were carried out [see M.A. Tonnelat (1971)] with negative results. To explain these, Lorentz was led to a further hypothesis, namely that the mass of a body must vary with its velocity with respect to the ether:

$$m = \frac{m_0}{\sqrt{1 - v^2/c^2}}. \tag{2.2}$$

Lorentz then showed that clocks would be *slowed* in the ratio

$$T = \frac{T_0}{\sqrt{1 - v^2/c^2}}, \tag{2.3}$$

$(T_0 = \text{clock period})$

which with the length contraction (2.1) led easily to a light velocity independent of that of the observer with respect to the ether. However, now that clocks were slowed and lengths contracted, the idea of simultaneity required delicate, not to say artificial analysis, which led to the *Lorentz transformations* independently found by Voigt and by Poincaré:

$$\begin{cases} \mathbf{x}' = \dfrac{\mathbf{x} - \mathbf{v}t}{\sqrt{1 - v^2/c^2}} \\[2ex] t' = \dfrac{t - \mathbf{v}.\mathbf{x}/c^2}{\sqrt{1 - v^2/c^2}}. \end{cases} \tag{2.4}$$

These transformations leave Maxwell's equations *invariant*, and the velocity of light constant. *Lorentz's theory thus explained the various experimental results.* On the other hand, by distinguishing the "true" time and "true" distances (i.e. those relative to the ether) from time and distance relative to the observer (moving with respect to the ether), Lorentz's theory contained a fundamental ambiguity, which once again could not be resolved by experiment, precisely because the ether remained *unobservable*! However, even though it was unobservable, it retained a central role, particularly in Lorentz's (1904) (and also Poincaré's) interpretation of the transformation (2.4), which was completely different from that of Einstein (1905).

It was the rejection of the *unobservable* ether and the critique of the notion of simultaneity, a notion which had been self-evident and unproblematical for prerelativistic physicists, which led Einstein to Special Relativity (1905). This theory was soon recognised [H. Minkowski (1908)] as a *theory of the mutual connection of space and time*.

The theory is based on the two postulates

  (i) In a vacuum, light propagates with constant velocity c in all Galilean reference frames, independently of the motion of the source and the observer.

  (ii) No internal physical experiment can demonstrate the motion of a Galilean reference frame with respect to another Galilean frame.

The first postulate – that of the constancy of the velocity of light – explained a whole series of experiments and measurements [see M.A. Tonnelat (1971)], which we shall not refer to here. The second postulate is the *extension* of the Galilean relativity principle to *all* physical phenomena, including electromagnetic ones, not simply mechanical systems. This fundamental postulate, whose *physical* aspect we have discussed, also has a *descriptive* aspect:

  (ii′) The laws of physics have the same form in all inertial frames.

  This descriptive aspect is easy to understand by noting that different forms in two inertial frames would imply a physical effect, and thus a means of detecting their relative motion. Thus, if Newton's second law $F = m\gamma$ holds in one inertial frame, and if in a second frame it takes the form

$$F = m\gamma + \Gamma,$$

  the term Γ could be interpreted as an extra force which could in principle be measured.

The most immediate consequence of these postulates is that the transformations from one inertial system to another must preserve the form of the equations of electromagnetism, i.e. Maxwell's equations, as optical experiments are far more accurate than those on mechanical systems. It will follow that the equations of Newtonian

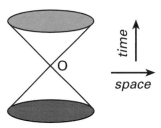

Fig. 2.1. The light cone in Minkowski space.

mechanics (invariant under the Galilei group rather than the Lorentz group) will not have the same form in different Galilean systems: they must therefore be modified so as to satisfy Einstein's relativity principle **(ii′)**. We must then check *experimentally* that the consequences of these new equations agree with what is observed in nature.

2.1   The space–time of special relativity

The existence of a universal velocity, that of light, must profoundly change the structure of space–time, as well as providing it with a *metric* (or more precisely a pseudo-metric[1]) defined by a quadratic form, and changing the nature of the *geometric objects* intrinsically attached to it. This four-dimensional space–time is called *Minkowski space–time* (1908), and we shall denote it by $\mathcal{M}$ (the designation is often shortened to *Minkowski space*).

Consider two inertial systems characterised by space–time coordinates $(t, \mathbf{x})$ and $(t', \mathbf{x}')$, and the propagation of a spherical light wave emitted from a point **O** of $\mathcal{M}$. We assume (without loss of generality) that at $t = t' = 0$ the two inertial systems coincide [i.e. the origin **O** is the same for both systems: $(t = 0, \mathbf{x} = \mathbf{0})$ and $(t' = 0, \mathbf{x}' = \mathbf{0})$ represent the same point of $\mathcal{M}$].

In the first reference system the light wave is at distance $|\mathbf{x}|$ from **O** at time t, i.e.

$$c^2 t^2 - \mathbf{x}^2 = 0. \tag{2.5}$$

In the second inertial system, the *same* light wave also appears spherical because of Einstein's relativity principle **(ii)** and, moreover, its velocity is also c by postulate **(i)**. It follows that in this system the distance travelled by the light wave after time t' satisfies

$$c^2 t'^2 - \mathbf{x}'^2 = 0. \tag{2.6}$$

What can we deduce from this? We note first that equation (2.5) [or (2.6)] defines a *three-dimensional* manifold in $\mathcal{M}$, i.e. a three-dimensional surface, which is a *cone* with origin **O**. This is called the *light cone* **(Fig. 2.1)** or *null cone*. The formal identity of (2.5) and (2.6) – one obtained from the other by coordinate changes associated with the change of inertial system – shows that the light cone stays the same, i.e. is *invariant*

---

[1] In fact – as we shall see below – the metric is defined by a quadratic form which is not positive definite. It would actually be better to speak of "providing space–time with a Lorentzian structure".

under changes of Galilean reference system. Two inertial observers will find that light propagates in the same way for each of them: the light cone is a geometric object of Minkowski space–time.

Further, as the properties of spatial homogeneity and uniformity of the flow of time survive, each point of $\mathcal{M}$ can be regarded as the vertex of a light cone.

> The light cone has two sheets: the *future light cone*, representing the expansion of a spherical light wave into space from **O**, and the *past light cone*, generated by the the the set of light rays reaching **O**. We can also regard the future light cone as generated by all possible light rays that can be emitted at **O**: the generators of the cone are the lines with equation
>
> $$\mathbf{x} = \mathbf{n}ct \qquad\qquad (2.7)$$
>
> where **n** is any unit vector ($\mathbf{n}^2 = 1$) of $\mathbf{R}^3$.

The light cones of space–time will play a considerable role in the development of relativity. In particular, they will allow us to give $\mathcal{M}$ a *metric*, as we shall see below. However, besides the light cones, there exists another class of geometric objects, representing inertial motions, i.e. *straight lines*. We shall, however, see that, unlike in the Newtonian case, not all straight lines correspond to observable inertial motions, but only those inside a light cone.

> If we consider cartesian coordinates $(t, \mathbf{x})$ in $\mathcal{M}$ (for clarity **Fig. 2.2** shows only one space dimension), the straight line $\mathbf{x} = 0$, i.e. the axis **O**t, represents an inertial motion $\{\mathbf{K}\}$. This line appears on **Fig. 2.2** as *a symmetry axis* of the cone with equation $c^2t^2 - \mathbf{x}^2 = 0$. Another inertial system $\{\mathbf{K}'\}$ moving with velocity v with respect to $\{\mathbf{K}\}$ is represented by a line which *on the figure* does not appear as a symmetry axis of the light cone. Does this violate the principle of relativity, as $\{\mathbf{K}\}$ and $\{\mathbf{K}'\}$ appear to have different properties in their relations to the null cone? Clearly it does not. It is our Euclidean representation of a non-Euclidean space which gives this appearance. This is inevitably inadequate and cannot convey *all* the properties of $\mathcal{M}$. In fact the line $\{\mathbf{K}\}$ as well as $\{\mathbf{K}'\}$, or indeed any other line through the vertex of the cone and lying in $\{\Gamma\}$ is a symmetry axis. For this reason, and by analogy with the case of the usual sphere, for which any line passing through the centre is a symmetry axis, the light cone is sometimes called a *pseudosphere* [J.L. Synge (1958)].

As a consequence of the principle of constancy of the velocity of light and Einstein's relativity principle, 3-planes do not appear. In Minkowski space the *only* geometric objects with a direct physical meaning, related neither to a definition nor a convention, are the light cones. Of course, we can foliate space–time with families of parallel 3-planes, all cutting a time axis. However, such a foliation is not only arbitrary but never necessary.

The most important consequence of this special structure of the space–time of Special Relativity is that it induces a *pseudo-Euclidean* metric, which can be written[2]

$$ds^2 = c^2dt^2 - d\mathbf{x}^2. \qquad\qquad (2.8)$$

---

[2] The convention $ds^2 = d\mathbf{x}^2 - c^2dt^2$ is also found. Setting $x_4 = ict$, we can write $ds^2 = \pm\left[dx_1 + dx_2{}^2 + dx_3{}^2 + dx_4{}^2\right]$, the sign depending on the convention adopted. Linear transformations such as $ds^2 = ds'^2$ are then formally rotations of $\mathbf{R}^4$, with one purely imaginary "angle".

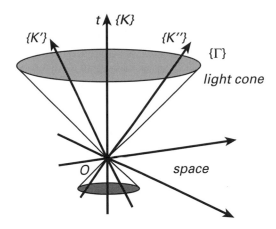

Fig. 2.2. Time axes of the light cone. The light cone has no privileged symmetry axis; any inertial motion – denoted by $\{K\}$ and $\{K'\}$ – can be regarded as a symmetry axis of the cone $\{\Gamma\}$.

This metric has the fundamental property of being *invariant* under changes of Galilean coordinates; put another way, if (t', x') denote new Cartesian coordinates attached to a new Galilean reference frame, we have

$$ds^2 = ds'^2 \tag{2.9}$$

or

$$c^2 dt^2 - d\mathbf{x}^2 = c^2 dt'^2 - d\mathbf{x}'^2. \tag{2.10}$$

The properties (2.9) and (2.10) hold for $ds^2 = 0$ as we have seen above ($ds^2 = ds'^2 = 0$ only expresses the constancy of the velocity of light). We now show that these relations are generally true. If we make a change of Cartesian coordinates $(t, \mathbf{x}) \rightarrow (t', \mathbf{x}')$ we find

$$ds^2 = f\left(t', \mathbf{x}'; \mathbf{v}\right) ds'^2 \tag{2.11}$$

where f is for the moment an arbitrary function of the space–time positions and relative velocity of the two inertial systems. In fact the homogeneity of space and the uniformity of the flow of time imply that f depends neither on t' nor on $\mathbf{x}'$. Further, if $\mathbf{v} = \mathbf{0}$ we must have $f(\mathbf{0}) = \text{const.} = 1$. If we now perform the inverse transformation, we must find

$$ds^2 = f(-\mathbf{v})f(+\mathbf{v})ds^2, \tag{2.12}$$

and we then have

$$f(-\mathbf{v})f(+\mathbf{v}) = 1. \tag{2.13}$$

But as f can only depend on $\mathbf{v}$ in the form $|\mathbf{v}|/c$, this implies that $f = 1$. In all of this reasoning we have implicitly assumed that all Galilean observers use the same units of time and length.

We note finally that the quadratic form (2.8) does not define a metric on $\mathcal{M}$ but only a pseudo-metric: ds^2 can have any sign, and moreover vanishes for *nonzero* dt^2 and $d\mathbf{x}^2$.

We shall also see below that the existence of this quadratic form, which is invariant under changes of inertial system, defines also the *causal structure* of $\mathcal{M}$. Moreover, it gives an absolute meaning to the *physical definition*, to Einstein's definition, of simultaneity. At each point, one can use electromagnetic waves only to observe events on the past light cone; and these events appear simultaneous.

2.2   The Lorentz transformation

There are many proofs of the Lorentz transformations, i.e. the formulæ used to pass from one inertial system to another while preserving the mathematical structure of space–time. The one we give here is not necessarily the simplest; it is however quite instructive.

We look for *linear* transformations leaving the light cones *invariant* [they must transform inertial motions into inertial motions; linearity can also be demonstrated using the homogeneity of space and the uniformity of time].

One can show that the most general transformations leaving the light cones invariant are the *conformal transformations*, which, besides the Lorentz transformations, include the following *nonlinear* transformations:

$$\begin{cases} t' = \left[t - \left(c^2 t^2 - \mathbf{x}^2 \right) a^0 \right] / N \\ \mathbf{x}' = \left[\mathbf{x} - \left(c^2 t^2 - \mathbf{x}^2 \right) \mathbf{a} \right] / N \\ N = 1 - 2 \left(c t a^0 - \mathbf{a} \cdot \mathbf{x} \right) + \left(a^{02} - \mathbf{a}^2 \right) \left(c^2 t^2 - \mathbf{x}^2 \right) \end{cases} \tag{2.14}$$

where $\mathbf{a}$ is a constant vector and where a^0 is a "constant"[3]. Similarly, *changes of scale* also preserve the light cone

$$t' = \lambda \mathbf{t}, \qquad \mathbf{x}' = \lambda \mathbf{x}. \tag{2.15}$$

All these transformations form a continuous group with 15 parameters.

Let us now set

$$X = \begin{vmatrix} ct \\ x \end{vmatrix}, \qquad X' = \begin{vmatrix} ct' \\ x' \end{vmatrix}, \tag{2.16}$$

(we consider henceforth a single space dimension, which greatly simplifies the calculations without altering the physics) and

$$G = \begin{vmatrix} 1 & 0 \\ 0 & -1 \end{vmatrix}. \tag{2.17}$$

We thus seek linear transformations L,

$$X' = LX, \tag{2.18}$$

where

$$L = \begin{vmatrix} \ell_{00} & \ell_{01} \\ \ell_{10} & \ell_{11} \end{vmatrix}, \tag{2.19}$$

---

[3] In fact it is the zeroth component of a four-vector (see below).

which preserve the light cone, whose equation [cf. Eqs. (2.5) and (2.6)], using the current notation, can be rewritten[4]

$$\begin{cases} X^T G X = 0, \\ X'^T G X' = 0. \end{cases} \tag{2.20}$$

The second of these equations can be rewritten again as

$$X^T L^T G L X = 0, \tag{2.21}$$

which must hold for all X satisfying the first of Eqs. (2.20). Under these conditions, it is not difficult to show that Eqs. (2.20) and (2.21) imply the matrix equation

$$L^T G L = \lambda G, \tag{2.22}$$

(where λ is a parameter specifying a change of units from one observer to another, and which we can always take equal to unity), characteristic of the transformations L that we seek; the latter relation, multiplied on the left by X^T and on the right by X, implies that

$$c^2 t^2 - x^2 = c^2 t'^2 - x'^2, \tag{2.23}$$

the equivalent equation to (2.9) [and (2.10)] implying the invariance of the metric on $\mathcal{M}$.

We note further that among the solutions of (2.22) we have

$$T = \begin{vmatrix} -1 & 0 \\ 0 & 1 \end{vmatrix}, \quad P = \begin{vmatrix} 1 & 0 \\ 0 & -1 \end{vmatrix} \tag{2.24}$$

representing respectively *time reversal* and *spatial reflection (parity)*.

The transformations of determinant $+1$ are called *proper* Lorentz transformations; the others are called *improper*. Similarly, those with $\ell_{00} > 0$ are called *orthochronous*, the others (i.e. those with $\ell_{00} < 0$) are called *antichronous*. As the identity is a proper transformation, all transformations continously related to the identity are also proper transformations. Orthochronous transformations have the property of leaving unchanged the direction of the flow of time as we pass from one inertial system to another.

We now limit ourselves to proper orthochronous transformations. To solve (2.22) (with $\lambda = 1$) we cast it in the form

$$L^T G = G L^{-1}, \tag{2.25}$$

which is a linear set of equations for the matrix elements of L, and can be solved easily, taking account of the fact that $\det L = +1$. The condition $\ell_{00} > 0$ allows us to put the matrices L into the form

$$L(\theta) = \begin{vmatrix} \text{ch } \theta & \text{sh } \theta \\ \text{sh } \theta & \text{ch } \theta \end{vmatrix}. \tag{2.26}$$

---

[4] Letting T denote the transpose.

The parameter θ must now be related to the relative velocity of the two inertial systems. We find

$$\mathrm{th}\,\theta = -v/c \tag{2.27}$$

The Galilean coordinate changes are

$$\begin{cases} t' = t\,\mathrm{ch}\,\theta + x/c\,\mathrm{sh}\,\theta \\ x' = x\,\mathrm{ch}\,\theta + ct\,\mathrm{sh}\,\theta, \end{cases} \tag{2.28}$$

so that a point at rest in the second inertial system, e.g. $x' = 0$, moves with velocity $v = x/t$ in the first system. This gives the relation (2.28).

Now inserting (2.27) into (2.26), we get

$$L(v) = \frac{1}{\sqrt{1 - v^2/c^2}} \begin{vmatrix} 1 & -v/c \\ -v/c & 1 \end{vmatrix}, \tag{2.29}$$

i.e. the "pure" or "restricted" Lorentz transformations (2.4) (the "boosts").

In the case above, where there was only one space dimension, the Lorentz group had only one parameter (v). In the general case, the Lorentz group has six parameters ($\mathbf{v}$ and three rotational parameters) and is non-Abelian (non-commutative).

2.3   Remarks

(1) Consider first two Lorentz transformations $L(\theta_1)$ and $L(\theta_2)$. We can easily show that

$$L(\theta_1) \cdot L(\theta_2) = L(\theta_1 + \theta_2) \tag{2.30}$$

using the addition formulæ for hyperbolic functions. Using (2.28) above to relate v and θ, we get

$$w = \frac{v_1 + v_2}{1 + v_1 v_2/c^2}, \tag{2.31}$$

which is the law of (parallel) *velocity addition* in Special Relativity (w corresponds to $\theta_1 + \theta_2$; v_1 to θ_1 and v_2 to θ_2). This relation shows that if one of the two velocities v_1 or v_2 is equal to the velocity of light, w is also equal to c, as expected *a priori* as just this constancy of the velocity of light was postulated *ab initio*.

Another consequence of the addition law (2.31) is that the velocity of light is the *limiting velocity* for an inertial system: if a Galilean reference system has an initial velocity $v < c$, it can never exceed c; whatever extra velocity it acquires, the resultant w [see Eq. (2.31)] always remains less than the velocity of light. This does not at all mean that relativity forbids velocities greater than that of light, [see E. Recami *et al.* (1974)] but only that the lines in $\mathcal{M}$ are divided into two categories, *timelike lines*[5] situated inside light cones (**Fig. 2.3**) and *spacelike lines*[6] lying outside these cones. These two families are totally disconnected, and one cannot pass from one to the other *via* a

---

[5] Representing uniform rectilinear motions with velocities $v < c$.
[6] Which would represent uniform rectilinear motions with velocities $v > c$.

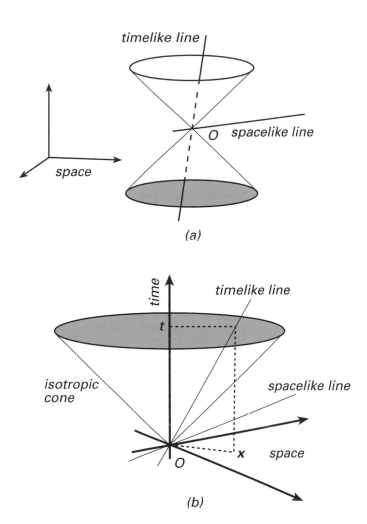

Fig. 2.3. Different types of straight lines in space–time. (a) A line is timelike if there exists a light cone which contains it entirely. A line is spacelike if it is outside any light cone whose vertex lies on it. A line is null (or lightlike) if it is a generator of a light cone. (b) The slope of a timelike line represents the inverse of the velocity of an associated inertial system relative to the inertial system linked to the coordinates (t, x). It is thus greater than c^{-1} (slope of the generators of the null cone), a timelike line thus represents an inertial motion at a velocity less than the velocity of light. By contrast a spacelike line would represent a motion at a velocity exceeding c.

Lorentz tranformation. *Only timelike lines correspond to inertial motions* according to experiment (and not for any theoretical reason). Spacelike lines correspond to uniform rectilinear motions at velocities exceeding that of light, which in practice are not observed.

(2) The existence of the *additive* parameter θ appearing in $L(\theta)$ now allows the explicit construction of a real (pseudo-)metric geometry [see I.M. Yaglom (1979)] in

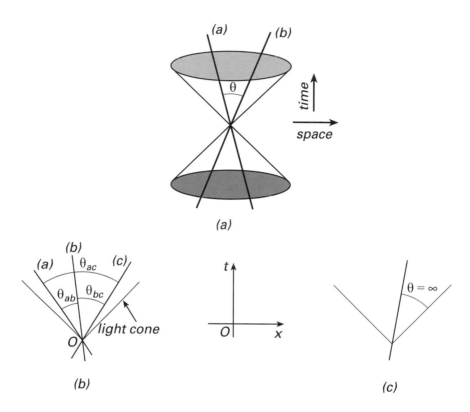

Fig. 2.4. Hyperbolic angle of two timelike lines. (a) The hyperbolic angle θ of the two lines (a) and (b) is uniquely related to the relative velocity of the two inertial motions associated by the relation $\theta = -th(v/c)$. (b) Passing from the inertial motion (a) to the inertial motion (c) can always be effected by use of a third inertial motion (b): the additivity of the hyperbolic angles ($\theta_{ac} = \theta_{ab} + \theta_{bc}$) gives the relativistic law of velocity addition. (c) The hyperbolic angle between any timelike line and a generator of the light cone is always infinite: all timelike lines make the same "angle" with the null cone and may thus be regarded as a symmetry axis for it.

$\mathcal{M}$. Thus, given two inertial systems A and B (**Fig. 2.4a**) moving with respect to each other with velocity v, we shall call the parameter θ associated with v the "angle" between the two lines representing them. This angle has an intrinsic meaning, in that it does not depend on the relative situation of the two lines. This is not true in classical space–time. Similarly, if we consider three inertial observers (a), (b) and (c), we have

$$\theta_{ac} = \theta_{ab} + \theta_{bc} \tag{2.32}$$

where the notation is obvious.

We can now see more easily why any inertial motion can be regarded as a symmetry axis of the light cone. The angle between any timelike line and the null cone is infinite: as $v \to c$, $\theta \to \infty$ [see Eq. (2.27)]. We shall return later to the geometry of $\mathcal{M}$.

(3) It is easy to see that when $v \ll c$, the Lorentz transformations (2.4) reduce to pure

Galilean transformations. This raises the question of what happens to the mathematical structures of Minkowski space, and in particular the light cones, in this limit.

Clearly, a null cone with vertex at Galilean coordinates $(t_0, \mathbf{x}_0)$ has the equation

$$c^2 (t - t_0)^2 - (\mathbf{x} - \mathbf{x}_0)^2 = 0, \tag{2.33}$$

so that as $c \to \infty$ the cone becomes more and more flat, reducing to the 3-plane $t = t_0$. In this limit, all the null cones reduce to the 3-planes $t = $ const. of classical space–time. The more the cones spread out, the more lines they include, so that in the limit every line passing through the vertex of a cone becomes a timelike line. In the classical limit, any line cutting a spatial plane is timelike, a result we can recover by passing to the limit $c \to \infty$. In this limit the only spacelike lines remaining are those contained in the 3-planes $t = $ const., corresponding to propagation at infinite velocity.

(4) Given two points $\mathbf{M}_1$ and $\mathbf{M}_2$ of $\mathscr{M}$, we can always form a *vector*, with four components, whose (pseudo-)length is given by

$$\mathbf{M}_1 \mathbf{M}_2 = c^2 (t_2 - t_1)^2 - (\mathbf{x}_2 - \mathbf{x}_1)^2 \tag{2.34}$$
$$= X^T G X \tag{2.35}$$

where X is the matrix representing $\mathbf{M}_1 \mathbf{M}_2$ in the Galilean reference frame. The vector $\mathbf{M}_1 \mathbf{M}_2$ is timelike if $\mathbf{M}_1 \mathbf{M}_2{}^2 > 0$, spacelike if $\mathbf{M}_1 \mathbf{M}_2{}^2 < 0$ and null if $\mathbf{M}_1 \mathbf{M}_2{}^2 = 0$. Put another way, $\mathbf{M}_1 \mathbf{M}_2$ is of the same type as the line passing through $\mathbf{M}_1$ and $\mathbf{M}_2$.

If $(t_1, \mathbf{x}_1)$ and $(t_2, \mathbf{x}_2)$ are the Galilean coordinates of the points $\mathbf{M}_1$ and $\mathbf{M}_2$ respectively, the vector $\mathbf{M}_1 \mathbf{M}_2$ has components $[(t_2 - t_1), (\mathbf{x}_2 - \mathbf{x}_1)]$.

As the quadratic form (2.34) is preserved under Galilean coordinate changes,

$$c^2 (t_2 - t_1)^2 - (\mathbf{x}_1 - \mathbf{x}_2)^2 = c^2 \left(t_2' - t_1' \right)^2 - \left(\mathbf{x}_2' - \mathbf{x}_1' \right)^2, \tag{2.36}$$

it follows that

$$c^2 t_1 \cdot t_2 - \mathbf{x}_2 \cdot \mathbf{x}_1 = c^2 \, t_1' \cdot t_2' - \mathbf{x}_1' \cdot \mathbf{x}_2'. \tag{2.37}$$

This allows us to define a (pseudo-)scalar product which is invariant under changes of inertial system, by

$$\mathbf{M}_1 \mathbf{M}_2 \cdot \mathbf{M}_1 \mathbf{M}_3 \ = X^T G Y. \tag{2.38}$$

In another inertial system we have

$$X' = LX \quad \text{and} \quad Y' = LY \tag{2.39}$$

and

$$X'^T G Y' = X^T L^T G L Y$$
$$= X^T G Y \tag{2.40}$$

using the relation (2.22). It follows that the pseudo-scalar product (2.38) is invariant under Lorentz transformations.

With this scalar product we can define in $\mathcal{M}$ the idea of *orthogonality* (actually pseudo-orthogonality) of two vectors by

$$X \perp Y \iff X^T G Y = 0, \tag{2.41}$$

where X and Y are the matrices representing $\mathbf{M_1 M_2}$ and $\mathbf{M_1 M_3}$.

2.4   Causality and simultaneity

We return to the geometrical and topological structure of Minkowski space. Each light cone allows us to divide space–time into three regions and two boundaries **(Fig. 2.5)**. There are two regions, one lying inside each of the two sheets of the null cone, and also the region outside the cone.

To characterise these three regions exactly – to give them a physical meaning – it is useful to fix an *orientation* of the light cones of space–time. To do this, consider an inertial observer; his trajectory in $\mathcal{M}$ is a timelike line (inside the null cone) *oriented in the direction that time is flowing*. We note, incidentally, that this orientation must be the same for all inertial systems whose space–time trajectories pass through the same point **O** (see **Fig. 2.5**); this is a minimal coherence condition (all Galilean observers must see time running in the same sense) which by construction is preserved by orthochronous Lorentz transformations. Γ^+ thus denotes the *future sheet*, i.e. that corresponding to a positive flow of time for every Galilean observer (whose trajectory passes through its vertex, (see **Fig. 2.5**). Γ^- denotes the *past sheet*, i.e. the other sheet. Once any one light cone is so oriented, all the others are if we require that this orientation is preserved by translations of space–time. It follows that, in all inertial systems, time flows in the same way.

We note that the pure Lorentz transformations (2.4), conserve the direction of time for different inertial observers in the same way as the various regions defined above. In any Galilean coordinate system $(t, \mathbf{x})$, we have

$$\left\{ \begin{array}{ll} \text{Int.}\Gamma^+ = \left\{ (t, \mathbf{x}) \quad : c^2 t^2 - \mathbf{x}^2 > 0, t > 0 \right\} \\ \text{Int.}\Gamma^- = \left\{ (t, \mathbf{x}) \quad : c^2 t^2 - \mathbf{x}^2 > 0, t < 0 \right\} \\ \text{Ext.}\Gamma \; = \left\{ (t, \mathbf{x}) \quad : c^2 t^2 - \mathbf{x}^2 < 0 \right\}, \end{array} \right.$$

now every Lorentz tranformation conserves the quadratic form $c^2 t^2 - \mathbf{x}^2$ [*see* Eq. (2.23)] and hence its sign.

The space–time points in the interior of $\Gamma^+(\mathbf{O})$, i.e. Int.Γ^+, lie in the *future* of **O**: events at **O** can influence or cause of those in the interior of Γ^+. Similarly, the space–time points in the interior of $\Gamma^-(\mathbf{O})$, i.e. Int.Γ^-, lie in the *past* of **O**: events here may have influenced or caused those at **O**. The third region, outside the light cone Γ, is composed of points which are completely inaccessible to **O**, as we do not observe signals propagating at speeds exceeding that of light.

In any Galilean system of coordinates, the condition

$$c^2 t^2 - \mathbf{x}^2 < 0$$

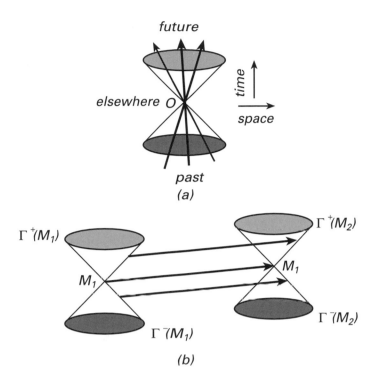

Fig. 2.5. (a) Each light cone divides space–time into three regions: the future, the interior of the sheet Γ^+; the past, the interior of the sheet Γ^-; and the exterior of the cone. (b) If one cone $\Gamma(M_1)$ is oriented, all the other null cones $\Gamma(M_2)$ are also, so that the homogeneity of space–time is preserved.

is equivalent to

$$\mathbf{x}^2/t^2 > c^2.$$

The exterior of the light cone is therefore inaccessible except by using physical phenomena which propagate *faster* than light.

We should note that the last point – the inaccessibility of the exterior of the light cone at **O** or, equivalently, the non-existence of effects propagating at velocities greater than light – constitutes an *additional postulate*, justified in the last analysis by experiment. We assume it holds here (in non-quantum situations).

Under these conditions, a Galilean observer (**Fig. 2.6**) can observe only a *limited* region of space–time: he must wait a certain time **OO′** (see figure) before gaining access to part of the exterior of **O**; in some sense at each moment he has a *horizon* (see also **Fig. 2.7**). In other words, some regions are too distant for light to have time to reach **O**.

Light cones oriented as above define a *causal structure* in Minkowski space $\mathcal{M}$, completely different from that of Newtonian space–time determined by the hyperplanes $t = \text{const}$. This causal structure determines a *partial ordering* on $\mathcal{M}$: we say that the

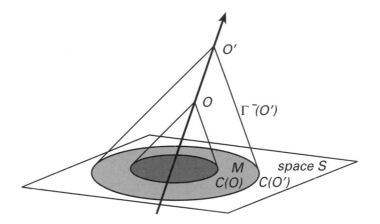

Fig. 2.6. The inacessibility of the global universe. An observer at **O** can only observe events in the interior of the sphere $\mathscr{C}(\mathbf{O})$; to have access to events outside $\mathscr{C}(\mathbf{O})$ in the spacelike plane **S**, he must wait until he reaches **O′**, i.e. until the time **OO′** has elapsed; thus **M** is unobservable from **O**, but is observable from **O′**.

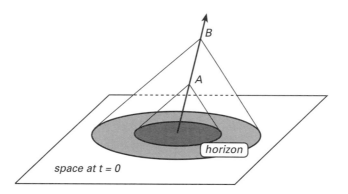

Fig. 2.7. A cosmological horizon. In the "big bang" cosmology, an observer has access only to a limited part of the Universe (shaded) and his horizon only increases with time.

point $\mathbf{M}_1$ *precedes* $\mathbf{M}_2$, and write this as

$$\mathbf{M}_1 \prec \mathbf{M}_2 \iff \mathbf{M}_1\mathbf{M}_2{}^2 \geq 0, \quad \mathbf{M}_1\mathbf{M}_2 \in \Gamma^+(\mathbf{M}_1),$$

if and only if the vector $\mathbf{M}_1\mathbf{M}_2$ is (i) timelike or possibly null, and (ii) oriented towards the future (**Fig. 2.8**). We can also show that the linear transformations preserving this partial ordering are precisely the Lorentz transformations [E.C. Zeeman (1964)]. In contrast, causality in classical space–time defines a *complete ordering*.

We come now to the notion of simultaneity. If we refer to the physical definition of simultaneity, that of Einstein, given in the preceding chapter, then at a given point of his local time a Galilean observer can regard as simultaneous all the events on his

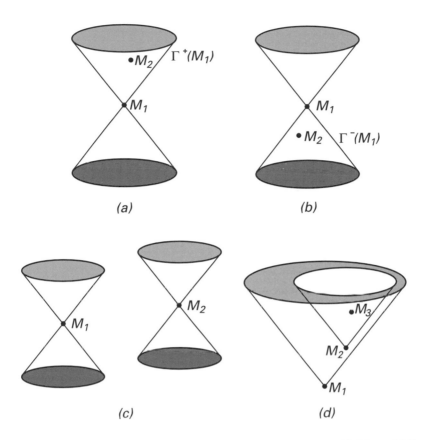

Fig. 2.8. **Causality in Minkowski space.** (a) $M_1 \prec M_2$; (b) $M_2 \prec M_1$; (c) M_1 and M_2 are not comparable and cannot be causally related; (d) $M_1 \prec M_2$ and $M_2 \prec M_3$ imply $M_1 \prec M_3$.

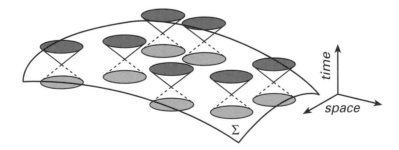

Fig. 2.9. **A spacelike surface.** Each point of the surface Σ lies outside the light cone of any other point of Σ.

past null cone Γ^-. Actually, only the points outside the light cone with no influence on the observer (at a given instant) constitute the analogue of simultaneous events in the classical case. However, while spatial hyperplanes in the Newtonian case only have three dimensions, the exterior of a light cone of a Galilean observer is a four-dimensional region. Thus, the notion of a three-dimensional space of simultaneous events is profoundly modified in special relativity, and is to some extent arbitrary.

If we wish to retain the three-dimensional character of space, the concept of simultaneity which emerges is that defined by *spacelike surfaces* (**Fig. 2.9**). Any spacelike surface Σ has the property that each of its points lies outside the light cone of each of its other points, so that two observers, Galilean or not, can regard the points of Σ as simultaneous (**Fig. 2.10**).

It follows that a Galilean observer can use whatever notion of simultaneity (i.e. choose whatever spacelike surfaces) he prefers, for example those defined by a *foliation* of space–time by a family of spacelike surfaces which do not intersect, specified by a real parameter (**Fig. 2.11**).

In reality, the usual notion of simultaneity, i.e. that *defined* by spacelike hypersurfaces orthogonal to a time axis (in the sense of the Minkowski metric) can *always* be used in special relativity, where it is probably the *most convenient*. We must, however, remember that this is arbitrary, and not required except for reasons of psychology and simplicity. However, in General Relativity, where the notion of inertial observer disappears and space–time is curved, there is no *a priori* foliation by spatial hyperplanes; thus any notion of simultaneity that remains is provided by spacelike surfaces whose arbitrariness can sometimes be limited by considering the symmetries of the problem under consideration.

As the simultaneity of two events, i.e. two points of space–time, has a certain arbitrariness, we can expect the same to hold for their spatial distance (**Fig. 2.12**). We examine this from an "operational" point of view, i.e. using a "thought experiment"[7] reproducing ideal experimental conditions. We thus consider two inertial observers $\{K\}$ and $\{K'\}$ (**Fig. 2.13**) and assume that $\{K\}$ wishes to measure his distance from $\{K'\}$. To this end, he sends an electromagnetic signal to $\{K'\}$ at time t_0, and receives an echo at time t_1 (such experiments have been performed using radar or lasers, with the aim of measuring the Earth–Moon distance, or the distances of the planets; see *Chapter 7*). If he employs the usual concept of simultaneity, i.e. that defined by orthogonal spacelike 3-planes (in the sense of the metric on $\mathcal{M}$) to his worldline (the line **(K)** on **Fig. 2.13**), he will define the time t_A when the reflection occurs at $\{K'\}$ by

$$t_A = \frac{t_0 + t_1}{2} \tag{2.42}$$

and so its distance to $\{K'\}$ at time t_A as

$$d_{\mathbf{KK'}} = \frac{c\,(t_1 - t_0)}{2}. \tag{2.43}$$

We note that any other notion of simultaneity defined by spacelike 3-planes leads to

---

[7] *"Gedankenexperiment"*.

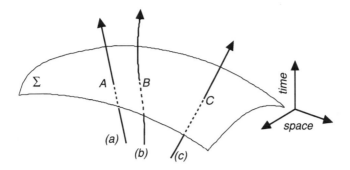

Fig. 2.10. Simultaneous events. Observers whose trajectories in $\mathcal{M}$ are denoted by (a), (b), (c), ...can regard the events A, B, C, ...as simultaneous, relative to the spacelike surface Σ. In the figure, (a) and (c) represent Galilean observers (straight lines), and (b) represents a non-Galilean observer.

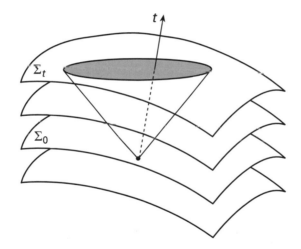

Fig. 2.11. Foliation of Minkowski space by a family of spacelike surfaces. The Galilean observer $\{\mathbf{O}\}$ can use any spacelike surface Σ_t to define a three-dimensional space of events which are simultaneous at time t. Further, he can use a family of such surfaces to obtain a convenient definition of physical space.

a time t_A such that [H. Reichenbach (1958)]

$$\begin{cases} t_A = t_0 + \varepsilon\,(t_1 - t_0) \\ 0\ < \varepsilon < \tfrac{1}{2} \end{cases} \tag{2.44}$$

and, as a result, to a distance $d_{\mathbf{KK'}}(\varepsilon) < d_{\mathbf{KK'}}\left(\tfrac{1}{2}\right)$.

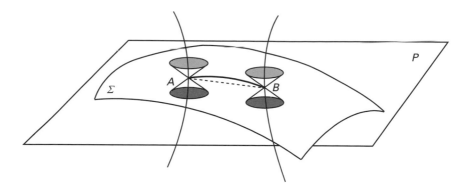

Fig. 2.12. Distance of two simultaneous events. The distance of the events A and B depends on the notion of simultaneity adopted. We can use the following convention: the distance from A to B is the shortest on a spacelike surface Σ used to define simultaneity. On the figure, P represents a spacelike 3-plane passing through A and B.

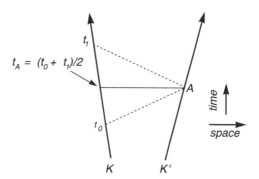

Fig. 2.13. Measurement of the distance between two inertial observers {K} and {K′}. {K} sends an electromagnetic signal to {K′} at the instant t_0 and receives its echo at t_1; the two numbers t_0 and t_1, together with a convention specifying the notion of simultaneity adopted, allow one to define a distance between {K} and {K′}. The usual convention of simultaneity using 3-planes orthogonal to the line {K}, leads to the definition that the instant when the reflection of the electromagnetic wave occurred [point A] is given by $(t_1 + t_0)/2$, so that the distance from {K} to {K′} at that instant is $d_{KK'} = c(t_1 - t_0)/2$.

2.5   Times and distances measured by inertial observers

Consider again two inertial observers {K} and {K′} using the same notion of simultaneity – that defined by spacelike hyperplanes orthogonal to their worldlines – and who have synchronised their clocks, for example at point **O** of **Fig. 2.14**.

Assume further that these observers measure a length ℓ_0 between two points at rest in {K}'s system (**Fig. 2.15a**). For {K} the measured length at time $t = 0$ is ℓ_0. For {K′}

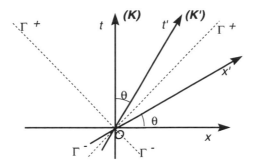

Fig. 2.14. Two conventional Galilean observers (i.e. using the same notion of simultaneity, that defined by spacelike 3-planes orthogonal to their respective wordlines {**K**} and {**K'**}). Note that in a Lorentz transformation of "angle" θ (which goes from {**K**} to {**K'**}), the relation of the two observers with respect to the light cone remains the same: the axis $O\Gamma^+$ (in one space dimension) is a symmetry axis for tOx as well as for $t'Ox'$; the Euclidean character of the figure plane does not allow a faithful representation of a non-Euclidean space.

at $t' = 0$, this same length is $\ell = \ell_0/\text{ch } \theta$, or

$$\ell = \ell_0\sqrt{1 - v^2/c^2}. \tag{2.45}$$

In other words the length measured by {**K'**} at rest relative to {**K**} appears shorter to him: this is the phenomenon of *apparent contraction*[8]. Far from being a *physical* effect of the ether, as Fitzgerald and Lorentz believed, this is an apparent effect [for a more detailed discussion, see M. Bunge (1967); M. Jammer (1979)] resulting essentially from the definitions and conventions used by the two observers in fixing their notion of simultaneity.

We now compare the two observer's clocks (**Fig. 2.15b**). A physical phenomenon – a clock – of duration **OA** in {**K**}, situated at x=0, appears to last longer in {**K'**}'s system. The point **A** (see figure) marking the end of the phenomenon has {**K'**}-coordinates (**OA'**, x) and we have

$$\mathbf{OA'} = \mathbf{OA}\text{ch } \theta, \tag{2.46}$$

and if $\mathbf{OA} = t_0$ and $\mathbf{OA'} = t'$, we get

$$t' = \frac{t_0}{\sqrt{1 - v^2/c^2}}; \tag{2.47}$$

effects in {**K**} appear to last longer for {**K'**} than for {**K**}: this is the apparent time dilation observed in all disintegrations of elementary particles.

---

[8] The visual appearance of a moving body has little to do with this contraction [see e.g. V.F. Weisskopf (1960)].

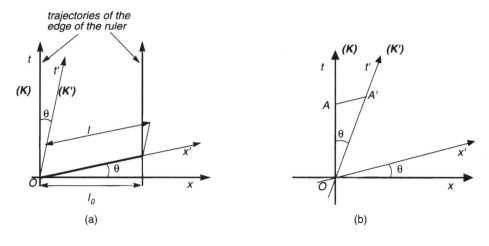

Fig. 2.15a. Apparent length contraction. $\ell_0 = \ell \; \text{ch} \; \theta$.
Fig. 2.15b. Apparent time dilation. The duration **OA** of an event in $\{\text{K}\}$ becomes **OA'** = **OA** $\text{ch} \; \theta >$ **OA** in $\{\text{K}'\}$.

2.6   Global properties of space–time

The properties we have just studied are in reality *local* properties in the sense that they are connected by the metric of Minkowski space to nearby events in space–time. The designation "local" can mean the atomic scale, or that of the laboratory, solar system, Galaxy, etc. Of course, this simply reflects the local character of the measurements and observations we can make.

However, it is by no means obvious from either a theoretical or observational viewpoint that space–time has the *topology* – i.e. the *global* structure – of Minkowski space.

To see this, consider a two-dimensional space–time (one space dimension). Imagine now that we *identify* the spatial points (see **Fig. 2.16**) to form a *spatially closed* two-dimensional surface. Locally this is still Minkowski space; but its global properties have been changed. The spacelike sections were open, but are now closed.

No local measurement can tell us about the topology of space–time. However, observation may give us an indication: there may exist galaxies (see **Fig. 2.17**) whose light reaches us from two completely different or even opposed directions. However, no such effect has yet been observed. Note that there are plenty of other possible global structures in which space would be bounded.

Another possible global structure for two-dimensional Minkowski space is one in which, contrary to the previous case, the points at infinite time are identified (**Fig. 2.18**). In general, such possibilities are excluded by causality: we could in principle influence our future. Actually, this argument is not as constraining as it at first appears. If the temporal periodicity is very great, the possibility of influencing our future does not really exist in practice.

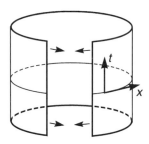

Fig. 2.16. The global structure of Minkowski space may be a cylinder (identifying points at spacelike infinity). In this case the spacelike sections are closed.

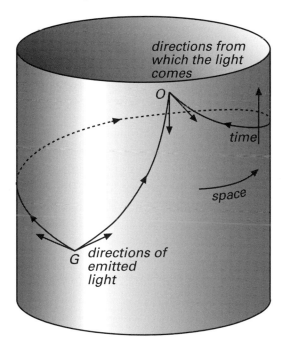

Fig. 2.17. The light emitted by galaxy G can be received by an observer O from various trajectories, so that G may be visible in several directions.

A study of the global properties of cosmological space–time can be found in M. Lachièze-Rey and J.P. Luminet (1995).

2.7 Experimental verification of Special Relativity

There have been many experimental tests of Special Relativity[9], which all confirm this theory to great accuracy (see Table 2.1). To these we can add the daily confirmations

---

[9] See the excellent article by M.P. Haugan and C.M. Will (1987).

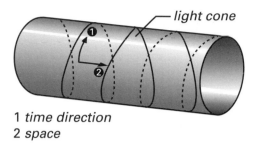

1 *time direction*
2 *space*

Fig. 2.18. If we identify points at timelike infinity in Minkowski space, causality is lost: an observer can influence both his past and his future!

of high-energy physics. It is not straightforward to compare the various results, and to understand their theoretical significance. It is clear, for example, that the two postulates stated at the beginning of the chapter can be divided into a multitude of sub-postulates and/or experimental "facts". This number is greatly increased by subtle questions of clock synchronisation, simultaneity and what part convention plays in these [see, for example, H. Reichenbach (1958)].

In each case, once the experimental reality of a given relativistic effect is established and its magnitude is measured, it is difficult to say quantitatively what leeway remains for an eventual alternative theory. At present there is no complete formalism for discovering this, although there have been interesting attempts [H.P. Robertson (1949); H.P. Robertson, T.W. Noonan (1968); R. Mansouri, R.U. Sexl (1977a, b, c)].

To take a simple example, let us consider the postulate of the constancy of the velocity of light [K. Brecher (1977)]. To model a possible violation, let us assume that the velocity, c', of light measured by an inertial observer is given by

$$c' = c + kv, \qquad (2.48)$$

where v is the source velocity and c the velocity of light in the source's reference system: k is a constant depending on the theory considered [$k = 0$ in Special Relativity; $k = 1$ in Newtonian theory; k has other values in various alternative theories]. $k \neq 0$ leads to [see W. De Sitter (1913)] many effects on binary star systems: apparent orbital eccentricity; the possibility of multiple images, inversion of the time-ordering of events, etc. Such effects are not observed [K. Brecher (1977); J.G. Fox (1962) (1965)] as the emitted light propagates in a dispersive medium (the interstellar medium) and there is a characteristic extinction distance

$$d_{\text{ext}} = \frac{\lambda}{2\pi(n-1)} \sim \frac{mc^2}{\lambda e^2 N}$$

(λ = wavelength, n = refractive index, N = electron density of the interstellar plasma) which is of order two light-years. At very short wavelengths (X-rays of 70 keV), the extinction distance is of order 20 kpc. Observations of binary X-ray sources (Her X-1,

Table 2.1. **Some recent measurements** *[from D. Newman et al. (1978)]*

authors	date	effect measured	precision
Brillet *et al.*	(1979)	ether wind	$\pm 2 \times 10^{-11}$
Mandelberg *et al.*	(1962)	transverse Doppler effect	5×10^{-2}
Brecher	(1977)	velocity of light	2×10^{-9}
Grove *et al.*	(1953)	relativistic mass	6×10^{-4}
Alvager *et al.*	(1964)	velocity of light	1.3×10^{-4}
Gviragossián *et al.*	(1975)	velocity of high energy electrons	2×10^{-7}
Bailey *et al.*	(1977)	muon g factor	2.7×10^{-7}
Van Dyck *et al.*	(1977)	electron g factor	3.5×10^{-9}
Wesley *et al.*	(1971)	electron g factor	3.5×10^{-9}

Cen X-3, SMC X-1) give a limit on k [K. Brecher (1977)]

$$k < 2 \times 10^{-9}.$$

Thus a theory *predicting* $k = 10^{-10}$ would be acceptable.

We turn now to the interesting treatment developed by H.P. Robertson (1949) (1968). Robertson imposes four *a priori* postulates which changes of inertial reference frames must obey, and deduces a general form for these generalised "Lorentz transformations", which depend on arbitrary parameters. These are the following:

(1) There exists a reference system in which space is isotropic.
(2) In this reference system, the velocity of light is c, independently of direction, motion of the source, wavelength, etc. ...
(3) There exists another reference system which moves with velocity v (along the x axis, for example) with respect to the first reference frame.
(4) The transformation between these two frames is linear.

Under these conditions it is not difficult to show that the coordinate transformation from the first system ($\{t, x, y, z\}$) to the second ($\{t', x', y', z'\}$) has the form

$$\left\{ \begin{array}{l} t' = a(v)t + \varepsilon x + \varepsilon' y + \varepsilon' z \\ x' = b(v)[x - vt] \\ y' = d(v)y \\ z' = d(v)z \end{array} \right. \tag{2.49}$$

where the functions $a(v), b(v), c(v)$ and $d(v)$ depend on the "theory of relativity" adopted, while ε and ε' are fixed by the procedure used to synchronise clocks. [H.P. Robertson (1949) (1968); R. Mansouri, R.U. Sexl (1977a, b, c)]. Using these transformations and Einstein's clock synchronisation procedure, the velocity of light is then given by

$$c/c(\theta) = 1 + \left(\beta + \delta - \frac{1}{2} \right) \left(\frac{v}{c} \right)^2 \sin^2 \theta + (\alpha - \beta + 1) \left(\frac{v}{c} \right)^2 + \cdots \tag{2.50}$$

where θ is the angle between the x' axis and the direction of propagation of a light

Table 2.2. **Experimental results for the parameters** α, β, δ *[from A. Brillet et al. (1982)]*

combination	experimental value	authors
α	$-1/2 \pm 10^{-7}$	G.R. Isaak (1970)
		D.C. Champeney et al. (1963)
$\beta + \delta$	$1/2 \pm 5 \times 10^{-7}$	A. Brillet et al. (1979)
$\alpha - \beta$	$-1.02 \pm 2 \times 10^{-2}$	R.J. Kennedy et al. (1932)

ray (in the second reference frame), and where the parameters α, β, δ represent the first deviations from Special Relativity, as the leading coefficients in the expansions of the functions $a(v), b(v), c(v)$ and $d(v)$ in powers of $(v/c)^2$:

$$\begin{cases} a(v) = 1 + \alpha(v/c)^2 + \cdots \\ b(v) = 1 + \beta(v/c)^2 + \cdots \\ c(v) = 1 + \delta(v/c)^2 + \cdots \end{cases} \tag{2.51}$$

Einstein's theory corresponds to $\alpha = -1/2, \beta = +1/2, \delta = 0$.

Experiments measure combinations of these parameters; thus we have the results shown in Table 2.2 [see A. Brillet, J.L. Hall (1982)].

Although this framework is very useful for testing Special Relativity, it can never give a final answer, as it is purely kinematic, and assumes an initial theoretical framework – Robertson's four postulates – which are arbitrary, however natural they may seem.

In closing this chapter, we remark that the Hughes–Drever experiment mentioned in *Chapter 1* can also be regarded as a high-precision test of local Lorentz invariance [see C.M. Will (1981)]. For this reason the experiment has been repeated several times [see e.g. T.E. Chupp et al. (1989); S.K. Lamoreaux et al. (1986); J.D. Prestage et al. (1985)].

Exercises

2.1   Consider a spherically symmetric star in *hydrostatic equilibrium*. Let $\rho(r)$ be its mass density at distance r and $m(r)$ the mass contained in a sphere of radius r. Finally, let $P(r)$ be the pressure at distance r and $P = P^*(\rho)$ the equation of state obeyed by the stellar matter.

(i) From the condition that a volume element of the star is in hydrostatic equilibrium (gravitational attraction exactly balanced by the pressure force) find an equation satisfied by dP/dr.

(ii) In the relativistic case, and with units such that $G = c = 1$, the analogous equation to that found in (i) is

$$\frac{dP(r)}{dr} = \frac{(\rho + P) \cdot (m + 4\pi r^3 P)}{r(r - 2m)}.$$

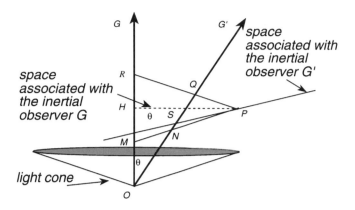

Fig. 2.19. Exercise 4.

    (a) Replace the factors G and c in this equation.

    (b) Verify that as $c \to \infty$, we recover the equation found in (i).

2.2   Starting from the relation $L^T G = G L^{-1}$, show that the orthochronous Lorentz transformations can also be written as

$$L(\theta) = \begin{vmatrix} \mathrm{ch}\,\theta & \mathrm{sh}\,\theta \\ \mathrm{sh}\,\theta & \mathrm{ch}\,\theta \end{vmatrix}.$$

Deduce that $L(\theta_1) L(\theta_2) = L(\theta_1 + \theta_2)$, and verify the velocity addition law.

2.3   Let

$$||L^{\nu}_{\mu}|| = \begin{vmatrix} 1 & 0 & 0 & 0 \\ 0 & \cos\theta & \sin\theta & 0 \\ 0 & -\sin\theta & \cos\theta & 0 \\ 0 & 0 & 0 & 1 \end{vmatrix}.$$

Show that this is a Lorentz transformation. Which?

2.4   In this problem Minkowski space has only two dimensions, one space and one time. The observers considered here are identical in the sense that they possess identical clocks, can send radar signals at the same time intervals, and can receive their echos with identical receivers. We choose units such that the velocity of light c is unity. See **Fig. 2.19**.

Part 1: *Longitudinal Doppler effect*

Let G and G' be two inertial observers moving with respect to each other with velocity v: the worldlines of G and G' cross at O. We assume that G and G' have synchronised their clocks at O $\{t = t' = 0\}$ and Γ is the light cone with vertex O. Let T_0 be the period of the radar signals from G to G' and T the period of these signals as received by G'.

(i) Is $T_0 = T$? (Justify your answer by a "Newtonian" argument.) Draw a space–time diagram corresponding to the transmission by G and reception by G' of periodic signals.

(ii) We set $k(v) = T/T_0$ and try to determine $k(v)$ by using the principle of relativity. If we denote by T' the period of radar signals emitted by G' and received by G, we can set $k'(v) = T/T_0$. Evaluate $k'(v)$ as a function of $k(v)$ (give a space–time diagram: how does this differ from that of question (i)?).

(iii) We now aim to calculate $k(v)$ explicitly, i.e. to prove the relation

$$k(v) = \left[\frac{1+v}{1-v}\right]^{1/2}.$$

At his time T_0, G sends a radar signal to G' which G' receives at time T and which he instantaneously reflects towards G. We denote by N and N'' the points of the line G where the signals are emitted and reflected. We denote by N' the point of the line G' where the signal emitted by G is reflected. We denote by S the middle of the line NN''.

(a) Draw a space–time diagram corresponding to the preceding experiment.
(b) Why is $ON'' = \{k(v)\}^2 T_0$?
(c) Use the diagram of question (a) to find the relation for $k(v)$.

Part 2: *Some applications*

(i) Let t be the time coordinate of N' for the observer G, and let t' be the time coordinate of N' for the observer G'. Show that

$$t = t' \left\{1 - v^2\right\}^{-1/2}$$
(slowing of moving clocks).

(ii) Consider now a third inertial observer G'', analogous to the preceding. G'' has velocity w with respect to G and velocity u with respect to G'. Using a space–time diagram like the earlier ones, show that

$$k(w) = k(v) \cdot k(u),$$

and deduce that

$$w = \frac{u+v}{1+uv}$$
(velocity addition).

Part 3: *The Lorentz transformation*

The distance of a point P of Minkowski space to an inertial observer is defined as c times half the time interval between emission of a radar signal and the reception of its echo at P (give a space–time diagram).

Let P be a point of space–time. G and G' attribute space and time coordinates (t, x) and (t', x') respectively to P. Show that (t, x) and (t', x') are related by the usual Lorentz transformations. Use the figure given, the principle of relativity, and the following notation: $OM = T_1$, $OQ = T_2'$, $OR = T_2$. [See H. Bondi (1965); D. Bohm (1965).]

2.5   Show that the Lorentz group reduces to the Galilei group as $c \to \infty$.

2.6   Let **x** be a timelike 4-vector and **y** an orthogonal 4-vector (using the Minkowski pseudo-metric). Show that **y** is spacelike.

2.7   Show that a 3-plane $\mathbf{k.x} = $ const. is spacelike if the vector **k** is timelike.

2.8   In Minkowski space $\mathcal{M}$, consider the 3-surface Σ defined by

$$\begin{cases} x^1 = a \, \mathrm{sh}\chi \cos \varphi \sin \theta \\ x^2 = a \, \mathrm{sh}\chi \sin \varphi \sin \theta \\ x^3 = a \, \mathrm{sh}\chi \cos \theta \\ x^0 = a \, \mathrm{ch}\chi \end{cases}$$

where a is a positive constant.

(i) Show that Σ is spacelike.

(ii) Calculate the metric (i.e. the element of length) on this surface.

(iii) What happens when the constant a is negative?

2.9   Verify that the causality relation $\prec$ is an ordering.

2.10   Verify that simultaneity is an equivalence relation.

2.11   Let P be a spacelike 3-plane and Γ^+ the future light cone. Find the intersection of P and Γ^+ (use the principle of relativity).

2.12   Let Σ be the 3-surface defined by the relation

$$(x^0)^2 - \vec{x}^2 = a^2$$

and let P be a spacelike 3-plane, with equation

$$k^0 x^0 - \vec{k}.\vec{x} = b.$$

What is the intersection of Σ and P?

3 The Relativistic Form of Physical Laws

There is no purely internal physical experiment which will demonstrate the relative motion of a Galilean reference system – this is the principle of relativity. It follows that *all* physical laws must necessarily have the *same* form in all inertial systems (see *Chapter 2*). To express this property in a simple analytic way, and to avoid the need to show explicitly in each case that the principle of relativity is satisfied, it is extremely useful to write the equations of physics in a form which is *manifestly covariant* under Lorentz transformations. Not only do we avoid the need for such proofs (often very delicate, and subject to error), but this also allows us to set physics in the context of space–time furnished with a pseudo-metric. Further, the need for manifest covariance of the analytic formulation of a physical phenomenon restricts its possible forms. For this reason we shall use various representations of the Lorentz group: spinors, tensors, etc.

3.1 Tensor formalism

(1) Unless otherwise stated, we shall only use cartesian coordinates which are orthogonal in the sense of the metric on $\mathcal{M}$. They thus have a timelike axis Ox^0 and three spacelike axes Ox^i ($i = 1, 2, 3$) orthogonal in pairs, in the sense of the metric of $\mathbf{R}^3$; these three spatial axes are orthogonal to the axis Ox^0, the time axis, in the sense of the metric of Minkowski space (**Fig. 3.1**).

In other words, if $\{\mathbf{e}_\mu\}$ ($\mu = 0, 1, 2, 3$) is a *tetrad* ("vierbein") of four unit vectors, collinear with the four axes Ox^μ and with origin O, we have by construction

$$< \mathbf{e}_\mu, \mathbf{e}_\nu >= \eta_{\mu\nu} \tag{3.1}$$

where $<,>$ denotes the (pseudo-) scalar product of Minkowski space, fixed by $\eta \equiv ||\eta_{\mu\nu}||$.

Remark on notation:

(i) In preceding sections we have denoted the scalar product by a point; thus in the earlier notation we would have

$$< \mathbf{e}_\mu, \ \mathbf{e}_\nu >= \mathbf{e}_\mu \cdot \mathbf{e}_\nu \tag{3.2}$$

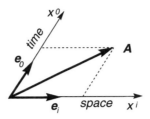

Fig. 3.1. Lorentz coordinate system and associated frame.

(ii) Similarly we denoted by G the matrix of the metric tensor, i.e.

$$G = \|\eta_{\mu\nu}\|. \tag{3.3}$$

From now on we shall also denote this by η, reserving

$$G = \|g_{\mu\nu}\|$$

for general relativity *or* special relativity in curvilinear coordinates: $\eta_{\mu\nu}$ is independent of the space–time event considered, whereas $g_{\mu\nu}$ depends on it.

(iii) We shall also use the following convention, which is common in the literature: *Greek indices run from 0 to 3; Latin indices from 1 to 3.*

(iv) We shall also use the *Einstein summation convention* for repeated indices. For example, $K^{\mu\nu i} \cdot A_{\mu i}$ means

$$K^{\mu\nu i} \cdot A_{\mu i} = \sum_{\mu=0}^{\mu=3} \sum_{i=1}^{i=3} K^{\mu\nu i} \cdot A_{\mu i}. \tag{3.4}$$

If $\mathbf{A}$ is any vector of M (i.e. the interval from the origin O to a point $\mathbf{M}$ of Minkowski space), we write

$$\mathbf{A} = \mathbf{e}_{\mu} \cdot A^{\mu}. \tag{3.5}$$

The A^{μ} are the *contravariant* components of $\mathbf{A}$. In (3.5) we have used the Einstein summation convention, and we have explicitly

$$\mathbf{e}_{\mu} \cdot A^{\mu} = \mathbf{e}_0 A^0 + \mathbf{e}_1 A^1 + \mathbf{e}_2 A^2 + \mathbf{e}_3 A^3. \tag{3.6}$$

The *covariant* components of $\mathbf{A}$, i.e. A_{μ}, are defined by

$$A_{\mu} = <\mathbf{A}, \ \mathbf{e}_{\mu}> \tag{3.7}$$

$$= A_{\nu} <\mathbf{e}_{\nu}, \ \mathbf{e}_{\mu}> \tag{3.8}$$

or

$$A_{\mu} = \eta_{\mu\nu} A^{\nu}. \tag{3.9}$$

Explicitly, we have

$$A^0 = A_0 \text{ and } A^i = -A_i. \tag{3.10}$$

The minus sign arises from the fact that the *metric tensor* $\eta_{\mu\nu}$ is not positive definite, unlike the Euclidean case. (In that case, in Cartesian coordinates we have a plus sign.)

We now set

$$\eta^{-1} = \|\eta^{\mu\nu}\| \tag{3.11}$$

so that, *numerically*

$$\|\eta_{\mu\nu}\| = \|\eta^{\mu\nu}\| = \begin{vmatrix} 1 & 0 & 0 & 0 \\ 0 & -1 & 0 & 0 \\ 0 & 0 & -1 & 0 \\ 0 & 0 & 0 & -1 \end{vmatrix}. \tag{3.12}$$

Mathematically, $\eta^{\mu\nu}$ is the (pseudo-)metric tensor of the dual space (i.e. the space of linear forms on the vectors **A** of $\mathscr{M}$); this is why the indices μ and v are raised.

We can invert (3.9) as

$$A^{\mu} = \eta^{\mu\nu} A_{\nu}. \tag{3.13}$$

Let **A** and **B** be two vectors at the same origin O; their (pseudo-)scalar product is then

$$< \mathbf{A}, \mathbf{B} > = < A^{\mu}\mathbf{e}_{\mu}, B^{\nu}\mathbf{e}_{\nu} > \tag{3.14}$$

$$= A^{\mu}B^{\nu} < \mathbf{e}_{\mu}, \mathbf{e}_{\nu} > \tag{3.15}$$

$$= \eta_{\mu\nu}A^{\mu} B^{\nu} \tag{3.16}$$

$$= A_{\mu} B^{\mu} \tag{3.17}$$

$$= A^{\mu}B_{\mu}, \tag{3.18}$$

where we have used the bilinearity of the scalar product [from (3.14) to (3.15)], the definition (3.1) of the metric tensor [from (3.15) to (3.16)] and the rule for lowering indices (3.13) [from (3.16) to (3.17) or (3.18)]. Explicitly, we have again

$$A^{\mu}B_{\mu} = A^{0}B^{0} - \sum_{i=1}^{i=3} A^{i} \cdot B^{i}.$$

We note finally that if $d\mathbf{M}$ is an infinitesimal interval between two events of $\mathscr{M}$, we have

$$ds^{2} = < d\mathbf{M}, d\mathbf{M} > = \eta_{\mu\nu} \, dx^{\mu} \, dx^{\nu}; \tag{3.19}$$

which is the space–time metric expressed using the summation and index conventions.

Using this notion of scalar product, we can easily see that the *sign* of the (pseudo-) norm of a vector **A** determines whether it is timelike, etc.

$$\begin{cases} A_{\mu}A^{\mu} > 0 \rightarrow \mathbf{A} \text{ is } timelike \\ A_{\mu}A^{\mu} < 0 \rightarrow \mathbf{A} \text{ is } spacelike \\ A_{\mu}A^{\mu} = 0 \rightarrow \mathbf{A} \text{ is } null \end{cases}$$

(see **Fig. 3.2**). We see that whether a vector **A** is spacelike etc. i.e. the sign of its (pseudo-)norm, is *invariant* under Lorentz transformations: these transformations, by construction, leave the scalar product in $\mathscr{M}$ invariant.

(2) We now study rather more precisely the action of Lorentz transformations on the

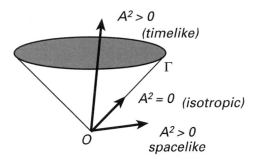

Fig. 3.2. Different types of 4-vectors. The nature of a 4-vector (spacelike, timelike, null) depends on its relation to the light cone.

components of 4-vectors in $\mathcal{M}$. A vector is a geometric object defined independently of any coordinate system. However, its components transform under coordinate changes. We limit ourselves here to Lorentzian coordinate changes (**Fig. 2.14**), general changes being studied later.

Let L be a Lorentz transformation, which thus satisfies (2.22), i.e.

$$L^T \eta L = \eta. \tag{3.20}$$

It transforms the tetrad $\{\mathbf{e}_\mu\}$ into the tetrad $\{\mathbf{e}'_\mu\}$ by the relation[1]

$$\mathbf{e}'_\mu = L_\mu{}^\nu \, \mathbf{e}_\nu, \tag{3.21}$$

leading immediately to

$$< \mathbf{e}'_\mu, \mathbf{e}'_\nu > = L_\mu{}^\alpha L_\nu{}^\beta < \mathbf{e}_\alpha, \mathbf{e}_\beta >$$
$$= \eta_{\alpha\beta} L_\mu{}^\alpha L_\nu{}^\beta; \tag{3.22}$$

but as the metric is invariant under Lorentz transformations [this is what (3.20) expresses],

$$< \mathbf{e}'_\mu, \mathbf{e}'_\nu > = < \mathbf{e}_\mu, \mathbf{e}_\nu > = \eta_{\mu\nu}, \tag{3.23}$$

we have

$$\eta_{\mu\nu} = \eta_{\alpha\beta} L_\mu{}^\alpha L_\nu{}^\beta. \tag{3.24}$$

This equation shows that if we invert the relation (3.21) as

$$\mathbf{e}_\mu = \left(L^{-1}\right)_\mu{}^\nu \, \mathbf{e}'_\nu, \tag{3.25}$$

then

$$\left(L^{-1}\right)_\mu{}^\nu = \eta_{\mu\alpha} \eta^{\nu\beta} L_\beta{}^\alpha. \tag{3.26}$$

---

[1] Above, we used L for the transformation $x' = Lx$. If L takes us from $\mathbf{e}$ to $\mathbf{e}'$, then $x' = L^{-1}x$. We shall use the latter definition from now on.

We note finally that under Lorentz transformations the coordinates x^μ (or more exactly, the components x^μ of the vector **OM**) transform so that

$$\mathbf{OM} = \mathbf{e}'_\mu x'^\mu = \mathbf{e}_\mu x^\mu \tag{3.27}$$

or

$$L_\mu{}^\nu \, \mathbf{e}_\nu x'^\mu = \mathbf{e}_\mu x^\mu; \tag{3.28}$$

i.e.

$$L_\mu{}^\nu x'^\mu = x^\nu. \tag{3.29}$$

Inverting this relation, we get

$$x'^\mu = \left(L^{-1}\right)_\nu{}^\mu x^\nu \tag{3.30}$$

or, by definition,

$$x'^\mu \underset{\text{def}}{\equiv} L^\mu{}_\nu x^\nu. \tag{3.31}$$

This definition of $L^\mu{}_\nu$ and the relation (3.24) then give

$$L^\mu{}_\nu = \eta_{\nu\alpha}\eta^{\mu\beta} L_\beta{}^\alpha \tag{3.32}$$

(*note* the position of the indices: see below).

(3) Quite generally, we call a *4-vector* a mathematical object with four components which behave under Lorentz transformations either as coordinates (contravariant components) or as the $\mathbf{e}_\mu$ (covariant components). In other words, **A** is a 4-vector if and only if

$$\begin{cases} A'^\mu = L^\mu{}_\nu A^\nu \\ A'_\mu = L_\mu{}^\nu A_\nu \end{cases} \tag{3.33}$$

under a change of basis $\{\mathbf{e}_\mu\} \to \{\mathbf{e}'_\mu\}$.

> At this point we should remark that the "definition" of a 4-vector given above is not a definition! It represents only a *practical criterion* for recognising if four physical quantities (e.g. charge density and electric current density) form a mathematical object which corresponds to physical reality: two Galilean observers measuring the same physical phenomenon – one obtaining the numbers (A^0, A^1, A^2, A^3) and the other the numbers (A'^0, A'^1, A'^2, A'^3) – must be able to reconcile their results with the relations (3.33). We add that this "definition" allows the practical use of *tensors* (see below) without the need for a definition which is rigorous but pointlessly complicated. Explicitly, this means that we are only interested in the sub-vector space of $\mathbf{R}^4$ which is *stable* under Lorentz transformations, the sub-space to which we can give physical meaning.

Thus, a physical law expressed in the analytic form

$$A^\mu = B^\mu,$$

is *manifestly covariant* under the Lorentz group, i.e. it has the *same form* in all Galilean frames, as in any other frame it becomes

$$A'^\mu = B'^\mu.$$

We note that the set of 4-vectors, subject to physical significance, form a vector space, and we can add them and multiply them by scalars: they are also vectors of $\mathbf{R}^4$. Finally, we note that 4-vectors of the same type (timelike, spacelike, etc.) do not form vector subspaces of $\mathbf{R}^4$ (see **Fig. 3.3**).

(4) As well as 4-vectors, there exist also *invariants*, of which we have already met several examples, such as

$$ds^2 = \eta_{\mu\nu}dx^\mu dx^\nu = dx_\mu dx^\mu$$
$$A^2 = A_\mu A^\mu,$$
$$< \mathbf{A}, \mathbf{B} > = A_\mu B^\mu.$$

We can also have scalar operators, such as the *Dalembertian*

$$\Box \equiv \frac{1}{c^2}\frac{\partial^2}{\partial x^{02}} - \Delta, \tag{3.34}$$

which can also be written as

$$\Box = \eta_{\mu\nu}\partial^\mu\partial^\nu = \partial_\mu\partial^\mu \equiv \partial_\mu{}^\mu \text{ etc.}, \tag{3.35}$$

where we have put

$$\partial_\mu \equiv \frac{\partial}{\partial x^\mu}. \tag{3.36}$$

The fact that the derivative with respect to (*contravariant*) coordinates constitutes a *covariant* 4-vector operator results from the fact that if $f(x)$ is an arbitrary invariant function of the x^μ, its differential df is also invariant, and

$$df(x) = \partial_\mu f(x)dx^\mu. \tag{3.37}$$

This expression cannot be invariant unless $\partial_\mu f$ is a covector (a covariant 4-vector).

(5) Besides invariants and 4-vectors we also find mathematical objects with several indices, and which we can add and multiply by scalars (provided that they have the same number of indices, arranged in the same way), which we call *tensors*. Under Lorentz transformations they transform as

$$\Lambda^{\alpha'\beta'\gamma'...}{}_{\mu'\nu'...} \cdots = L^{\alpha'}{}_\alpha L^{\beta'}{}_\beta L^{\gamma'}{}_\gamma \cdots \left(L^{-1}\right)^\mu{}_{\mu'} \left(L^{-1}\right)^\nu{}_{\nu'} \cdots \Lambda^{\alpha\beta\gamma...}{}_{\mu\nu...} \cdots \tag{3.38}$$

Each tensor index possesses the transformation properties of a 4-vector: the components of a tensor transform like a product of components (contravariant or covariant) of 4-vectors [see *Appendix A*].

From tensors of the same type, e.g. $U^{\mu\nu}{}_{\alpha\beta\gamma}$ and $V^{\mu\nu}{}_{\alpha\beta\gamma}$, we can form other tensors, such as

$$\begin{cases} a\ U^{\mu\nu}{}_{\alpha\beta\gamma} + b\ V^{\mu\nu}{}_{\alpha\beta\gamma} \\ U^{\mu\nu}{}_{\alpha\beta\gamma} \cdot V^{\lambda\sigma}{}_{\tau\rho\delta}. \end{cases}$$

A tensor of rank n (with n tensor indices) can be *contracted* over two of its indices (one covariant, one contravariant) to give a tensor of rank $n-2$; for example,

$$\begin{array}{ccc} U^{\mu\nu}{}_{\alpha\beta\gamma} & \rightarrow & U^{\mu\nu}{}_{\nu\beta\gamma} & \rightarrow & U^{\mu\nu}{}_{\nu\mu\gamma} \\ \text{(rank 5)} & & \text{(rank 3)} & & \text{(rank 1)} \end{array}$$

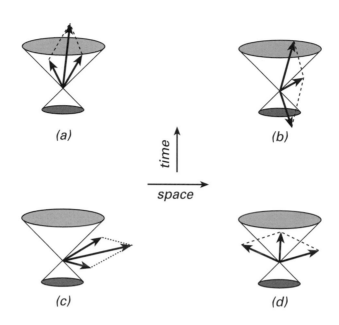

Fig. 3.3. **The addition of various types of 4-vectors (timelike, spacelike, null) does not in general conserve their type.** Conversely, a given 4-vector can be expressed as the sum of any types of 4-vectors.

or

$$U^{\mu\nu}{}_{\alpha\beta\gamma} \to U^{\mu\nu}{}_{\alpha\nu\gamma}, \quad \text{etc...}$$

Differentiation gives a tensor of rank $n+1$:

$$U^{\mu\nu}{}_{\alpha\beta\gamma} \quad \to \quad \partial_\lambda U^{\mu\nu}{}_{\alpha\beta\gamma}$$
$$\text{(rank 5)} \qquad\qquad \text{(rank 6)}$$

Transpositions of indices produce other tensors of the same rank, for example

$$U^{\mu\nu}{}_{\alpha\beta\gamma} \to U^{\nu\mu}{}_{\gamma\beta\alpha}.$$

Lorentz transformations conserve the *rank* of a tensor, and also any *symmetry* or *antisymmetry* properties it may have under permutations of indices of the same type (i.e. contravariant or covariant).

(6) A particularly useful tensor is the totally antisymmetric Levi–Civita symbol, defined by

$$\varepsilon^{\alpha\beta\gamma\delta} = \begin{cases} +1 & \text{if } \alpha\beta\gamma\delta \text{ is a cyclic permutation of 0123} \\ -1 & \text{if } \alpha\beta\gamma\delta \text{ is an anticyclic permutation of 0123} \\ 0 & \text{otherwise.} \end{cases}$$

We can easily see that ε is a tensor *if we limit ourselves to proper Lorentz transformations*; if not, the transformed tensor changes sign. Given a tensor of components $T^\mu{}_\nu$, its determinant is given by

$$\det \mathbf{T} = \frac{1}{4!}\varepsilon^{\gamma\iota\mu\rho}\varepsilon^{\delta\kappa\nu\sigma}T_{\gamma\delta}T_{\iota\kappa}T_{\mu\nu}T_{\rho\sigma}$$

Using the metric tensor η, we can also form (pseudo-)tensors $\varepsilon_{\mu\nu\alpha\beta}$, $\varepsilon^{\mu\nu}{}_{\alpha\beta}$, etc. The term "pseudo-" refers to the change of sign of the components of ε under improper Lorentz transformations.

(7) The *manifest covariance* of the equations of physics, i.e. their *form* invariance under Lorentz transformations is expressed in the form of tensor equations (or spinor equations, which we have not mentioned here). We shall meet many examples.

3.2 The Doppler effect and aberration

An application of the manifest covariance of physical laws is given by the relativistic Doppler effect and the related phenomenon of aberration: the frequency of an electromagnetic wave appears different depending on whether it is observed in the rest-frame of the emitter or in a Galilean frame moving with respect to this. Further, the propagation direction of the wave differs for two such observers. We write ν_{obs} and ν_{em} for the frequencies respectively observed in a Galilean frame $\{O_{\text{obs}}, x, y, z\}$ and emitted by a source at rest in another Galilean frame $\{O_{\text{em}}, x', y', z'\}$ moving with respect to the first frame with velocity $\mathbf{v}$ parallel to Ox. We call θ_{obs} the angle between the $O_{\text{obs}}x$ axis and the observed light ray, and θ_{em} that between the emitted ray and the axis $O_{\text{em}}x'$. We then have to relate ν_{obs} to ν_{em} and θ_{obs} to θ_{em}.

We note first that an electromagnetic wave propagating *in vacuo* obeys the wave equation $\Box\varphi = 0$, where φ is the electric or magnetic field of the wave. This equation has plane wave solutions of the form

$$\varphi(x) = A(k)e^{-ik\cdot x},$$

where k is an arbitrary 4-vector, necessarily so because of the relativistic invariance of the wave equation. The latter property can also be found by noting that the wave 4-vector points along the generator of the null cone of vertex O_{em} in Minkowski space, i.e. along a light ray from O_{em}. The 4-vector k is therefore null, i.e. $k^2 = 0$, which has to hold if φ is to be a solution of the wave equation.

The condition that the phase $k \cdot x$ of the wave φ should be Lorentz invariant is

$$k_\mu x^\mu = k'_\mu x'^\mu;$$

or

$$k_0 x^0 - \mathbf{k} \cdot \mathbf{x} = k'_0 x'^0 - \mathbf{k}' \cdot \mathbf{x}', \tag{3.39}$$

giving finally

$$k_0[t - \mathbf{n} \cdot \mathbf{x}] = k'_0 [t' - \mathbf{n}' \cdot \mathbf{x}'] \tag{3.40}$$

with $\mathbf{n} \equiv \mathbf{k}/k_0$, and where we have used the fact that the 4-vector k is null, i.e. the relation $k^{02} = \mathbf{k}^2$. The vector $\mathbf{k}$ (or $\mathbf{k}'$) can be chosen in the plane $xO_{\mathrm{obs}}y$ (or in the plane $x'O_{\mathrm{em}}y'$ respectively), so that θ_{obs} and θ_{em} are then given by[2] (see **Fig. 3.4**).

$$\mathbf{k} \cdot \mathbf{x} = kx \cos \theta_{\mathrm{obs}} + ky \sin \theta_{\mathrm{obs}}$$

$$\mathbf{k}' \cdot \mathbf{x}' = k'x' \cos \theta_{\mathrm{em}} + k'y' \sin \theta_{\mathrm{em}}.$$

Thus (3.40) can be rewritten as

$$k_0[t - \mathbf{n} \cdot \mathbf{x}] = k_0' \left[t' - \mathbf{n}' \cdot \mathbf{x}' \right]$$

$$= tk_0' \left[\frac{1}{\sqrt{1 - v^2}} + \frac{v \cos \theta_{\mathrm{em}}}{\sqrt{1 - v^2}} \right]$$

$$-xk_0' \left[\frac{v}{\sqrt{1 - v^2}} + \frac{\cos \theta_{\mathrm{em}}}{\sqrt{1 - v^2}} \right]$$

$$-yk_0' \sin \theta_{\mathrm{em}}$$

$$= k_0 t - kx \cos \theta_{\mathrm{obs}} - yk_0 \sin \theta_{\mathrm{obs}}.$$

Now, identifying the coefficients of t, x and y in this expression, we get

$$k_0 \sin \theta_{\mathrm{obs}} = k_0' \sin \theta_{\mathrm{em}}$$

$$k_0 = \frac{k_0'}{\sqrt{1 - v^2}} [1 + v \cos \theta_{\mathrm{em}}]$$

$$k_0 \cos \theta_{\mathrm{obs}} = k_0' \frac{v + \cos \theta_{\mathrm{em}}}{\sqrt{1 - v^2}}.$$

These equations immediately give the expressions we seek

$$\tan \theta_{\mathrm{obs}} = \frac{\sin \theta_{\mathrm{em}} \cdot \sqrt{1 - v^2}}{v + \cos \theta_{\mathrm{em}}} \tag{3.41a}$$

$$v_{\mathrm{obs}} = \frac{v_{\mathrm{em}}}{\sqrt{1 - v^2}} [1 + v \cos \theta_{\mathrm{em}}] \; ; \tag{3.41b}$$

analogous relations giving the source quantities in terms of the observed quantities can be found by making the substitutions obs $\longleftrightarrow$ em, $v \longleftrightarrow -v$.

In the non-relativistic limit, i.e. $v \ll 1$, (3.41b) reduces to

$$v_{\mathrm{obs}} \approx v_{\mathrm{em}} [1 + v \cos \theta_{\mathrm{em}}],$$

which is the usual non-relativistic Doppler effect formula. The relation (3.41a) demonstrates the *aberration effect*, while (3.41b), giving the *relativistic Doppler effect*, has several important special cases. If $\theta_{\mathrm{obs}} = 0$ or $\theta_{\mathrm{obs}} = \pi$, the Doppler effect is purely *longitudinal*, the source either moving towards or away from the observer. In the former case we have

$$v_{\mathrm{obs}} = v_{\mathrm{em}} \left[\frac{1 + v}{1 - v} \right]^{1/2} > v_{\mathrm{em}},$$

---

[2] In the rest of this paragraph x and k denote the norms of 3-vectors rather than full 4-vectors. Also, the light velocity is taken as unit.

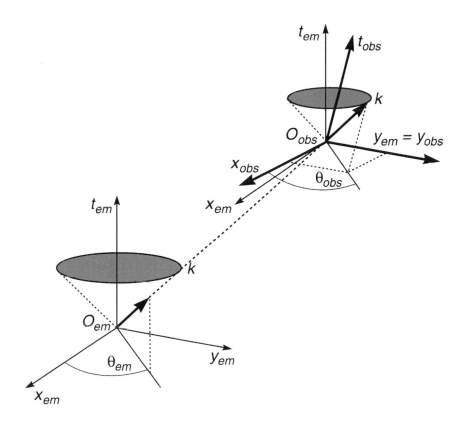

Fig. 3.4. The relativistic Doppler effect and aberration (space–time diagram).

i.e. an apparent *blue shift* of the source. In the latter case, where the source moves away from the observer, we get

$$v_{\text{obs}} = v_{\text{em}} \left[\frac{1-v}{1+v} \right]^{1/2} < v_{\text{em}},$$

i.e. a *red shift*. If instead $\theta_{\text{obs}} = \pi/2$, i.e. the line of sight is perpendicular to the relative motion, we have

$$v_{\text{obs}} = \frac{v_{\text{em}}}{\sqrt{1-v^2}},$$

a specifically relativistic effect, with no classical analogue: the *transverse* Doppler effect.

3.3 The kinematic description of particle motion

The space–time motion of a particle subject to a given force is completely described by giving a curve, its space–time trajectory or *worldline*. This is precisely the same in Minkowski space as in Newtonian space–time.

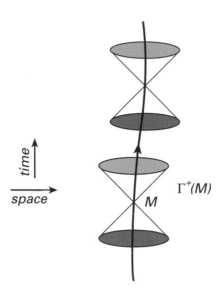

Fig. 3.5. A timelike curve. A timelike curve is contained in every light cone whose vertex lies on it.

A space–time trajectory (see **Fig. 3.5**) can be written parametrically as

$$x^\mu = x^\mu(\xi),$$

where ξ is an *arbitrary* parameter. Of course, we can always choose $\xi = x^0/c = t$, in which case we recover the usual form

$$\begin{cases} x^0 = ct \\ \mathbf{x} = \mathbf{x}(t) \end{cases}$$

however, contrary to the Newtonian case, where all space–time trajectories are allowed, – provided of course that time increases along them (see **Fig. 3.6**) – relativistic trajectories are constrained to be *timelike* (**Fig. 3.7**). This means that at each point **M** of the space–time trajectory the future trajectory must lie entirely inside the future light cone $\Gamma(\mathbf{M})$.

This restriction expresses the requirements of *causality*. Consider a point **M′** very close to **M** (**Fig. 3.7**). The average speed with which the particle moves from **M** to **M′** is $\Delta \mathbf{x}/\Delta t$, and must be less than that of light. This implies that the 4-vector **MM′** is always inside $\Gamma(\mathbf{M})$ and is thus timelike.

The fact that all possible trajectories must be timelike means that the 4-vectors tangent to the trajectory must also be timelike. Such a 4-vector is given by

$$u^\mu(\xi) = \frac{dx^\mu}{d\xi},$$

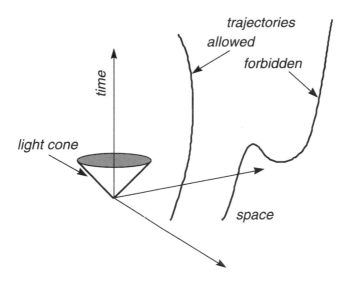

Fig. 3.6. Causality of trajectories. This is shown by the fact that they must be timelike everywhere.

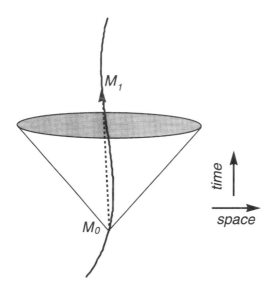

Fig. 3.7. Timelike trajectory. The average speed to go from any point of the trajectory to any other must be less than that of light. Thus the 4-velocity (limiting case where the two points are very close) must be timelike.

and must obey the condition

$$u^{\mu}(\xi)u_{\mu}(\xi) = \frac{dx_{\mu}}{d\xi}\frac{dx^{\mu}}{d\xi} = \frac{ds^2}{d\xi^2} > 0.$$

As it is tangent to the trajectory, the 4-vector $u^{\mu}(\xi)$ will by definition be a *4-velocity*.

In analogous fashion we can define a *4-acceleration* by

$$\gamma^{\mu}(\xi) = \frac{du^{\mu}(\xi)}{d\xi} = \frac{d^2x^{\mu}(\xi)}{d\xi^2}. \tag{3.42}$$

A priori the parameter ξ has no direct physical significance, although we could regard it as some kind of timescale. Among all possible parametrisations of a space–time trajectory, there is one which has an immediate physical significance. This is parametrisation by the (pseudo-)length of the trajectory, whose infinitesimal element is ds^2. More precisely, we use the parametrisation defined by ds^2/c^2, which has the dimensions of time, and set

$$d\tau^2 = ds^2/c^2. \tag{3.43}$$

The parameter τ is called the *proper time*. This is an invariant which therefore has the same numerical value in all inertial frames, and represents the time of an observer who moves with the particle. In the rest of this book we will use most generally *a system of units in which the velocity of light is the unit of velocity*; then there is no need to distinguish between the parameters τ and s, or between x^0 and t, etc...

With parametrisation defined by proper time, the 4-vector is a *unit* vector

$$\begin{cases} u^{\mu} = \dfrac{dx^{\mu}}{d\tau} \\ u^{\mu}u_{\mu} = 1. \end{cases} \tag{3.44}$$

Differentiating the second equality with respect to τ, we get

$$u_{\mu}\frac{du^{\mu}}{d\tau} = u_{\mu}\gamma^{\mu} = 0; \tag{3.45}$$

or, the 4-acceleration is always orthogonal to the 4-velocity. Further, since u^{μ} is a timelike 4-vector, the 4-vector γ^{μ} orthogonal to it is a spacelike 4-vector (**Fig. 3.8**).

> **Example 1:** Consider a free particle. Its motion is described by a timelike straight line in Minkowski space, with equation
>
> $$x^{\mu}(\tau) = u^{\mu}.(\tau - \tau_0) + x^{\mu}{}_0 \tag{3.46}$$
>
> in an arbitrary Lorentzian coordinate system. In (3.46), $x^{\mu}{}_0$ is an arbitrary 4-vector and τ_0 is a proper time, such that
>
> $$x^{\mu}(\tau_0) = x^{\mu}{}_0,$$
>
> while u^{μ} is an arbitrary timelike 4-vector.
>
> **Example 2:** The motion described by
>
> $$x^0(\xi) = \tau_0 \text{sh}\xi, \qquad x^1 = \tau_0 \text{ch}\xi, \qquad x^2 = x^3 = 0,$$
>
> is timelike as the corresponding 4-velocity
>
> $$u^0(\xi) = \tau_0 \text{ch}\xi, \qquad u^1(\xi) = \tau_0 \text{sh}\xi, \qquad u^2 = u^3 = 0,$$

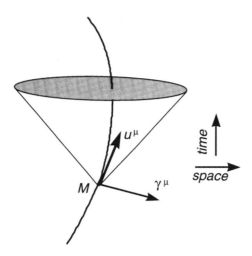

Fig. 3.8. Nature of kinematic quantities. The 4-velocity is timelike (causality), while the 4-acceleration, being orthogonal to it, is spacelike.

is timelike: $u^\mu(\xi)u_\mu(\xi) = \tau_0^2 > 0$. Similarly, the 4-acceleration is given by

$$\gamma^0(\xi) = \tau_0 \text{sh}\xi = x^0(\xi),$$
$$\gamma^1(\xi) = \tau_0 \text{ch}\xi = x^1(\xi)$$
$$\gamma^2 = \gamma^3 = 0.$$

We can easily verify that $\gamma_\mu u^\mu = 0$.

Returning to the proper time, (3.44) gives

$$dt = u^0 d\tau. \tag{3.47}$$

Since [Eq. (3.43)]

$$\frac{d\tau^2}{dt^2} = 1 - \frac{d\mathbf{x}^2}{dt^2} = 1 - \mathbf{v}^2, \tag{3.48}$$

we have

$$u^0 = \frac{1}{\sqrt{1 - \mathbf{v}^2}}. \tag{3.49}$$

Similarly we find

$$\mathbf{u} = \frac{\mathbf{v}}{\sqrt{1 - \mathbf{v}^2}} \tag{3.50}$$

(in a system of units where $c \neq 1$, we have to replace the factor $\mathbf{v}$ by $\mathbf{v}/c$).

In the non-relativistic limit $v \ll c$ we get

$$\begin{cases} u^0 \approx 1, & \mathbf{u} \approx \mathbf{0}. \\ dt \approx d\tau \end{cases}$$

If we take a limit where u^μ has the dimension of a velocity we find[3]

$$\mathbf{u} \approx \mathbf{v}.$$

3.4   Relativistic dynamics: $E = mc^2$

So far we have defined a new kinematics by imposing changes of inertial systems preserving the laws of electromagnetism [we still have to verify (see *Chapter 5*) that Maxwell's equations can actually be put into manifestly covariant form and thus satisfy the principle of relativity]. Under these conditions, the laws of Newtonian dynamics, which are covariant under the Galilei group, will not be covariant under the Lorentz group. They will therefore not satisfy Einstein's relativity principle and hence must be modified.

In Newtonian physics, Newton's second law constitutes an *additional postulate* (see *Chapter 1*) which we have to add to kinematic postulates such as the principle of inertia or the principle of relativity. In relativity things are obviously similar: the analogue of Newton's second law is a postulate which we could adopt *a priori*.

To find the relativistic form of Newton's second law we must use several properties: the kinematic quantities available to describe the space–time trajectory, the principle of relativity in writing manifestly covariant relations, the need to recover the Newtonian results in the limit of small velocities, and simplicity.

The first question is how to generalise the notions of momentum and energy. In non-relativistic physics these are *numbers which are conserved in collisions* of particles, or first integrals of the motion. We know [see H. Goldstein (1980), or L. Landau and E. Lifschitz (1960)] that these conservation laws result from the *homogeneity of space and the uniformity of time*, properties which also hold in Minkowski space. However, the conservation laws *cannot* simply be taken over; neither the kinetic energy of a free particle $1/2mv^2$ nor its momentum $m\mathbf{v}$ have the same form *after* a Lorentz transformation. We must start by changing the relations between momentum and velocity, and between energy and velocity. Now, the only mathematical object which does not involve higher than a first derivative is the 4-velocity u^μ. Similarly, the *simplest* generalisation of the Newtonian 4-vector (under the Galilei group) given by $(E, \mathbf{p})$ is a 4-momentum which we denote by p^μ. Thus *the only* possibility of connecting p^μ and u^μ is a proportionality

$$p^\mu = m_0 u^\mu \tag{3.51}$$

which we still have both to interpret and verify experimentally.

Consider first the space components $\mathbf{p}$. We have

$$\mathbf{p} = m_0 \mathbf{u} = \frac{m_0 \mathbf{v}}{\sqrt{1 - \mathbf{v}^2}} \tag{3.52}$$

---

[3] The 0 component of u^μ is singular in the limit and need not be considered.

$$= m_0\mathbf{v} + \frac{m_0\mathbf{v}v^2}{2c^2} + m_0\mathbf{v}O\left(v^2/c^2\right), \tag{3.53}$$

where we have replaced the factors c in the last equation, from which we can see that for $v^2 \ll c^2$ we get the usual Newtonian limit $\mathbf{p} = m_0\mathbf{v}$: m_0 thus represents the mass.

Now examine the time component p^0; we have

$$p^0 = m_0 u^0 = \frac{m_0}{\sqrt{1-v^2}} \tag{3.54}$$

$$= m_0 + \frac{m_0 v^2}{2} + m_0 O\left(v^4\right) \tag{3.55}$$

or, replacing the factors c and multiplying each side by c^2,

$$p^0 c^2 = m_0 c^2 + \frac{1}{2} m_0 v^2 + m_0 c^2 O\left(v^4/c^4\right). \tag{3.56}$$

In the non-relativistic approximation, when $v^2 \ll c^2$, the last expression reduces to

$$p^0 c^2 \approx m_0 c^2 + \frac{1}{2} m_0 v^2 \tag{3.57}$$

showing that p^0 (in units with $c = 1$, or $p^0 c^2$ in units with $c \neq 1$) must be interpreted as an energy, *the energy of a free particle.* Further, (3.56) clearly confirms the interpretation of the constant m_0 as the particle mass. In the relativistic case [Eq. (3.54)], if we are in the particle's *rest frame*, i.e. if $\mathbf{v} = \mathbf{0}$, then p^0 reduces to

$$p^0 = E = m_0 c^2. \tag{3.58}$$

Unlike the classical case, where the zero of the energy of a free particle is arbitrary, in relativity its value is fixed. However (3.58) represents much more than a convention fixing the zero of the energy: it shows that *the mass of a body has a real energy content.* This equivalence of mass and energy will fundamentally change the laws of gravitation. The relation (3.58) [or (3.59) below] is well verified experimentally by a multitude of physical phenomena: nuclear reactions, elementary particle collisions, etc.

Relations (3.52) and (3.54) can also be written as

$$\begin{cases} \mathbf{p} = m\mathbf{v} \\ E = mc^2 \end{cases}$$

where we have put

$$m = \frac{m_0}{\sqrt{1-v^2}} \tag{3.59}$$

which represents the *relativistic mass* of a free particle. While m_0 is the rest-mass – the matter content of the particle – m is the moving mass, with $m \geq m_0$.

From the 4-vector p^μ we can form the invariant $p_\mu p^\mu$, whose value is easily found to be $m_0{}^2$,

$$p_\mu p^\mu = m_0{}^2, \tag{3.60}$$

which can be written explicitly as

$$p^\mu p_\mu = E^2 - \mathbf{p}^2 = m_0{}^2 c^4, \tag{3.61}$$

giving the relativistic relation between energy and momentum

$$E = \sqrt{m_0{}^2 + \mathbf{p}^2} \tag{3.62}$$

which is very different from the Newtonian relation $E = \mathbf{p}^2/2m_0$, to which it reduces when $c \to \infty$ *and* the rest-mass energy is removed.

The equation of motion of a free particle is thus

$$m_0 \frac{du^\mu}{d\tau} = \frac{dp^\mu}{d\tau} = 0. \tag{3.63}$$

If we are dealing with a point *collision* of two particles of 4-momentum $p^\mu{}_1$ and $p^\mu{}_2$, the quantity $p^\mu{}_1 + p^\mu{}_2$ is conserved and we have[4]

$$p_1^\mu + p^\mu{}_2 = p'^\mu{}_1 + p'^\mu{}_2 \tag{3.64}$$

where the p'^μ represent the 4-momenta after the collision.

We now come to real relativistic dynamics, i.e. to forces and interactions. As in the Newtonian case we assume – it is an empirical fact – that we need only know the initial position and velocity of a particle to determine its motion completely, i.e. its space–time trajectory. Mathematically, the *simplest* law of relativistic dynamics will be given by a second order differential equation. Further, if we demand, as in the Newtonian case, that the force generates changes of momentum, we must set

$$\frac{dp^\mu(\tau)}{d\tau} = f^\mu\left(x^\lambda, p^\lambda\right) \tag{3.65}$$

where $f^\mu\left(x^\lambda, p^\lambda\right)$ is the *4-force*. This is the only manifestly covariant equation that we could write, given the hypotheses. We still have to interpret f^μ. It is easy to verify that the spatial components of (3.65) give Newton's second law in the non-relativistic approximation $\mathbf{v}^2 \ll c^2$. The 0 component reduces in this approximation to

$$\frac{dE}{d\tau} = f^0, \tag{3.66}$$

showing that f^0 should be interpreted as the *power*.

Differentiating each side of Eq. (3.60), we get

$$p_\mu \frac{dp^\mu}{d\tau} = 0, \tag{3.67}$$

so that multiplying the relativistic dynamical law (3.65) by p^μ, we find

$$p_\mu f^\mu = 0. \tag{3.68}$$

---

[4] The kinematics of collisions between elementary particles can be found in the complementary treatments of R. Hagedorn (1964), and E. Byckling and K. Kajantie (1973).

This equation relates f^0 to the other components f^i of the 4-force,

$$f^0 = \frac{\mathbf{f} \cdot \mathbf{p}}{E} \equiv \mathbf{f} \cdot \mathbf{v}. \tag{3.69}$$

Moreover, this shows that in relativity *forces always depend on velocities*, contrary to the Newtonian case.

> An example of a relativistic force is the *Lorentz force* acting on a charged particle in an electromagnetic field: this has the general form
>
> $$f^\mu = \varphi^\mu{}_\nu p^\nu, \tag{3.70}$$
>
> where $\varphi^\mu{}_\nu$ is an antisymmetric tensor directly related to the electromagnetic field (see *Chapter 5*). We can easily check that (3.68) is satisfied.

We should add that the latter property (orthogonality of the 4-force and the momentum) does not hold if the particle has variable mass.

The case of interacting relativistic particles is more complex, as the actions of the forces must propagate at velocities less than light. These give systems of non-local integro-differential equations, or differential systems where the interaction is mediated by classical *fields*, which themselves obey further equations [see J. Rzewuski (1958); A.O. Barut (1965)]. In fact, relativistic dynamics only takes a truly satisfying form in the framework of quantum field theory [see e.g. C. Itzykson and J.P. Zuber (1980)].

3.5 Minkowski space in curvilinear coordinates

Up to now we have used Lorentzian coordinates (i.e. rectilinear and orthogonal with respect to the Minkowski metric) to describe geometric objects representing physical variables. However, these objects are defined independently of any coordinate system, Lorentzian or not. Clearly, in Special Relativity Lorentzian coordinates have a special significance as corresponding *physically* to the time and space of Galilean observers, and it is in principle possible to use only these coordinates. However, it is useful to introduce arbitrary curvilinear coordinates, not least because this brings in techniques extremely similar to those used in General Relativity, thus making the transition to the general theory easier.

(1) Let $\mathbf{M}$ be a point of Minkowski space and $\{x^\mu\}_{\mu=0,1,2,3}$ arbitrary curvilinear coordinates. $\{\mathbf{e}_\mu\}$ denotes a set of four linearly independent 4-vectors tangent to the coordinate lines. The *basis* formed by the $\mathbf{e}_\mu$ is not necessarily orthogonal and the $\mathbf{e}_\mu$ are not necessarily unit vectors. The basis $\{\mathbf{e}_\mu\}$ consists of one timelike 4-vector and three spacelike 4-vectors. This is not strictly necessary, but is required if we wish to give a physical significance to the coordinate lines: the timelike coordinate line then represents the worldline of an *accelerated observer* (see **Fig. 3.9**).

The problem we wish to solve is as follows: given two spacetime points $\mathbf{M}$ and $\mathbf{M}'$, to which are attached two frames $\{\mathbf{e}_\mu\}$ and $\{\mathbf{e}'_\mu\}$ and coordinate systems $\{x^\mu\}$ and $\{x'^\mu\}$, we wish to *transform the components* of the associated tensors from one frame to the other. This problem has several linked aspects. We need to know how to change

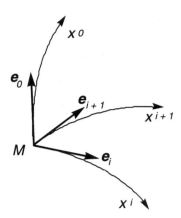

Fig. 3.9. Curvilinear coordinates and the associated basis.

coordinates at a given point **M**, and also to know how to compare (*or transport*) a tensor (actually its components) from **M** to **M**′. Also, we need to know how to write down the laws of physics in such a way that they are manifestly covariant under arbitrary coordinate changes.

(2) We start by considering the problem of coordinate changes at **M**. We pass from the basis $\{\mathbf{e}_\mu\}$ to the basis $\{\mathbf{e}'_\mu\}$ at **M** using relations like

$$\mathbf{e}'_\mu = a_\mu{}^\nu \mathbf{e}_\nu \tag{3.71}$$

$$\mathbf{e}_\mu = a'_\mu{}^\nu \mathbf{e}'_\nu, \tag{3.72}$$

which immediately imply the constraint

$$a_\mu{}^\nu a'_\nu{}^\lambda = \delta_\mu{}^\lambda. \tag{3.73}$$

Now considering an infinitesimally neighbouring point $\mathbf{M} + d\mathbf{M}$, we have

$$d\mathbf{M} = \mathbf{e}_\mu dx^\mu \tag{3.74}$$

$$= \mathbf{e}'_\mu dx'^\mu \tag{3.75}$$

with $\mathbf{e}_\mu = \partial_\mu \mathbf{M}$ and $\mathbf{e}'_\mu = \partial'_\mu \mathbf{M}$. Comparing these two relations shows that

$$\mathbf{e}'_\mu = \partial_\nu \mathbf{M} \frac{\partial x^\nu}{\partial x'^\mu} = \mathbf{e}_\nu \frac{\partial x^\nu}{\partial x'^\mu}, \tag{3.76}$$

giving

$$a_\mu{}^\nu = \frac{\partial x^\nu}{\partial x'^\mu} \tag{3.77}$$

$$a'_\mu{}^\nu = \frac{\partial x'^\nu}{\partial x^\mu} \tag{3.78}$$

and thus

$$\frac{\partial x^\nu}{\partial x'^\mu} \cdot \frac{\partial x'^\mu}{\partial x^\lambda} = \delta_\lambda{}^\mu. \tag{3.79}$$

At this point we introduce a convention which will be very convenient in the following. To avoid the need to specify the new variable (coordinate) on each occasion, we use *primed indices*. Thus, instead of writing $a_\mu{}^\nu$ and $a'_\mu{}^\nu$ we shall write $a_{\mu'}{}^\nu$ and $a_\mu{}^{\nu'}$. With this convention we can, for example, write the relation (3.71) as

$$\mathbf{e}_\mu' \equiv \mathbf{e}_{\mu'} = a_{\mu'}{}^\nu \mathbf{e}_\nu, \tag{3.80}$$

and (3.73) as

$$a_\mu{}^{\nu'} a_{\nu'}{}^\lambda = \delta_\mu{}^\lambda. \tag{3.81}$$

Starting from the invariant $ds^2 = d\mathbf{M}^2 \equiv d\mathbf{M} \cdot d\mathbf{M}$, we find that

$$ds^2 = \mathbf{e}_\mu dx^\mu \cdot \mathbf{e}_\nu dx^\nu \tag{3.82}$$
$$= \mathbf{e}_\mu \cdot \mathbf{e}_\nu dx^\mu dx^\nu, \tag{3.83}$$

and a similar relation for the new coordinates and basis vectors. As by definition the metric tensor $g_{\mu\nu}$ is given by

$$g_{\mu\nu} = \mathbf{e}_\mu \cdot \mathbf{e}_\nu, \tag{3.84}$$

it follows that this tensor transforms as

$$g_{\mu\nu} = \frac{\partial x^{\lambda'}}{\partial x^\mu} \cdot \frac{\partial x^{\sigma'}}{\partial x^\nu} g_{\lambda'\sigma'} \tag{3.85}$$
$$g_{\mu'\nu'} = \frac{\partial x^\lambda}{\partial x^{\mu'}} \cdot \frac{\partial x^\sigma}{\partial x^{\nu'}} g_{\lambda\sigma} \tag{3.86}$$

The components $g_{\mu\nu}$ of the metric tensor η represent the geometry of Minkowski space in a given coordinate system; this can vary, and we must *always* find a coordinate system in which it takes the form $\eta_{\mu\nu}$.

(3) By *definition*, in a coordinate change the dx^μ transform as a *contravariant* 4-vector:

$$dx^{\nu'} = \frac{\partial x^{\nu'}}{\partial x^\mu} dx^\mu, \tag{3.87}$$

where the transformation coefficients reduce to Lorentz transformations if we use Lorentzian coordinates.

Similarly, if $\varphi(x)$ is a scalar function, $\partial_\mu \varphi(x)$ transforms like the components of a *covariant* 4-vector. Thus

$$\partial_\mu \varphi(x) = \frac{\partial \varphi\{x(x')\}}{\partial x^{\prime\nu}} \frac{\partial x^{\prime\nu}}{\partial x^\mu} \tag{3.88}$$
$$= \frac{\partial x^{\nu'}}{\partial x^\mu} \partial_{\nu'} \varphi'(x'); \tag{3.89}$$

i.e.

$$\partial_{\mu'} \varphi' = \partial_\nu \varphi \frac{\partial x^\nu}{\partial x^{\mu'}}. \tag{3.90}$$

The metric tensor thus behaves as a covariant tensor of the second rank under these transformations. A 4-vector $\mathbf{A}$ has contra- and co-variant components A^{μ} and A_{ν} in the basis $\{\mathbf{e}_{\mu}\}$:

$$\mathbf{A} = A^{\mu} \cdot \mathbf{e}_{\mu} \tag{3.91}$$

$$A_{\mu} = \mathbf{A} \cdot \mathbf{e}_{\mu}, \tag{3.92}$$

exactly as in the usual Lorentzian case. We should note that the foregoing relations hold only at a given point. Given any two 4-vectors $\mathbf{A}$ and $\mathbf{B}$, their scalar product is given by

$$\mathbf{A} \cdot \mathbf{B} = \mathbf{e}_{\mu} \cdot \mathbf{e}_{\nu} A^{\mu} B^{\nu} \tag{3.93}$$

$$= g_{\mu\nu} A^{\mu} B^{\nu}, \tag{3.94}$$

as before. We also have

$$A^{\mu} = g^{\mu\nu} A_{\nu} \tag{3.95}$$

$$A_{\mu} = g_{\mu\nu} A^{\nu} \tag{3.96}$$

where $g^{\mu\nu}$ is the "inverse" of $g_{\mu\nu}$:

$$g^{\mu\nu} = (-1)^{\mu+\nu} \frac{\text{minor of } g_{\mu\nu}}{\text{determinant of } g_{\mu\nu}} \tag{3.97}$$

with

$$g_{\mu\alpha} \cdot g^{\alpha\nu} = \delta_{\mu}{}^{\nu}. \tag{3.98}$$

The manifest covariance of the laws of physics can thus be expressed by relations between quantities behaving as tensors under transformations between general coordinates systems.

We have seen above that the gradient of a scalar function is a 4-vector. We can ask if this holds more generally: i.e. is the gradient of a 4-vector a tensor, etc.? To discover this we consider the differential of a 4-vector $\mathbf{A}$:

$$d\mathbf{A} = d\{A^{\mu}\mathbf{e}_{\mu}\} \tag{3.99}$$

or

$$d\mathbf{A} = dA^{\mu} \cdot \mathbf{e}_{\mu} + A^{\mu} d\mathbf{e}_{\mu} \tag{3.100}$$

$$= \partial_{\nu} A^{\mu} dx^{\nu} \mathbf{e}_{\mu} + A^{\mu} d\mathbf{e}_{\mu}. \tag{3.101}$$

For the case of Lorentzian coordinates, $d\mathbf{e}_{\mu} = 0$ and $\partial_{\nu} A^{\mu}$ behaves as a second rank tensor. But for general curvilinear coordinates $d\mathbf{e}_{\mu} \neq 0$, and this extra term means that $\partial_{\nu} A^{\mu}$ does not behave as a tensor. Thus, in a new coordinate system,

$$\partial_{\nu} A^{\mu} = \frac{\partial}{\partial x^{\nu}} \left\{ \frac{\partial x^{\mu}}{\partial x'^{\lambda}} A^{\lambda} \left(x\{x'\} \right) \right\} \tag{3.102}$$

$$= \frac{\partial x'^{\alpha}}{\partial x^{\nu}} \frac{\partial}{\partial x'^{\alpha}} \left\{ \frac{\partial x^{\mu}}{\partial x'^{\lambda}} A^{\lambda} \left(x\{x'\} \right) \right\}. \tag{3.103}$$

We thus get an extra term containing

$$\frac{\partial^2 x^\mu}{\partial x'^\alpha \partial x'^\lambda},$$ (3.104)

which vanishes for Lorentz transformations.

What causes this? We note that the change in A^μ contains not only the change resulting from transporting the 4-vector **A** from **M** to **M** + d**M**, but also the change of $\{\mathbf{e}_\mu\}$. It is therefore convenient to define a covariant type of derivation which measures the *absolute* change of a vector or tensor.

(4) We define the *covariant derivative* ∇_μ of a 4-vector A^μ using the relation

$$d\mathbf{A} \underset{\text{def}}{\equiv} \mathbf{e}_\mu \nabla_\nu A^\mu dx^\nu.$$ (3.105)

The structure of this relation shows immediately that

$$\nabla_\nu A^\mu = \frac{\partial x^{\nu'}}{\partial x^\nu} \cdot \frac{\partial x^\mu}{\partial x^{\mu'}} \nabla_{\nu'} A^{\mu'},$$ (3.106)

i.e. that $\nabla_\nu A^\mu$ is a tensor. The following notation is very convenient:

$$\nabla_\nu A^\mu \equiv A^\mu{}_{;\nu}$$ (3.107)

$$\partial_\nu A^\mu \equiv A^\mu{}_{,\nu}.$$ (3.108)

To evaluate the covariant derivative ∇_ν we need to find the change of $d\mathbf{e}_\mu$ in the basis $\{\mathbf{e}_\mu\}$, i.e.

$$d\mathbf{e}_\mu = \omega_\mu{}^\nu \mathbf{e}_\nu$$ (3.109)

where $\omega_\mu{}^\nu$ is linear in dx^α,

$$\omega_\mu{}^\nu = \Gamma^\nu_{\mu\alpha} dx^\alpha.$$ (3.110)

$\Gamma^\nu_{\mu\alpha}$ is the connection – or rather the *connection symbol* – allowing a comparison of directions at infinitesimally neighbouring points, and thus a comparison of neighbouring bases.

Inserting these relations into the definition (3.101) we get

$$d\mathbf{A} = \partial_\nu A^\mu dx^\nu \mathbf{e}_\mu + A^\mu \Gamma^\nu_{\mu\alpha} dx^\alpha \mathbf{e}_\nu$$ (3.111)

$$= \left\{ \partial_\nu A^\mu + \Gamma^\mu_{\lambda\nu} A^\lambda \right\} \mathbf{e}_\mu dx^\nu,$$ (3.112)

where the second equality follows from the first by the index substitutions $\nu \to \mu, \alpha \to \nu, \mu \to \lambda$. Thus

$$\nabla_\nu A^\mu = \partial_\nu A^\mu + \Gamma^\mu_{\lambda\nu} A^\lambda.$$ (3.113)

We have finally to evaluate the connection symbols $\Gamma^\mu_{\nu\lambda}$. We note first that the connection symbols are *symmetrical* with respect to the two lower indices:

$$\Gamma^\mu_{\nu\lambda} = \Gamma^\mu_{\lambda\nu}.$$ (3.114)

This follows from the definition (3.110) of the $\Gamma^\mu_{\alpha\beta}$, the fact that $\partial_\mu M = e_\mu$ and from Schwarz's theorem $\partial_\mu \partial_\nu M = \partial_\nu \partial_\mu M$. We thus have

$$\partial_\mu \partial_\nu M = \Gamma^\alpha_{\mu\nu} e_\alpha \tag{3.115}$$

$$\partial_\nu \partial_\mu M = \Gamma^\alpha_{\nu\mu} e_\alpha, \tag{3.116}$$

which shows the symmetry of the Γ. To calculate the Γ explicitly we evaluate the derivative of the metric tensor

$$dg_{\mu\nu} = d\left(e_\mu \cdot e_\nu\right) \tag{3.117}$$

$$= \left\{\Gamma^\lambda_{\mu\rho} g_{\lambda\nu} + \Gamma^\lambda_{\nu\rho} g_{\mu\lambda}\right\} dx^\rho \tag{3.118}$$

$$= \partial_\rho g_{\mu\nu} dx^\rho; \tag{3.119}$$

which implies

$$\partial_\rho g_{\mu\nu} = \Gamma^\lambda_{\mu\rho} g_{\lambda\nu} + \Gamma^\lambda_{\nu\rho} g_{\mu\lambda}. \tag{3.120}$$

Changing indices $\mu \to \rho$ and $\nu \to \rho$ gives

$$\Gamma^\lambda_{\rho\mu} g_{\lambda\nu} + \Gamma^\lambda_{\nu\mu} g_{\rho\lambda} = \partial_\mu g_{\rho\nu} \tag{3.121}$$

$$\Gamma^\lambda_{\mu\nu} g_{\lambda\rho} + \Gamma^\lambda_{\rho\nu} g_{\mu\lambda} = \partial_\nu g_{\mu\rho}. \tag{3.122}$$

Adding the first two relations and subtracting the third, the symmetry of the Γ shows that

$$\Gamma^\lambda_{\mu\nu} g_{\lambda\rho} = \frac{1}{2}\left\{\partial_\mu g_{\nu\rho} + \partial_\nu g_{\mu\rho} - \partial_\rho g_{\mu\nu}\right\}$$

$$= \Gamma_{\mu\nu,\rho}. \tag{3.123}$$

Finally, contracting this equation with $g^{\alpha\rho}$, we get

$$\Gamma^\alpha_{\mu\nu} = \frac{1}{2} g^{\alpha\rho}\left\{\partial_\mu g_{\nu\rho} + \partial_\nu g_{\mu\rho} - \partial_\rho g_{\mu\nu}\right\}. \tag{3.124}$$

The $\Gamma_{\mu\nu,\rho}$ and $\Gamma^\lambda_{\mu\nu}$ are called the *Christoffel symbols* of the first and second kind respectively. They are sometimes also written as

$$\Gamma_{\mu\nu,\rho} = [\mu\nu, \rho] \tag{3.125}$$

$$\Gamma^\lambda_{\mu\nu} = \left\{\begin{matrix} \lambda \\ \mu\nu \end{matrix}\right\}. \tag{3.126}$$

We define the *absolute differential* of the components A^μ of a 4-vector A by comparing the differential

$$dA = d\left(e_\mu A^\mu\right) \tag{3.127}$$

and $e_\mu \nabla_\nu A^\mu dx^\nu$, i.e. by

$$\nabla A^\mu \equiv \nabla_\nu A^\mu dx^\nu. \tag{3.128}$$

The above applies to contravectors. For covectors we get very similar results. Thus

$$\nabla_\nu A_\mu = \partial_\nu A_\mu - \Gamma^\lambda_{\mu\nu} A_\lambda. \tag{3.129}$$

For any tensor **T** with components $T^{...}_{..}$, we get

$$\nabla_\nu T^{...}_{..} = \partial_\nu T^{...}_{..} + \Gamma^{.}_{\nu\lambda} T^{\lambda..}_{..} + \Gamma^{.}_{\nu\lambda} T^{.\lambda..}_{..} + \cdots$$
$$- \Gamma^\lambda_{\nu.} T^{...}_{\lambda.} - \Gamma^\lambda_{.\nu} T^{...}_{.\lambda} - \cdots; \qquad (3.130)$$

which is not surprising, as in its transformation properties a tensor can be regarded as a (tensor) product of 4-vectors.

For the metric tensor $g_{\mu\nu}$ we thus find

$$\nabla_\rho g_{\mu\nu} = \partial_\rho g_{\mu\nu} - \Gamma^\sigma_{\mu\rho} g_{\nu\sigma} - \Gamma^\sigma_{\nu\rho} g_{\mu\sigma} \equiv 0, \qquad (3.131)$$

using the explicit equation for the Christoffel symbols of the second kind. This is *Ricci's theorem*.

(5) We return now to the manifest covariance of the equations of physics under arbitrary changes of space–time coordinates. In practice, the Lorentzian indices are simply interpreted as general indices and any partial derivatives as covariant derivatives. For example, Maxwell's equations will become

$$\nabla_\nu F^{\mu\nu} = 4\pi J^\mu \qquad (3.132)$$
$$\nabla_\nu{}^* F^{\mu\nu} = 0, \qquad (3.133)$$

while the basic equations of relativistic hydrodynamics will be written as

$$\nabla_\mu J^\mu = 0 \qquad (3.134)$$
$$\nabla_\nu T^{\mu\nu}_{\text{tot}} = 0 \qquad (3.135)$$

(see *Chapter 5*).

(6) We apply these results to *inertial* motions, i.e. those of Galilean observers. We could start from the equation of motion

$$\frac{d^2 x^\mu}{d\tau^2} = 0, \qquad (3.136)$$

and pass to arbitrary curvilinear coordinates $\{x'^\mu\}$. However, it is more instructive to start from the definition of these motions as *geodesics* in Minkowski space. We thus use

$$\delta \int_A^B d\tau = 0 \qquad (3.137)$$

with $d\tau = \{g_{\mu\nu} dx^\mu dx^\nu\}^{1/2}$, in arbitrary coordinates. If ξ is any parameter along a geodesic (here a timelike line), we have

$$\delta \int_A^B \frac{d\tau}{d\xi} d\xi = 0 \qquad (3.138)$$

with

$$\frac{d\tau}{d\xi} = \left\{ g_{\mu\nu} \frac{dx^\mu}{d\xi} \frac{dx^\nu}{d\xi} \right\}^{1/2}. \qquad (3.139)$$

We then get

$$
\delta \int_A^B d\tau = \int_A^B d\xi \left\{ \frac{1}{2} \frac{d\xi}{d\tau} \partial_\sigma g_{\mu\nu} \frac{dx^\mu}{d\xi} \frac{dx^\nu}{d\xi} - \frac{d}{d\xi} \left(\frac{d\xi}{d\tau} g_{\mu\sigma} \frac{dx^\mu}{d\xi} \right) \right\} \delta x^\sigma
$$
$$
+ \frac{d\xi}{d\tau} g_{\mu\nu} \frac{dx^\mu}{d\xi} \delta x^\nu \Bigg|_B^A = 0. \tag{3.140}
$$

The last term comes from integrating by parts and vanishes, remembering that the points A and B are *fixed* (we study geodesics from A to B). From (3.140) and the fact that the δx^μ are arbitrary, we have

$$
\frac{d^2 x^\mu}{d\xi^2} + \Gamma^\mu_{\alpha\beta} \frac{dx^\alpha}{d\xi} \frac{dx^\beta}{d\xi} - \frac{d\tau}{d\xi} \frac{d^2\xi}{d\tau^2} \frac{dx^\mu}{d\xi} = 0; \tag{3.141}
$$

and now choosing a parametrisation of the form $a\xi + b = \tau$, we have

$$
\frac{d^2 x^\mu}{d\tau^2} + \Gamma^\mu_{\alpha\beta} \frac{dx^\beta}{d\tau} \frac{dx^\beta}{d\tau} \equiv \frac{\nabla u^\mu}{d\tau} = 0, \tag{3.142}
$$

i.e. the *geodesic equation*. In this equation $\nabla u^\mu / d\tau$ is a covariant derivative along the motion.

Exercises

3.1   Let $\mathbf{K}$ and $\mathbf{K}'$ be two Galilean frames with relative velocity $\mathbf{v} = (v, 0, 0, 0)$. Let $T^{\mu\nu}$ be the components of a tensor in $\mathbf{K}$. Give the components T'^{01} and T'^{23} in $\mathbf{K}'$. Give also the components $T'^0{}_1$ and $T'^2{}_3$.

3.2   Let $A^{\mu\nu}$ be a second rank tensor. Show that $A^\mu{}_\mu = A_\mu{}^\mu$.

3.3   Let A^μ and B^μ be two 4-vectors. Verify that the following equations are not covariant: (i) $A^\mu = 1/B^\mu$, (ii) $A^0 = B^1$, (iii) $A^0 B^1 = A^1 B^2$.

3.4   Let $A^{\mu\nu}$ and $B^{\mu\nu\alpha}$ be two tensors. Show that $A_{\mu\nu} B^{\mu\nu\alpha}$ is a 4-vector.

3.5   Let $T^{\mu\nu}$ be a symmetric (or antisymmetric) tensor. Show that $T_{\mu\nu}$ has the same symmetry properties.

3.6   Let $S_{\mu\nu}$ and $A^{\mu\nu\alpha}$ be tensors respectively symmetric and antisymmetric with respect to the indices μ and ν. Show that $S^{\mu\nu} A_{\mu\nu}{}^\alpha = 0$.

3.7   Consider in $\mathbf{R}^3$ the tensor $\Delta_{ij} = \delta_{ij} - k_i k_j$, with $k_i k^i = 1$. Show that Δ_{ij} is a projector[5]. On which subspace does it operate?

3.8   Let u^μ and n^μ be two 4-vectors with $u^\mu n_\mu = 0$ and $u^\mu u_\mu = +1$. Which is timelike? Spacelike? Let $\Pi^{\mu\nu} = a u^\mu u^\nu + b n^\mu n^\nu + c \eta^{\mu\nu} + d [u^\mu n^\nu + u^\nu n^\mu]$. Under what conditions is this tensor a projector? On what subspace?

3.9   In $\mathbf{R}^n$, let two tensors $S^{\mu\nu}$ and $A^{\mu\nu}$ be symmetric and antisymmetric respectively. How many independent components do they have?

3.10   Let $K_{\mu\nu\alpha\beta}{}^\lambda$ be a tensor which is symmetric in $(\mu\alpha)$ and antisymmetric in $(\nu\beta)$. How many independent components does it have?

---

[5] A linear hermitian operator such that $P^2 = P$.

3.11   Let $R_{\mu\nu\alpha\beta}$ be symmetric under the interchange $(\mu\nu) \leftrightarrow (\alpha\beta)$ and antisymmetric under the interchange $\mu \leftrightarrow \nu$ or $\alpha \leftrightarrow \beta$, and satisfying $\sum R_{\mu(\nu\alpha\beta)} = 0$, where the sum is over cyclic permutations of the indices (ν, α, β). Indices μ, ν, α, β vary from 1 to n.

(i) Let $z_{\mu\nu}$ be a symmetric tensor. Verify that the tensor $\{z_{\mu\alpha} z_{\nu\beta} - z_{\mu\beta} z_{\nu\alpha}\}$ has the same symmetries as $R_{\mu\nu\alpha\beta}$.

(ii) Calculate the number of independent components of the tensor $R_{\mu\nu\alpha\beta}$.

(iii) Give the most general form of the tensor $R_{\mu\nu\alpha\beta}$ given that space is isotropic and homogeneous.

3.12   Prove the relation $\left(L^{-1}\right)_\mu{}^\nu = \eta_{\mu\alpha}\eta^{\nu\beta}L_\beta{}^\alpha$.

3.13   Show that the Levi–Civita tensor $\varepsilon^{\mu\nu\alpha\beta}$ indeed transforms as a tensor when we are limited to orthochronous Lorentz transformations. How does it behave under the transformations T and/or P?

3.14   In $\mathbf{R}^3$ calculate $\varepsilon_{ijk}a^j b^k$ where a^j and b^k are 3-vectors.

3.15   Let A^μ and B^μ be two 4-vectors and let C^μ be a pseudo 4-vector. Find all the symmetric tensors of rank two formed using A^μ, B^μ, C^μ, etc.

(i) The same as (i), but for antisymmetric tensors.

3.16   In the Euclidean plane $\mathbf{R}^2(O, x, y; \delta_{ij})$, consider polar coordinates

$$\begin{cases} x = r\cos\phi \\ y = r\sin\phi. \end{cases}$$

(i) Find the coordinate curves and give the $\mathbf{e}_r$, $\mathbf{e}_\phi$ corresponding to $x^r \equiv r$ and $x^\phi \equiv \phi$.

(ii) Find the components of the metric tensor $g_{ij}(r, \phi)$ in this coordinate system.

(iii) Deduce the Christoffel symbols Γ^k_{ij}.

(iv) Calculate $\nabla_i \nabla^i A$ in this coordinate system, where A is a scalar field.

(v) Calculate $(\nabla_i \nabla_j - \nabla_j \nabla_i)A$.

(vi) The same questions as (iv) and (v) but where A is a vector field with components A^i.

3.17   The same as 16, but for spherical polar coordinates in $\mathbf{R}^3$:

$$\begin{cases} x = r\cos\phi\sin\theta \\ y = r\sin\phi\sin\theta \\ z = r\cos\theta. \end{cases}$$

3.18   Same questions as 16, but for two-dimensional Minkowski space in relativistic polar coordinates:

$$\begin{cases} x^1 = \tau\cos\phi\,\mathrm{sh}\chi \\ x^2 = \tau\sin\phi\,\mathrm{sh}\chi \\ x^0 = \tau\,\mathrm{ch}\chi \end{cases}$$

(consider only points inside the light-cone $\Gamma^+(0)$).

3.19   Consider in $\mathbf{R}^3$ a cylinder of radius a and axis Oz. On this cylinder we use the coordinates (ϕ, z) defined by

$$\begin{cases} x = a\cos\phi \\ y = a\sin\phi \\ z = z. \end{cases}$$

(i)–(v) Same as in question 16.

(vi) Write down the equation for the geodesics.

3.20   Consider a sphere of radius a and centre O in $\mathbf{R}^3$.

(i)–(v) Same as in question 16.

(vi) Same as in question 16 but using the polar coordinates defined in 3.18 with $a = r$.

(vii) What are the geodesics on the sphere? Write down the equation of a geodesic and solve it.

(viii) Same questions as in the last exercise but for the "pseudo-sphere" with equation $x^{02} - \mathbf{x}^2 = a^2$ ($a =$ const.).

3.21   In Minkowski space, consider the coordinates

$$\begin{cases} \xi = t - x \\ \eta = t + x \\ y = y \\ z = z. \end{cases}$$

Calculate the metric tensor in this new coordinate system. Calculate the new basis vectors as functions of the old. Are the 4-vectors $\mathbf{e}_\xi$ and $\mathbf{e}_\eta$ timelike, spacelike or null? Are they orthogonal?

3.22   Show that the geodesic equation can be rewritten as

$$\frac{d}{d\tau}\left[g_{\mu\nu}\frac{dx^\mu}{d\tau}\right] - \frac{1}{2}\partial_\mu g_{\alpha\beta}\frac{dx^\alpha}{d\tau}\frac{dx^\beta}{d\tau} = 0.$$

4 Gravitation and Special Relativity

Gravitation plays a decisive role in all astronomical phenomena. Even though it is much weaker than electrostatic forces, gravitation dominates these completely at large distances: matter is electrically neutral and electrostatic forces are screened[1]. By contrast, there is no negative mass, and gravity acts at arbitrarily large distances.

We can thus see that in cosmology, gravitational forces will largely determine the evolution of the Universe, and among other things the scale factor $R(t)$ which characterises it. Of course, the state of matter (its motion, energy, composition etc.) also affects this evolution, particularly in influencing the gravitational forces themselves.

This basic point is worth explaining in detail. Consider a body of mass M and characteristic size R. Using the gravitational constant G we can form a characteristic velocity

$$v \sim \left(\frac{GM}{R} \right)^{1/2}, \tag{4.1}$$

which we can interpret as e.g. the escape velocity from its surface. We can imagine that if this velocity becomes of order c, the velocity of light,

$$v \sim c \quad \text{or} \quad \frac{GM}{Rc^2} \sim 1, \tag{4.2}$$

relativistic effects will play an important role in the theoretical description of the phenomena considered. The *gravitational parameter* GM/Rc^2 can also be interpreted as the ratio of gravitational potential energy to rest-mass energy of the body:

$$\frac{GM}{Rc^2} = \frac{\text{potential energy}}{\text{rest-mass energy}} = \frac{GM^2}{R} \cdot \frac{1}{Mc^2}. \tag{4.3}$$

Here again we see the necessity for a *relativistic theory of gravity* if we want to consider phenomena for which the gravitational parameter is not negligible. Equation (4.3) also shows that various energetic contributions to the mass M (particularly the physical state of the system, including its internal motions) will play an important part in such a relativistic theory.

Table 4.1 gives the order of magnitude of the gravitational parameter in a few cases. We see that astrophysical objects for which a relativistic gravitational theory is

---

[1] For planets, it can happen that electrostatic forces are comparable with gravity [see the interesting article by J.M. Lévy–Leblond (1969) and the references therein].

Table 4.1. *Some values of the gravitational parameter GM/Rc²*

object	mass	size	GM/Rc²
	g	cm	
Nucleus	10^{-23}	10^{-13}	10^{-38}
Atom	10^{-23}	10^{-8}	10^{-43}
Man	10^{5}	10^{2}	10^{-25}
Earth	6×10^{27}	6×10^{8}	10^{-9}
Sun	2×10^{33}	7×10^{10}	10^{-6}
Galaxy	10^{44}	10^{23}	10^{-7}
White dwarf	2×10^{33}g	10^{9}	3×10^{-4}
Neutron star	2×10^{33}	10^{6}	0.3
Universe			$\lesssim 1$
Black holes			1

needed are *neutron stars, the Universe,* and possibly existing *black holes*. The study of such objects constitutes *Relativistic Astrophysics*. We should add that all relativistic theories of gravity predict the existence of *gravitational waves*, which, although they have not yet been observed directly, are also part of relativistic astrophysics.

Among the various theoretical possibilities for a relativistic description of gravity, Einstein's General Relativity is the first (which was compatible with experiment) and also the simplest. There are a large number of rival theories, many of which have already been eliminated either by increased accuracy of the classical tests or by new tests (particularly those provided by the *binary pulsar* PSR 1913+16). Moreover, many theoretical arguments lend support to General Relativity.

In this chapter we shall briefly examine what the combination of Special Relativity and Newtonian gravity might reveal.

An essential element of Special Relativity is that each energy defines an associated mass ($E = mc^2$) and *vice versa*. From this it follows that *everything has weight*! Newton's law of universal gravitation,

$$\mathbf{F} = -G\frac{mm'}{r^3}\mathbf{r}, \qquad (4.4)$$

thus implies the gravitational attraction of *all* forms of energy, whether this corresponds to the usual Newtonian mass, to electromagnetic energy, or to gravitation itself. The consequences of this new effect are very profound, and will be studied in *Chapter 7* (*The Principle of Equivalence*). We shall, however, consider several possible consequences of the relativistic equivalence of mass and energy *and* of gravitation, using approximate reasoning. These approximate arguments demonstrate the basic physical processes of General Relativity in a simple way. We shall thus examine in turn (i) the gravitational redshift, (ii) the bending of light rays, (iii) the advance of the perihelion of Mercury, and (iv) the need for nonlinear equations to describe relativistic gravity.

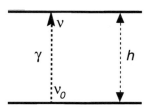

Fig. 4.1. Redshift of the frequency of a photon in a gravitational field. The photon does work against gravity to climb to height h.

4.1   The gravitational redshift

Consider a photon emitted in a gravitational field, e.g. from the surface of a star, and imagine that we observe it at height z_1 above the point of emission, which itself has height z_0 (see **Fig. 4.1**). The photon is emitted with frequency v_0. This frequency v_0 corresponds to an energy $h v_0$ (where h is the Planck's constant) and thus to a mass $m = h v_0 / c^2$.

To reach height z_1 the photon has to do work

$$\Delta W = m \Delta U, \tag{4.5}$$

against gravity, where ΔU is the change in the gravitational potential between emission and observation of the photon,

$$\Delta U = \frac{GM}{R + z_0} - \frac{GM}{R + z_1} \sim -\frac{GM}{R^2}(z_1 - z_0) \tag{4.6}$$

(M is here the mass of the star). At the height where it is observed the photon thus has energy E,

$$E = E_0 - \Delta W, \tag{4.7}$$

where $E_0 = h v_0$ and $E = h v$. We thus have

$$v = v_0 \left[1 - \frac{|\Delta U|}{h} \right] < v_0; \tag{4.8}$$

the photon is thus *redshifted*. The redshift z is given by

$$z = \frac{\Delta v}{v} = \frac{\Delta U}{c^2} = \frac{GM}{R^2 c^2} \cdot (z_1 - z_0). \tag{4.9}$$

Thus, the redshift of a photon emitted at the surface of the star, compared with the case where gravity is absent, is

$$z = \frac{GM}{R c^2}.$$

Of course this reasoning is not at all rigorous: the photon is massless, although everything happens *as if* its energy$/c^2$ were able to interact gravitationally. We see the appearance of the famous gravitational parameter GM/Rc^2 as expected. This redshift has actually been measured, first on Earth [R.V. Pound and G.A. Rebka (1960)], then

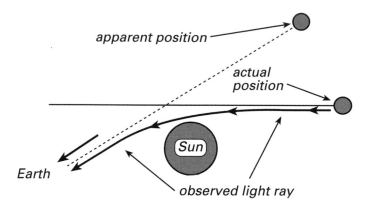

Fig. 4.2. Bending of light rays by the Sun.

in the radiation emitted from the centre of the Sun's disc [R.V. Pound and J.L. Snider (1964)], and finally in white dwarfs [J.L. Greenstein, J.B. Oke and H.L. Shipman (1971)]. All these measurements confirm the relation above, which can also be deduced in General Relativity. We shall see that the redshift can also be used as an argument [A. Schild (1967)] for curved space–time.

4.2   Light bending

Another consequence of the equivalence of mass and energy is that the photon has mass, and thus cannot propagate exactly in a straight line. It is sensitive to the presence of large masses like that of the Sun.

Figure 4.2 shows how light emitted by a star is deviated by the Sun when it passes sufficiently close to it. Thus, by measuring the position of a star at an epoch when its light is deflected by the Sun and another when it is not, we can detect a change of its angular position. This is actually observed (the first observations, comparing photographic plates, date from 1919) [see J. Eisenstaedt (1986)].

We shall estimate the order of magnitude of this effect by imagining the photon to have a mass $h\nu/c^2$ and a momentum $h\nu/c$. This is then a simple Kepler problem (see **Fig. 4.3** for the notation) and the hyperbolic trajectory of the photon subject to the pull of the Sun is given in polar coordinates by [see e.g. L. Landau and E. Lifschitz (1960) or H. Goldstein (1980)]

$$\varphi = \int_r^\infty \frac{\frac{L}{r^2}\,dr}{\left[2m_\gamma\left[E - \frac{GM_\odot m_\gamma}{r}\right] - \frac{L^2}{r^2}\right]^{1/2}}, \tag{4.10}$$

from which we can immediately deduce the deviation $\delta\varphi$ (see **Fig. 4.3**) of a photon

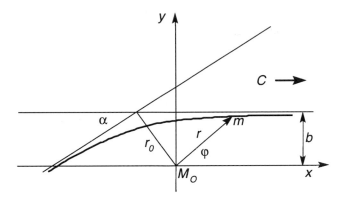

Fig. 4.3. Bending of light by the Sun. Notation used in the text.

which grazes the Sun's surface:

$$\delta\varphi = 2\varphi_0 - \pi = \int_{R_\odot}^{\infty} \frac{\frac{L}{r^2} dr}{\left[2m_\gamma \left[E - \frac{GM_\odot m_\gamma}{r}\right] - \frac{L^2}{r^2}\right]^{1/2}}, \qquad (4.11)$$

where L is the angular momentum[2] of the photon, $L = m_\gamma R_\odot c$, and φ_0 is the angle φ at the photon's minimum distance from the Sun. As the energy E is conserved, we have

$$E = \frac{m_\gamma}{2}\left[\left(\frac{dr}{dt}\right)^2 + r^2\left(\frac{d\varphi}{dt}\right)^2\right] - \frac{GM_\odot m_\gamma}{r}$$

$$= \frac{m_\gamma b^2 c^2}{2R_\odot^2} - \frac{GM_\odot m_\gamma}{R_\odot}. \qquad (4.12)$$

Equations (4.10) to (4.12) finally give

$$\delta\varphi = \frac{2GM_\odot}{bc^2} \sim \frac{2GM_\odot}{R_\odot c^2}.$$

In fact General Relativity gives twice this value, in exact agreement with what is observed: we shall see that this arises from the need to respect the relativistic character of the photon...

4.3   The advance of the perihelion of Mercury

At the end of the 19th century, observations showed a residual advance of its perihelion by 43 arc-seconds per century. Here "residual" means the advance left unexplained

---

[2] The angular momentum of the photon is $L_\odot = bm_\gamma c$, where b is the impact parameter. Strictly speaking, $b \neq R_\odot$, but as the deviation is very small, b is little different from $R_\odot$. Remember that the deviation shown in **Fig. 4.2** is greatly exaggerated.

by Newtonian theory (see *Chapter 1*). An advance or retard of the perihelion requires only that the law of attraction should differ – even very little – from a $1/r^2$ law: the only central force laws giving closed trajectories are $1/r^2$ and r. It follows that *any* modification of Newtonian theory implies a variation of the perihelion.

Relativity implies such a modification. The usual relativistic corrections (i.e. of order v^2/c^2), and those resulting from the energy of the gravitational field explain both the qualitative effect and the order of magnitude of the perihelion advance of Mercury. We denote by M_{0P} and M_P the rest-mass and relativistic mass of a planet, so that

$$M_P = \frac{M_{0P}}{\sqrt{1 - v^2/c^2}}, \tag{4.13}$$

giving the first correction. The gravitational field inside a sphere of radius r centred on the Sun has energy and thus *mass*, which must be added to that of the Sun. The energy density ρ_G of this gravitational field is given by

$$\rho_G = -\frac{(\nabla V)^2}{8\pi G} \tag{4.14}$$

$$= -\frac{M^2 G}{8\pi r^4} \tag{4.15}$$

by analogy with the electromagnetic case (note the minus sign in these two equations, which express the fact that the gravitational interaction between two masses is attractive, whereas the electrostatic force between two like charges is repulsive). The planet thus feels not only the mass of the Sun but also that resulting from the gravitational field, i.e.

$$M(r) = M_\odot + \frac{M_\odot^2 G}{2rc^2}. \tag{4.16}$$

The calculation is straightforward [see R. Sexl and H. Sexl (1979)]; we write the equation for conservation of energy, keeping only terms up to order v^2/c^2, and estimating v^2 for nearly circular orbits:

$$\frac{1}{2} M_P v^2 - M_P \frac{GM(r)}{r} = E \tag{4.17}$$

(*energy conservation*)

$$M_P \approx M_{0P} + \frac{1}{2} M_{0P} v^2/c^2 + \cdots \tag{4.18}$$

(*order* v^2/c^2)

$$\frac{M_{0P} v^2}{r} \approx \frac{GM_{0P} M_\odot}{r^2} \longrightarrow v^2 \approx \frac{GM_\odot}{r}. \tag{4.19}$$

(*circular orbit*)

We can see that the corrections caused by the energy of the gravitational field are of the same order as those in v^2/c^2. This is not surprising as $v^2/c^2 \approx GM_\odot/rc^2$. Using these results in the equation of energy conservation (4.17) we get

$$E = \frac{1}{2}M_{0P}v^2 - G\frac{M_{0P}M_\odot}{r} - \frac{3G^2 M_{0P}M_\odot^2}{4r^2c^2} \tag{4.20}$$

which shows that we must add an extra potential $V_S(r)$

$$V_S(r) = -\frac{3}{4}\frac{GM_\odot}{Rc^2}\frac{R}{r}\frac{GM_\odot M_{0P}}{r} \tag{4.21}$$

to the usual energy. The perihelion advance $\delta\Psi$ of the planet is essentially of order the ratio between the Newtonian potential and that caused by the relativistic corrections, i.e.

$$\frac{\delta\Psi}{2\pi} \approx \frac{V_S}{V_N} = \frac{3}{4}\frac{GM_\odot}{Rc^2}. \tag{4.22}$$

General Relativity gives an effect four times larger: nevertheless, this crude reasoning[3] gives both the order of magnitude and the sign (advance not retard) of the effect.

4.4   The need for nonlinear equations for gravitation

The classical gravitational potential V is given by Poisson's equation as a function of the mass density ρ,

$$\Delta V = 4\pi G\rho. \tag{4.23}$$

Since energy in all its forms is equivalent to mass, we must include in ρ all the corresponding mass densities, *including that of the gravitational field ρ_G*,

$$\rho_G = -\frac{(\nabla V)^2}{8\pi Gc^2}. \tag{4.24}$$

Poisson's equation thus becomes

$$\Delta V + \frac{(\nabla V)^2}{2c^2} = 4\pi G\rho_{\text{mat}}. \tag{4.25}$$

This equation in *nonlinear*, expressing the fact that *the gravitational field is coupled to itself*. Of course, this equation is neither manifestly covariant, nor even relativistic: it simply gives us an idea of what to expect. If we try to find a relativistic equation, we must first make the substitution $\Delta \rightarrow \Box$ and also know the relativistic form of V. It can be shown that if V is chosen as a *scalar*, one gets a *retard* of the perihelion of the planets. If V is the fourth component of a 4-vector, one finds a theory in which gravitation is *repulsive*. On the other hand, and this is the aim of the following chapters, a second-rank tensor produces a sensible theory, even though this is not the only possibility.

---

[3] A full calculation is given in L. Landau and E. Lifschitz (1960).

Exercises

4.1   Using the perturbing potential found in this chapter:

$$V_S(r) = -\frac{3}{4}\frac{GM_\odot}{Rc^2}\frac{R}{r}\frac{GM_\odot M_{0P}}{r},$$

calculate the perihelion advance for Mercury exactly.
[Reference: L. Landau and E. Lifschitz (1960); exercise no. 10; chapter 1.]

4.2   Explain (4.16): what mass do we actually measure? Explain your answer.

5 Electromagnetism and Relativistic Hydrodynamics

We still have two important questions. We have first to establish that electromagnetic theory actually does exhibit the Lorentz invariance required by Einstein's relativity principle, and in particular find a manifestly covariant form of Maxwell's equations. Also we need some elementary notions of relativistic hydrodynamics in view of various applications to relativistic astrophysics.

5.1 Densities and currents

Consider a system of N particles, with the space–time trajectory of the ith particle denoted by $x_i^\mu(\tau)$. The expression

$$J^\mu(x) = \sum_{i=1}^{i=N} \int \delta^{(4)} \left[x^\mu - x_i^\mu(\tau) \right] u_i^\mu(\tau) d\tau \tag{5.1}$$

is a 4-vector, the particle *4-current*. We verify that it does indeed correspond to a particle current in terms of its timelike and spacelike components. We show this for the 0 component, as the proof is similar for the other components. We have[1]

$$J^0(x) = \sum_{i=1}^{i=N} \int \delta \left[t - t_i(\tau) \right] \cdot \delta^{(3)} \left[\mathbf{x} - \mathbf{x}_i(\tau) \right] \frac{dt_i}{d\tau} d\tau$$

$$= \sum_{i=1}^{i=N} \int \delta \left[t - t_i(t) \right] \cdot \delta^{(3)} \left[\mathbf{x} - \mathbf{x}_i(t) \right] dt_i$$

$$= \sum_{i=1}^{i=N} \delta^{(3)} \left[\mathbf{x} - \mathbf{x}_i(t) \right],$$

where the second line follows from the first by using the same parametrisation for the various particle trajectories, i.e. the usual time coordinate, and then integrating over t.

---

[1] The integral is a line integral along the trajectory of the ith particle.

The last expression is just the usual particle density. Similarly, we find

$$\mathbf{J}(x) = \sum_{i=1}^{i=N} \mathbf{v}_i(t) \delta^{(3)} \left[\mathbf{x} - \mathbf{x}_i(t)\right] \tag{5.2}$$

where $\mathbf{v}$ is the usual velocity: $\mathbf{v} \equiv d\mathbf{x}/dt$.

Multiplying both sides of the definition of $J^\mu(x)$ by the common charge of these particles – assuming that they are charged – we see that the four quantities "charge density" and "electric current" form a 4-vector.

What meaning can we give to the invariant $\left[J^\mu(x) \cdot J_\mu(x)\right]^{1/2}$? Similarly, how do we interpret the (4-) direction of $J^\mu(x)$? To see this, and to simplify the notation, we denote the first invariant by $n(x)$ and let $v^\mu(x)$ be the unit vector in the direction of $J^\mu(x)$:

$$J^\mu(x) = n(x) \cdot v^\mu(x). \tag{5.3}$$

We note first that J^μ and v^μ are both timelike 4-vectors[2]; they are linear combinations of timelike 4-vectors with positive time components: the 4-velocities of particles have these properties and the δ are certainly positive. We can thus use a frame in which v^μ reduces to its zeroth component: $v^\mu = (1, \mathbf{0})$. In this frame J^μ becomes $J^\mu = (n, \mathbf{0})$; in other words, in this frame the 3-current vanishes and n is the particle density.

The total number of particles in the system, or for charged particles, the total charge, can be found by evaluating the flux of the 4-vector across a spacelike 3-surface (see **Fig. 5.1**):

$$N = \int_\Sigma d\Sigma_\mu J^\mu(x), \tag{5.4}$$

where $d\Sigma_\mu$ is the differential form "surface element" (see *Appendix B*):

$$d\Sigma_\mu = \frac{1}{3!} \varepsilon_{\mu\nu\alpha\beta} dx^\nu \wedge dx^\alpha \wedge dx^\beta. \tag{5.5}$$

This makes intuitive sense. We need to integrate a 4-vector in such a way as to obtain a scalar: this requires us to integrate *covariantly* (and so to contract J^μ with the only 4-vector we have, i.e. $d\Sigma_\mu$). Moreover, the total particle number is equal to the number of intersections of their worldlines with the 3-surface Σ. It is easy to check that if the spacelike 3-surface is chosen as $t = $ const., then $d\Sigma_\mu$ reduces to (see *Appendix B*)

$$d\Sigma_\mu = (d^3x, 0, 0, 0)$$

and we have

$$\int d\Sigma_\mu J^\mu(x) = \int d^3x J^0(x) = N. \tag{5.6}$$

The preceding properties hold quite generally, even for a continuous system. To understand why a scalar quantity (like particle number or total charge) is associated with a 4-vector density, we can use the following heuristic argument. With total charge

---

[2] J^μ is timelike for like particles; if we deal with a mixture of particles bearing both signs of charge, then J^μ can be spacelike or even null.

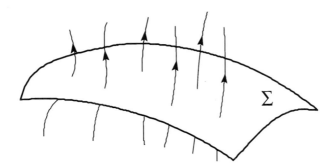

Fig. 5.1. Particle flux across a spacelike surface Σ. The total number of particles in the system is equal to the number of intersections of their worldlines with Σ.

N_0, for example, we can associate the density $n = N_0/V_0$, where V_0 is the volume containing the system. Under a Lorentz transformation, the volume V_0 contracts to $V = V_0(1 - v^2)^{1/2}$; in other words the density becomes $n/(1 - v^2)^{1/2}$. Thus, it behaves like the fourth component of a 4-vector. The same argument holds for any scalar quantity.

This argument is quite general: *any tensorial quantity of type* **T** *corresponds to a density with an additional Lorentz index* $\mathcal{T}^\mu$. Hence, the energy–momentum P^μ of a physical system corresponds to an energy–momentum density called the *energy–momentum tensor* $T^{\mu\nu}$, which we shall study later. The total energy–momentum is given by

$$P^\mu = \int_\Sigma d\Sigma_\nu \, T^{\mu\nu}. \tag{5.7}$$

Similarly, the spin corresponds to a tensor with three indices.

We consider the particle 4-current further. If we choose two arbitrary spacelike 3-surfaces Σ and Σ', it is clear that they have the same number of intersections with the worldlines of the particles: this number is constant, at least provided that we are dealing with stable particles. To express this conservation of particle number (or charge, or other scalar quantity) we consider the space–time volume bounded by the two surfaces Σ and Σ' and the surface S generated by a current tube (see **Fig. 5.2**). Stokes' theorem (see *Appendix B*) gives

$$\int d^4x \, \partial_\mu J^\mu(x) = \int_\Sigma d\Sigma_\mu J^\mu(x) - \int_{\Sigma'} d\Sigma_\mu J^\mu(x)$$
$$+ \int_S d\Sigma_\mu J^\mu(x) = 0, \tag{5.8}$$

which expresses the conservation of particle number (the last integral vanishes as J^μ is orthogonal to $d\Sigma_\mu$, in the sense of the metric of $\mathcal{M}$; the second integral has a minus sign, as $d\Sigma_\mu$ is parallel to the normal to Σ, and this normal is oriented towards the past on Σ', and towards the future on Σ).

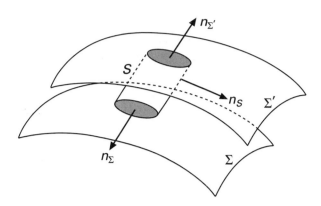

Fig. 5.2. Conservation of particle number. Stokes' theorem is applied to the volume bounded by a current tube with surface S, and two spacelike surfaces Σ and $\Sigma' \cdot n_S$ is the normal to S, while n_Σ ($n_{\Sigma'}$) is the normal to Σ (Σ'). Here the outward normal to the surface is chosen.

As the integral (5.8) vanishes (conservation of N implies that the number of intersections of worldlines with Σ and with Σ' are equal: see **Fig. 5.1**) for any integration volume, it follows that the integrand vanishes (*modulo* some mathematical conditions which always hold in practice), and we have

$$\partial_\mu J^\mu(x) = 0. \tag{5.9}$$

In the usual coordinates this becomes

$$\frac{\partial n}{\partial t} + \nabla \cdot \mathbf{J} = 0; \tag{5.10}$$

which is the customary continuity equation.

5.2   The equations of electromagnetism

To find the form of changes of inertial frame which leave the velocity of light constant, we assumed that the laws of electromagnetism retained the same forms in all Galilean frames. In particular, a spherical wave remains a spherical wave. This property allowed us to find the Lorentz transformations. However, we still have to *verify* that the equations of electromagnetism (i.e. Maxwell's equations) satisfy this condition, and that in particular they can be written in *manifestly covariant* form. To show this, we take Maxwell's equations in the form

$$\nabla \cdot \mathbf{E} = 4\pi n \tag{5.11}$$

$$\nabla \wedge \mathbf{B} - \frac{\partial \mathbf{E}}{\partial t} = 4\pi \mathbf{J}, \tag{5.12}$$

$$\nabla \cdot \mathbf{B} = 0 \tag{5.13}$$

$$\nabla \wedge \mathbf{E} + \frac{\partial \mathbf{B}}{\partial t} = 0, \tag{5.14}$$

where we have paired the equations with nonzero right hand sides and the homogeneous ones, and where we have set $c = 1$. These equations are *linear* in **E** and **B**, and first order in the various partial derivatives ∂_μ. As n and **J** can be grouped as a 4-vector J^μ, it *must* be possible to cast the first pair of Maxwell's equations in the form

$$\partial_\mu F^{\mu\nu} = 4\pi J^\nu. \tag{5.15}$$

In other words, the fields **E** and **B** should form the components of a second rank tensor $F^{\mu\nu}$. However, although **E** and **B** have 6 independent components, the tensor **F** has 16 *a priori*. This tensor must therefore possess particular symmetry properties: we note that an *antisymmetric* second rank tensor in Minkowski space has 6 independent components. The conservation of charge

$$\partial_\mu J^\mu = 0, \tag{5.16}$$

requires

$$\partial_{\mu\nu} F^{\mu\nu} = 0, \tag{5.17}$$

whatever **F** is. It follows that this tensor is antisymmetric as $\partial_{\mu\nu}$ is symmetric. We can identify the components of **F** in terms of **E** and **B** by identifying the first pair of Maxwell's equations with the covariant equation (5.15). We then find

$$\|F^{\mu\nu}\| = \begin{vmatrix} 0 & -E_1 & -E_2 & -E_3 \\ E_1 & 0 & -B_3 & B_2 \\ E_2 & B_3 & 0 & -B_1 \\ E_3 & -B_2 & B_1 & 0 \end{vmatrix} \tag{5.18}$$

We find the components $F_{\mu\nu}$ and $F^{\mu\nu}$ easily by raising and lowering the indices of $F^{\mu\nu}$ using the metric tensor $\eta_{\mu\nu}$.

The second pair of Maxwell's equations, the *constraint* equations for the electromagnetic field, can be found in a similar way[3] and reads

$$\partial_\mu \,^*F^{\mu\nu} = 0, \tag{5.19}$$

where $^*F^{\mu\nu}$ is given by

$$^*F^{\mu\nu} = \frac{1}{2}\varepsilon^{\mu\nu\alpha\beta} F_{\alpha\beta}. \tag{5.20}$$

(***F** is the *dual* tensor to **F**.) In writing the preceding expression for $^*F^{\mu\nu}$, it is easy to see that the second pair of Maxwell's equations can be written as

$$\partial_\mu F_{\nu\lambda} + \partial_\nu F_{\lambda\mu} + \partial_\lambda F_{\mu\nu} = 0. \tag{5.21}$$

This second pair imply[4] that the fields **E** and **B** can be derived from *electromagnetic potentials* V and **A** using the relations

$$\mathbf{B} = \nabla \wedge \mathbf{A}, \qquad \mathbf{E} = -\nabla V - \frac{\partial \mathbf{A}}{\partial t}, \tag{5.22}$$

---

[3] This second pair of equations (i) is linear in the derivatives ∂_μ, (ii) is linear in $F^{\mu\nu}$ and (iii) can only be written in a manifestly covariant way using the tensors $\eta_{\mu\nu}$, $F_{\mu\nu}$ and $\varepsilon_{\mu\nu\alpha\beta}$.
[4] See *Appendix B*.

which can be rewritten in the manifestly covariant form

$$F_{\mu\nu} = \partial_\mu A_\nu - \partial_\nu A_\mu, \tag{5.23}$$

where we have set $A^\mu = (V, \mathbf{A})$. The 4-vector character of A_μ is manifest when we replace $\mathbf{F}$ by these potentials in Maxwell's equation (5.15): we get

$$\Box A^\mu - \partial^\mu [\partial_\nu A^\nu] = 4\pi J^\mu; \tag{5.24}$$

showing the tensorial nature of $(V, \mathbf{A})$. This equation could also have been found from the non-manifestly-covariant form (5.11)–(5.14) of Maxwell's equations, the definition (5.22) of the potentials, and grouping the components of J^μ. Equation (5.24) is indeed covariant; the operators $\eta_{\mu\nu}$ and $\partial^\mu \partial_\nu$ have the transformation properties indicated by their indices.

If we apply to the 4-potential A^μ a *gauge transformation*

$$A^\mu \longrightarrow A^\mu + \partial^\mu \Lambda, \tag{5.25}$$

where Λ is an arbitrary function, we note that the electromagnetic field $F^{\mu\nu}$ does not change. The definition of the potentials leaves a certain arbitrariness, which can be lifted by *imposing* supplementary conditions[5] called *gauge conditions*. An example is the *Lorentz condition*

$$\partial_\mu A^\mu = 0; \tag{5.26}$$

this condition does not determine the 4-potential uniquely, as a gauge transformation whose function Λ satisfies

$$\Box \Lambda = 0 \tag{5.27}$$

gives the same electromagnetic field. Of course, other gauge conditions are possible.

From the electromagnetic field tensor $\mathbf{F}$ and its dual $^*\mathbf{F}$ one can form two *invariants*

$$F^{\mu\nu} \cdot F_{\mu\nu} \quad \text{and} \quad {}^*F_{\mu\nu} \cdot F^{\mu\nu}. \tag{5.28}$$

In terms of the fields $\mathbf{E}$ and $\mathbf{B}$, these two invariants are

$$F^{\mu\nu} \cdot F_{\mu\nu} = 2[\mathbf{B}^2 - \mathbf{E}^2] \tag{5.29}$$

$$^*F_{\mu\nu} \cdot F^{\mu\nu} = 4\mathbf{E} \cdot \mathbf{B}. \tag{5.30}$$

For an electromagnetic wave propagating in vacuum, these two invariants vanish: the fields $\mathbf{E}$ and $\mathbf{B}$ are orthogonal and the wave has as much magnetic as electric energy.

---

[5] In a sense, once A^μ is assumed to satisfy a gauge condition which is not manifestly covariant, such as the Lorentz condition, it is no longer really a 4-vector: in changing the system of inertia, one has also to make a gauge transformation.

5.3   The energy–momentum tensor

Above, we considered a system of identical particles, and the corresponding density of energy and momentum is the energy–momentum tensor,

$$T^{\mu\nu}_{\text{part}} = \sum \int d\tau p^\mu u^\nu_i \delta[x - x_i(\tau)]. \tag{5.31}$$

In the non-relativistic approximation this reduces to

$$T^{00}_{\text{part}} = \sum p^0_i \delta[\mathbf{x} - \mathbf{x}_i(t)], \tag{5.32}$$

where p^0 reduces to the usual kinetic energy if we neglect the rest-mass energy of the particles. If, moreover, $p^\mu = u^\mu$, T^{00} is indeed the energy density of the system. Similarly, we can verify that T^{i0} is a momentum density, that T^{0i} is an energy flux, that T^{ij} represents the *stress tensor*, and finally that $T^{\mu\nu}$ is *symmetric*.

We examine this more closely. We start with the spatial components of $T^{\mu\nu}$. The element of 3-momentum is

$$dP^i = T^{ij} d\Sigma_j$$
$$= T^{ij} \frac{1}{3!} \varepsilon_{jk\lambda\mu} dx^\kappa \wedge dx^\lambda \wedge dx^\mu$$

or

$$dP^i \sim T^{ij} dx^0 dx^k dx^l.$$

dP^i/dx^0 is the force in the direction i and $dP^i/dx^0 \times 1/(dS = dx^k dx^l)$ is the force (in direction i) applied to a surface element oriented parallel to the plane (k, l); this is a tension.

As the equations of motion of a system of particles can often be derived from a Lagrangian, including the relativistic case [A.O. Barut (1965); G. Kalman (1961); A. Peres and N. Rosen, (1960)], we can show that, despite the different definition, the energy–momentum tensor of the fields and particles can be given the same physical interpretation (see *Appendix C*).

In general, a physical system is made up of fields (see *Appendix C*) and particles, and the total energy–momentum tensor in the absence of external fields is

$$T^{\mu\nu}_{\text{tot}} = T^{\mu\nu}_{\text{part}} + T^{\mu\nu}_{\text{field}}. \tag{5.33}$$

For identical charged particles the conservation of *total* energy and momentum of the system are

$$\partial_\mu T^{\mu\nu}_{\text{tot}} = 0, \tag{5.34}$$

giving the conservation law

$$\partial_\mu T^{\mu\nu}_{\text{part}} = 4\pi F^{\mu\nu} J_\mu; \tag{5.35}$$

the right hand side of this equation represents the *coupling* between the electromagnetic fields and the charges of the system.

5.4   Relativistic hydrodynamics

We make here some brief remarks on relativistic hydrodynamics, which are needed not only for handling problems in General Relativity (cosmology, dense stars and black holes) but also problems in high-density matter (heavy ion collisions etc.).

Just as in Newtonian theory, relativistic hydrodynamics is based on the explicit form of the matter current J^μ (or other possible currents) and the total energy–momentum tensor $T^{\mu\nu}$, and particularly the *conservation equations*

$$\begin{cases} \partial_\mu J^\mu &= 0 \\ \partial_\mu T^{\mu\nu} &= 0. \end{cases} \tag{5.36}$$

A physically interesting case is given by a *perfect fluid*. Here there is only one 4-vector, denoted u^μ. This 4-vector is then necessarily equal to the mean 4-velocity of the fluid, as the current can only be proportional to this vector, which we assume to be a unit vector (although this is not an important restriction) thus u^μ must be timelike, as all 4-currents are. Then the most general form of the energy–momentum tensor is

$$T^{\mu\nu} = Au^\mu u^\nu + B\eta^{\mu\nu} \tag{5.37}$$

$$= (A + B)u^\mu u^\nu - B\Delta^{\mu\nu}(u), \tag{5.38}$$

where $\Delta^{\mu\nu}(u)$ is the projection tensor orthogonal to u^μ, i.e. onto the rest-space of the time-direction defined[6] by u^μ:

$$\Delta_{\mu\nu}(u) \equiv u^\mu u^\nu - \eta^{\mu\nu}. \tag{5.39}$$

The *only* symmetric tensors we can construct from the metric tensor $\eta^{\mu\nu}$ and the 4-vector u^μ are exactly $\eta^{\mu\nu}$ and $u^\mu u^\nu$: thus $T^{\mu\nu}$, being a symmetric tensor, must therefore be a linear combination of these tensors[7]. Let us now interpret the arbitrary functions A and B. In a reference frame comoving with the fluid, u^μ reduces to

$$u^\mu = (1, \mathbf{0}), \tag{5.40}$$

and the only non-vanishing components of $T^{\mu\nu}$ are

$$T^{00} = A, \quad T^{ii} = B \qquad (i = 1, 2, 3). \tag{5.41}$$

A thus represents the *energy density* ρ of the system, while T^{ij} represents the *stresses* of the medium: these stresses are the same in all directions i, the medium is isotropic, and B is just the usual *pressure* P. Thus the most general form of the energy–momentum tensor for a perfect fluid is

$$T^{\mu\nu} = (\rho + P)u^\mu u^\nu - P\eta^{\mu\nu}. \tag{5.42}$$

In the expression "perfect fluid", the word "perfect" refers to the *absence of transport processes* – dissipative phenomena – like heat conduction, viscosity, or diffusion. This

---

[6] One also finds the definition $\Delta^{\mu\nu} = \eta^{\mu\nu} - u^\mu u^\nu$.
[7] We note that u^μ is the timelike eigenvector of $T^{\mu\nu}$; the three other spacelike ones are completely arbitrary.

is shown by the fact that the energy–momentum tensor contains *no gradients* of macroscopic quantities (temperature, velocity, density, etc.).

The hydrodynamics problem thus has *six* unknown functions [P, ρ, n (where n is the particle density) and the three independent components of u^μ] which obey only *five* equations, the five conservation equations (5.36). We thus need another equation: giving an *equation of state* of the type

$$P = P(\rho), \qquad n = n(\rho) \tag{5.43}$$

or

$$P = P(n), \qquad \rho = \rho(n), \tag{5.44}$$

which closes the system, provided that the equation of state does not itself introduce further parameters. For example, if the equation of state depends on temperature, we also need an equation governing the transport of heat, and so on.

One also needs *relativistic thermodynamics*. One can show that the usual relations hold in a local frame [see e.g. J. Ehlers (1971) or C.W. Misner, K.S. Thorne and J.A. Wheeler (1973)] and we shall return to this when necessary.

Exercises

5.1 Let $F^{\mu\nu}$ be an antisymmetric tensor. Calculate $F^\mu{}_\nu$ and $F_{\mu\nu}$. Calculate $*F^{\mu\nu}$.

5.2 Let P be a spacelike 3-plane with equation $k_\mu x^\mu = $ const. Calculate the 3-surface element of $d\Sigma_\mu$.

5.3 Calculate the surface element $d\Sigma_\mu$ of the surface $x_\mu x^\mu = a^2$.

5.4 Consider the motion of a particle of charge e and mass m in an external electromagnetic field $A^\mu(x)$. The action is given by

$$S = \int_a^b \{[\eta_{\mu\nu} dx^\mu dx^\nu]^{1/2} + eA_\mu(x)dx^\mu\}.$$

By requiring $\delta S = 0$, find the corresponding equations of motion.

5.5 Consider a system of N particles, with coordinates $x_i^\mu(\tau)$ in Minkowski space, a 4-velocity $u_i^\mu(\tau)$; $i = 1, 2, \ldots, N$; $\tau = $ proper time. The particles are identical, of charge e and mass m. Their dynamics are defined by

$$m \frac{d^2 x_i^\mu(\tau)}{d\tau^2} = eF^{\mu\nu}(x_i)u_{i\nu}(\tau), \qquad i = 1, 2, \ldots, N,$$

where $F_{\mu\nu}$ is the electromagnetic field. We set

$$R(x, u) = \sum_{i=1}^{i=n} \int d\tau \, \delta^{(4)}[x - x_i(\tau)] \cdot \delta^{(4)}[u - u_i(\tau)]$$

$$[x \equiv x^0, x^1, x^2, x^3, \ldots].$$

(i) Show that the system's 4-current is given by

$$J^\mu(x) = e \int d^4u\, R(x, u) u^\mu.$$

(ii) Show that the energy–momentum tensor of the particles is given by

$$T^{\mu\nu}_{\text{part}} = \int d^4u R(x, u) m u^\mu u^\nu.$$

(iii) Show that $R(x, u)$ obeys the equation

$$u^\mu \partial_\mu R(x, u) + \frac{e}{m} F^{\mu\nu}(x) u_\nu \frac{\partial}{\partial u^\mu} R(x, u) = 0.$$

(iv) By integration of this relation, deduce that

$$\partial_\mu J^\mu(x) = 0.$$

(v) In an analogous fashion, show that

$$\partial_\nu T^{\mu\nu}_{\text{part}} = J_\nu(x) F^{\mu\nu}(x).$$

5.6   Let a^μ, b^μ, c^μ be three 4-vectors of $\mathcal{M}$. Write down the most general second-rank tensor which is (i) symmetric, (ii) antisymmetric.

5.7   Let $F_{\mu\nu}$ be the electromagnetic field tensor. Let u^μ and h^μ be two 4-vectors such that

$$u^\mu u_\mu = 1, \quad h^\mu h_\mu = h^2, \quad u^\mu h_\mu = 0.$$

(i) Calculate $F_{\mu\nu} F^{\mu\nu}$ and $F_{\mu\nu}{}^* F^{\mu\nu}$.

(ii) How should we interpret h^μ and u^μ if $F_{\mu\nu} = -\varepsilon_{\mu\nu\alpha\beta} h^\alpha u^\beta$?

(iii) Then find

$$h^\mu F_{\mu\nu}, \quad u^\mu F_{\mu\nu}, \quad u^{\mu*} F_{\mu\nu}, \quad h^{\mu*} F_{\mu\nu}.$$

Interpret your results.

(iv) Calculate the energy–momentum tensor of the electromagnetic field as a function of u^μ and h^μ.

(v) Is h^μ really a 4-vector?

5.8   Suppose we try to write a covariant form of Ohm's law, $\mathbf{J} = \sigma \mathbf{E}$. In the relativistic case we thus write an expression of the form

$$J^\mu = \Lambda^{\mu\nu\beta} F_{\alpha\beta},$$

where J^μ is the 4-current and $F_{\alpha\beta}$ is the electromagnetic field tensor. In the following we shall assume that all the quantities appearing are actually Fourier transforms.

(i) How do we write the conservation of charge and what does this imply for the tensor $\Lambda^{\mu\alpha\beta}$?

(ii) What symmetries does the tensor $\Lambda^{\mu\alpha\beta}$ have? How many independent components does it have?

(iii) Give the most general form of $\Lambda^{\mu\alpha\beta}$, given that it can only be formed from k^μ, the wavevector, and u^μ, a timelike 4-vector parallel to J^μ.

5.9 Let $A^\mu(x)$ be the electromagnetic 4-potential. We try to write an equation for it using the following postulates: (i) manifest covariance: (ii) the equation should be second-order, (iii) linearity. We further assume that the 4-current J^μ, the source term for A^μ, should appear on its own on the right hand side of the equation.

(i) Find the most general form of the equation satisfied by $A^\mu(x)$.

(ii) Show that this equation is totally determined if we require gauge invariance and that $\partial_\mu J^\mu = 0$.

(iii) Show that the equation is completely determined if we impose that plane waves of the type

$$A^\mu(x) = A^\mu(k)e^{ik\cdot x}$$

are solutions (with $J^\mu \equiv 0$), with $k^2 = 0$.

5.10 Let $F^{\mu\nu}$ be an electromagnetic field tensor such that $\Gamma_{\mu\nu}{}^*F^{\mu\nu} = 0$. Show that if $F_{\mu\nu}F^{\mu\nu} > 0$, there exists a Galilean reference system in which $F_{\mu\nu}$ reduces to a pure magnetic field. What happens if $F_{\mu\nu}F^{\mu\nu} < 0$?

5.11 By hypothesis, the vacuum is homogeneous (in the sense of Minkowski space) and isotropic. If we try to treat it as a material medium, (i) give the most general form of the associated energy–momentum tensor, and (ii) deduce the form of the corresponding "equation of state".

5.12 The relativistic Maxwell–Boltzmann distribution is

$$f(p^\mu) = 2A\,\theta(p^0)\,\delta(p^2 - m^2)\,\exp\{-\beta u^\mu p_\mu\} \tag{1}$$

and is normalized by

$$J^\mu = \int d^4p\,\frac{p^\mu}{m}\,f(p^\lambda) \tag{2}$$

where J^μ is the particle 4-current, and $\theta(p^0)$ is the Heaviside step function.

(i) Write the expression (2) in the form of a triple integral. Using a change of variables of the type $p^0 = mch\,\chi$, evaluate J^μ as a function of A and u^μ, then deduce A.

(ii) Explain the form (1) for the relativistic Maxwell–Boltzmann distribution. Show that for $\beta m \gg 1$, the function

$$f(p) = A\exp\{-\beta u^\mu p_\mu\}$$

reduces to the ordinary Maxwell distribution.

(iii) We set

$$T^{\mu\nu} = \int d^4p \frac{p^\mu p^\nu}{m} f(p).$$

Interpret the various components of this tensor. Show that this is indeed the energy–momentum tensor (of a perfect gas).

(iv) Verify that $T^{\mu\nu}$ must have the form

$$T^{\mu\nu} = (\rho + P)u^\mu u^\nu - P\eta^{\mu\nu}.$$

Thus calculate the energy density ρ and the pressure P. Verify that $P = n\beta^{-1}$, with $n^2 = J^\mu J_\mu$.

(v) Is energy in equipartition among these degrees of freedom?

(vi) Calculate $T^{\mu\nu}$ for $m \to 0$, and deduce that $P = \frac{1}{3}\rho$ for photons or neutrinos. *You may use the results*:

$$K_n(x) = \int d\chi \operatorname{ch} n\chi \exp\{-x \operatorname{ch}\chi\}$$

$$= \frac{x^n}{1.3.5\ldots 2n-1} \int d\chi \operatorname{sh}^{2n}\chi \exp\{-x \operatorname{ch}\chi\}$$

$$K_{n+1}(x) - K_{n-1}(x) = 2n \frac{K_n(x)}{x}$$

(K_n *are MacDonald or Kelvin functions, modified Bessel functions.*)

$$K_n(x) = \sqrt{\frac{\pi}{2x}}e^{-x}\left\{1 + \frac{(4n^2-1)}{1!8x} + \frac{(4n^2-1)(4n^2-3^2)}{2!(8x)^3} + \cdots\right\}$$

$$\text{for} \quad x \gg 1$$

$$K_n(x) = \frac{1}{2}(n-1)!\left[\frac{2}{x}\right]^n + \cdots$$

$$K_0(x) = -0.5772 - \log\frac{x}{2} + \cdots$$

$$\text{for} \quad x \ll 1$$

$$K'_n(x) = \frac{n}{x}K_n(x) - K_{n+1}(x)$$

5.13   How do we write the relativistic Fermi–Dirac distribution? How is it normalised? Calculate the pressure and density of a completely degenerate Fermi gas at $T = 0K$. Use the result

$$f(x) \equiv \int_0^x t^2[1+t^2]^{1/2}dt$$

$$= \frac{1}{4}\left\{x(1+x^2)^{3/2} - \frac{1}{2}[x(1+x^2)^{1/2} - \log(x+(1+x^2)^{1/2})]\right\}.$$

[References for exercises **12** and **13**: J.L. Synge (1957); S.R. de Groot *et al.* (1980).]

5.14 Starting from the conservation of energy and momentum, $\partial_\mu T^{\mu\nu} = 0$, for a perfect fluid, show that

$$dU + P dV = 0.$$

Justify the fact that a perfect fluid is non-dissipative.

6   What is Curved Space?

We shall see at the end of this chapter that relativistic gravitation will require the introduction of curved space–time. We should ask what "curved space" actually is. Our intuition, based on surfaces in $\mathbf{R}^3$, can be extended to spaces of dimension larger than two. We can deduce the essentials from simple examples [like the sphere] of surfaces in $\mathbf{R}^3$. We shall do this, first by studying some geometric properties of known surfaces, and then comparing them with corresponding properties of the plane $\mathbf{R}^2$. We then define the Riemann curvature and finally give arguments leading to curved space–time.

6.1   Some manifestations of curvature

Here we consider only a sphere of radius R embedded in $\mathbf{R}^3$: clearly this is a curved surface. We shall try to construct elementary geometrical figures whose properties we compare with the analogous plane figure.

(1) Geodesic triangle (**Fig. 6.1**): In the plane, a triangle is formed by the intersection of three non-parallel straight lines. On a sphere, arcs of great circles play the role of straight lines: a straight line in the plane $\mathbf{R}^2$, is the shortest path (geodesic) between two points, while for the sphere S_2 the geodesics are arcs of great circles. We thus can construct a triangle on the sphere S_2; between two points A and B there is an arc of a great circle (exactly one, if the distance AB is to be a minimum and A, B are not at poles). We now choose point C so that $\widehat{CAB}$ equals $\pi/2$. We could also choose C so that $\widehat{CBA} = \pi/2$ (see **Fig. 6.1**). We thus get *a triangle the sum of whose angles is greater than* π.

This is generally true: in a curved space, the sum of the angles of a geodesic triangle is usually different from π.

(2) Geodesic circle (**Fig. 6.2**): Consider now a circle of radius a with centre at point O (arbitrary) of a sphere. This means that, starting from point O, we trace out all the geodesic arcs of length a.

If θ (see **Fig. 6.2**) is the angle subtended at the centre of the sphere by one one of the radii of the geodesic circle, and r_0 the radius of the circle *in* $\mathbf{R}^3$, we have

$$a = R\theta \qquad \text{and} \quad r_0 = R \sin \theta \tag{6.1}$$

with $0 < \theta < \pi/2$, i.e. $0 < \sin \theta < 1$, where R is the radius of the sphere. Under these

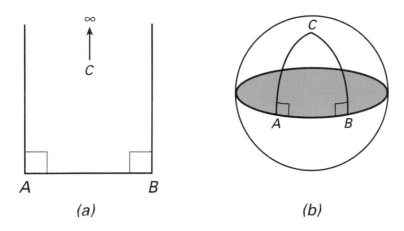

Fig. 6.1. Triangles with two base angles $\pi/2$ in the plane and on the sphere. On the sphere the sum of the angles of the triangle is greater than π.

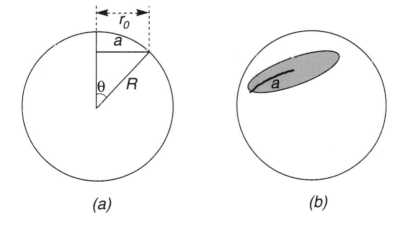

Fig. 6.2. Geodesic circle (b) and geometric quantities (a) used to calculate its perimeter and area.

conditions, the circumference of the geodesic circle is $\ell = 2\pi r_0$, while its area is the area of the spherical cap which it subtends, i.e. $S = 2\pi R^2(1 - \cos\theta)$. Thus

$$\ell = 2\pi R \sin\theta = 2\pi R \sin\frac{a}{R} \sim 2\pi a \left[1 - \frac{a^2}{6R^2} + \cdots\right], \tag{6.2}$$

$$S = 2\pi R^2(1 - \cos\theta) = 2\pi R^2 \cos\frac{a}{R} \sim \pi a^2 \left[1 - \frac{a^2}{12R^2} + \cdots\right]. \tag{6.3}$$

The above expressions for ℓ and S should be compared with the usual values $2\pi a$ and πa^2, respectively. We note first that if R is very large the values are very similar: a sphere of large radius is locally very similar to a plane. Next, if $R \ll a$, ℓ and S

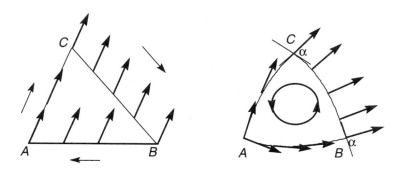

Fig. 6.3. Parallel transport (a) in the Euclidean case, and (b) for a sphere S_2.

differ strongly from their usual Euclidean values. This is sometimes rather incorrectly rendered by saying that *in curved space, the value of π is no longer 3.14159·· · .*

(3) Parallel transport of a vector (**Fig. 6.3**): Consider again two geodesic triangles, one in the plane $\mathbf{R}^2$, the other on the sphere S_2. Let $\vec{X}$ be a vector in the plane, and transport it parallel to itself along the closed path defined by the triangle. After completing one turn, the vector $\vec{X}$ coincides exactly with the vector $\vec{X}$ with which we started; moreover this is true for any triangle. What happens in S_2? For simplicity we choose a birectangular triangle (**Fig. 6.3**). We must however be clear about what is meant by "parallel transport". In the Euclidean case, the vector $\vec{X}$ is always parallel to itself; also this means that the angle it makes with each side of the triangle ABC is constant. For the sphere the comparison of the successive directions of $\vec{X}$ is much more delicate, as is the comparison with the tangent vectors forming the curvilinear triangle ABC. Thus let us take the vector $\vec{X}$ at point A as the tangent vector to the side AC. Starting from C up to the point B, the vector $\vec{X}$ makes an angle α with CB; then, from B to A, it makes an angle $\alpha + \pi/2$. Arriving at A, the vector $\vec{X}$ now makes an angle α with the vector which began the journey, and thus no longer coincides with it, unlike the Euclidean case. This is also a manifestation of the curvature, as one can easily calculate by evaluating the angle α as a function of R and the geodesic distance AB.

(4) Geodesic deviation (**Fig. 6.4**): Consider now two geodesics of the sphere S_2, meeting at a point O at a small angle α. Let A and B be two points on the two geodesics at a distance a from O and let χ be the distance between A and B measured *on* the sphere.

The distance χ from A to B can be found easily: we have

$$\chi = \alpha R \sin \theta = \alpha R \sin \frac{a}{R} \tag{6.4}$$

$$\sim \alpha a. \left[1 - \frac{1}{6} \frac{a^2}{R^2} + \cdots \right] \tag{6.5}$$

showing that the *geodesic deviation* χ obeys the differential equation (geodesic deviation

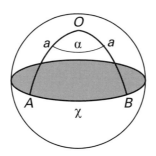

Fig. 6.4. Geodesic deviation.

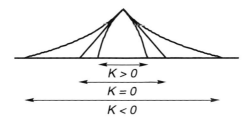

Fig. 6.5. Geodesic deviation as a function of the curvature K.

equation; see *Chapter 7*)

$$\frac{d^2\chi}{da^2} = -\frac{1}{R^2}\chi = -K\chi \tag{6.6}$$

Figure 6.5 gives a qualitative idea of the difference of the Euclidean deviation $K = 0$ and that for the sphere $K > 0$ (K is the *curvature*); we have added the case where $K < 0$, which we discuss below. Geodesic deviation plays an important role in General Relativity (the angular diameter of galaxies in cosmology; antennae for the possible detection of gravitational waves, etc.) and we shall see the covariant form of it in the next chapter.

(5) Geodesic rectangle (**Fig. 6.6**): We give a final example of the effects of curvature. We draw a curvilinear rectangle $ABCD$ in S_2, i.e. a closed curve with four right angles. We can easily verify that while the sides AD and BC have the same length, this is not true of AB and CD. Moreover, the arc DC is *not* a geodesic, and if we draw the geodesic of length equal to AB we end at a point C' such that $ABC'D$ no longer has four right angles, nor two pairs of opposite equal sides. This is another illustration of curvature.

6.2   Curvature of two-dimensional surfaces

It is particularly important to understand two-dimensional curvature because its definition is used to construct that of more general curved spaces with n dimensions. We define this curvature by using the standard notion of the curvature of a plane curve.

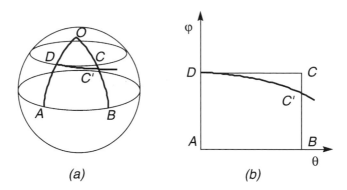

Fig. 6.6. Geodesic rectangle. (a) The quadrilateral $ABCD$ is a rectangle; however, if AB, AD and BC are geodesic arcs, CD is not. The geodesic arc through D orthogonal to AD ends at $C' \neq C$: the quadrilateral cannot be closed. (b) The same situation expressed in polar coordinates (θ, φ) which allows us to describe the sphere S_2 locally.

From the preceding examples, which all concerned the sphere S_2, we might be tempted to define the curvature of a surface in one of several ways. The first possibility is as the excess (positive or negative) length of the perimeter of a geodesic circle drawn on the surface considered, compared with that of a Euclidean circle of the same radius, i.e.

$$K = \lim_{a \to 0} \frac{3}{\pi} \left\{ \frac{2\pi a - \ell}{a^3} \right\}. \tag{6.7}$$

The second possibility is to define the curvature as the excess (positive or negative) area of the circle (**Fig. 6.7**). One might also use some other property in which the curvature manifests itself:

$$K = \lim_{a \to 0} \frac{12}{\pi} \left\{ \frac{\pi a^2 - S}{a^4} \right\}. \tag{6.8}$$

However, this type of definition suffers from a serious defect: it uses the symmetry properties of the sphere (isotropy) and is therefore of no use for a surface without any symmetries.

If we look at the way the curvature showed itself in the previous examples, it is clear that all the examples used *metric properties* (angles, parallelism, lengths, areas, etc.), which are thus linked to the surface metric.

A priori the metric defined by $g_{ij}(\xi^1, \xi^2)$ on a two-dimensional surface S embedded in $\mathbf{R}^3$ can be written in the form

$$d\sigma^2 = g_{ij}\left(\xi^1, \xi^2\right) d\xi^i d\xi^j \qquad (i, j = 1, 2), \tag{6.9}$$

where the ξ^i are two arbitrary parameters describing the surface in a neighbourhood

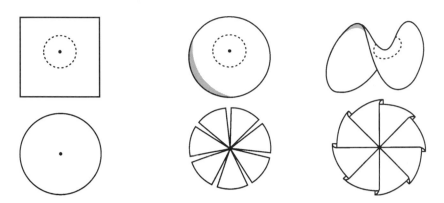

Fig. 6.7. Excess (or shortfall) of the area of a geodesic circle drawn on a surface S and "flattened" on a Euclidean plane $\mathbf{R}^2$, for $K = 0$, $K > 0$ or $K < 0$ [from W. Rindler (1977)].

of the given point. The surface S of $\mathbf{R}^3$ is defined parametrically by

$$\begin{cases} x = x(\xi^1, \xi^2) \equiv x^1 \\ y = y(\xi^1, \xi^2) \equiv x^2 \\ z = z(\xi^1, \xi^2) \equiv x^3 \end{cases} \tag{6.10}$$

so that we can always write the squared distance $d\sigma^2$ *on S* as

$$\begin{aligned} d\sigma^2 &= dx^2 + dy^2 + dz^2 = \delta_{ij} dx^i dx^j \\ &= d\left[x(\xi^1, \xi^2)\right]^2 + d\left[y(\xi^1, \xi^2)\right]^2 + d\left[z(\xi^1, \xi^2)\right]^2 \\ &= \delta_{i\ell} \frac{\partial x^i}{\partial \xi^j}(\xi^1, \xi^2) \frac{\partial x^\ell}{\partial \xi^k}(\xi^1, \xi^2) d\xi^j d\xi^k, \end{aligned} \tag{6.11}$$

which immediately gives

$$g_{ij}(\xi^1, \xi^2) = \delta_{k\ell} \frac{\partial x^k}{\partial \xi^i}(\xi^1, \xi^2) \cdot \frac{\partial x^\ell}{\partial \xi^j}(\xi^1, \xi^2). \tag{6.12}$$

We note that g_{ij} contains *a priori* three functions of the ξ^i (i.e. g_{11}, g_{22} and g_{12}), but as the *coordinates* of the points of the surface S (or rather of an open part of S) are arbitrary, it is always possible *for a two-dimensional surface* to find a local parametrization $\{\alpha^i\}_{i=1,2}$ in which the metric takes the form

$$d\sigma^2 = G(\alpha^1, \alpha^2)\left[(d\alpha^1)^2 + (d\alpha^2)^2\right], \tag{6.13}$$

where G is a known function for a given S. Under these conditions the curvature of a surface S of $\mathbf{R}^2$ is directly related to the function G. Both the pseudo-definitions (6.7) and (6.8), as well as the appearance of the curvature in the geodesic deviation equation (6.6), show that the *second derivatives of the metric tensor* (or of G) define the curvature. If $G(\alpha^1, \alpha^2)$ is constant the metric (6.13) can be put in the *Euclidean form*

$$d\sigma^2 = (d\alpha^1)^2 + (d\alpha^2)^2; \tag{6.14}$$

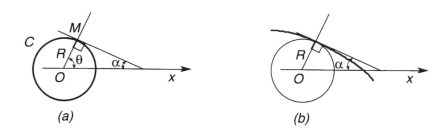

Fig. 6.8. Curvature of a plane curve. (a) For a circle we have $R = ds/d\alpha$, which we can extend (b) to an arbitrary curve.

the space (surface) with this metric is then *locally Euclidean*, or locally flat. For example, a point of a *cylinder* with axis Oz and radius R, represented in $\mathbf{R}^3$ using cylindrical polar coordinates

$$\begin{cases} x & = r\cos\theta \\ y & = r\sin\theta \\ z & = z, \end{cases} \tag{6.15}$$

is defined by the pair $(\xi^1 \equiv \theta, \xi^2 \equiv z)$ and has the metric

$$d\sigma^2 = dz^2 + R^2 d\theta^2 \equiv (d\xi^2)^2 + R^2(d\xi^1)^2 \tag{6.16}$$
$$= (d\alpha^1)^2 + (d\alpha^2)^2,$$

with $\alpha^1 \equiv \xi^2$ and $\alpha^2 \equiv R\xi^1$.

Before we come to the curvature of two-dimensional surfaces, we recall the definition of the curvature of a plane curve (**Fig. 6.8**). Consider a circle C with centre O and radius R, and let Ox be a straight line through O. We produce the tangent from a point M of C; it cuts the axis Ox at an angle α. If the point M is displaced to $M + dM$, it moves on C through a small arc $ds = R d\theta = d\alpha$ (see the figure) and we thus have

$$R = \frac{ds}{d\alpha}. \tag{6.17}$$

For any plane curve, *the radius of curvature R can be defined in an analogous fashion* – i.e. by the radius of the osculating circle at the point considered. For a plane curve represented by an equation of the form $y = y(x)$, we have

$$R = \frac{ds}{d\alpha} = \frac{\left[1 + y'^2\right]^{3/2}}{y''}, \tag{6.18}$$

and for a curve defined parametrically by $\{x = x(t), y = y(t)\}$, we have

$$R = \frac{ds}{d\alpha} = \frac{\left[\dot{x}^2 + \dot{y}^2\right]^{3/2}}{[\dot{x}\ddot{y} - \ddot{x}\dot{y}]}. \tag{6.19}$$

Now consider a surface S of $\mathbf{R}^3$ represented at least locally by an equation of the

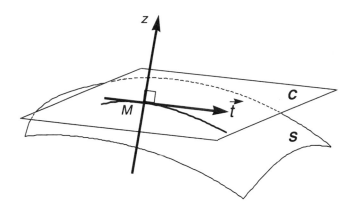

Fig. 6.9. The arbitrary vector $\vec{t}$ in the tangent plane at the point M of the surface S defines with the Mz axis (normal to S) a plane orthogonal to S at M. This plane cuts S in a curve C which depends on the chosen direction of $\vec{t}$, and whose curvature is the curvature of S in this direction.

form $z = f(\xi^1, \xi^2)$: a point M of S is then given by *two* coordinates ξ^1 and ξ^2. In a neighbourhood of a point $\left[\xi_0^1, \xi_0^2, z_0 \equiv f\left(\xi_0^1, \xi_0^2\right)\right]$ of S we can write

$$z = z_0 + \left(\xi_1^1 - \xi_0^1\right) \left.\frac{\partial f}{\partial \xi^1}\right|_{\xi^1 = \xi_0^i} + \left(\xi^2 - \xi_0^2\right) \left.\frac{\partial f}{\partial \xi^2}\right|_{\xi^2 = \xi_0^i}$$

$$+ \frac{1}{2} \left(\xi^1 - \xi_0^1\right) \left(\xi^2 - \xi_0^2\right) \left.\frac{\partial^2 f}{\partial \xi^1 \partial \xi^2}\right|_{\xi^i = \xi_0^i} + \cdots, \tag{6.20}$$

which reduces to

$$z = \frac{1}{2} f_{ij} \xi^i \xi^j + O(\xi^3), \tag{6.21}$$

if we choose the point M at the origin of coordinates and choose the tangent plane at M parallel to the plane (ξ^1, ξ^2) (in which case $\xi_0^i = z_0 = 0$ and $\partial f / \partial \xi^i|_{\xi^i = 0} = 0$). Put another way, near the point (ξ_0^1, ξ_0^2) the surface S is approximated by the anisotropic paraboloid (6.21). In (6.21) we clearly have

$$f_{ij} \equiv \left.\frac{\partial^2 f(\xi^1, \xi^2)}{\partial \xi^i . \partial \xi^j}\right|_{M}; \tag{6.22}$$

reflecting the anisotropy of S at M if $f_{ij} \neq \text{const.} \times \delta_{ij}$.

We now consider a point M in the tangent plane to S, and let $\vec{t}$ be an arbitrary direction in this plane. The plane defined by $\vec{t}$ and the z axis (see **Fig. 6.9**) is obviously orthogonal to S at M; it intersects the surface S in a curve C whose curvature of course depends on the chosen direction $\vec{t}$. In the tangent plane to S at M the curve C projects to $\xi^i = t^i \alpha$, where α is an arbitrary parameter, so that *in the plane* $(Mz, \vec{t})$ it has the equation

$$z = z(\alpha) \sim \frac{1}{2} f_{ij} t^i t^j \alpha^2, \tag{6.23}$$

which immediately provides its curvature as [cf. Eq. (6.19)]

$$K = f_{ij}t^i t^j. \tag{6.24}$$

As it must, K depends on the direction $\vec{t}$ considered. However, an acceptable definition of the curvature of the surface S cannot depend on $\vec{t}$, even though it must be related to the curvatures of the various interesecting curves. The definition can only depend on the *invariants* which can be formed from the matrix $\|f_{ij}\|$, i.e. its *trace* and its *determinant*. If $K_i\,(i = 1, 2)$ are the eigenvalues[1] of the matrix $\|f_{ij}\|$, we have

$$\text{Det}\,\|f_{ij}\| = K_1.K_2 \equiv K_G \tag{6.25}$$

(*Gaussian curvature*)

$$\frac{1}{2}\text{Tr}\,\|f_{ij}\| = \frac{1}{2}(K_1 + K_2) \equiv K_M \tag{6.26}$$

(*mean curvature*).

The two eigenvalues K_1 and K_2 are the *principal curvatures* of S, and the corresponding eigenvectors (for $K_1 \neq K_2$) are the *principal directions* of S at the point M. The Gaussian curvature[1] or *intrinsic curvature* of the surface, is the only one which will concern us in relativistic gravity, as it depends only on S itself and not on the way it is embedded in a larger space (this is illustrated in the next paragraph); in contrast the mean curvature[2], depends not only on S but also on the way it is embedded in $\mathbf{R}^3$, and is therefore not intrinsic.

To see this, consider a sphere of radius R and then a cylinder with axis Oz. For the sphere we have

$$z = \sqrt{R^2 - x^2 - y^2}$$
$$\sim R.\left\{1 - \frac{1}{2}\frac{x^2 + y^2}{R^2} + \cdots\right\},$$

so, considering the inward-pointing normal,

$$f_{ij} = \frac{1}{R}\delta_{ij}, \tag{6.27}$$

and in consequence, $K_1 = K_2 = 1/R$, or

$$K_G = \frac{1}{R^2}, K_M = \frac{1}{R}. \tag{6.28}$$

For the cylinder we can consider two perpendicular directions and find the curvature of the corresponding curves: if we take the generator of a circle we have $K_1 = 0$ and $K_2 = 1/R$, so that

$$K_G = 0 \quad \text{and} \quad K_M = 1/2R. \tag{6.29}$$

---

[1] Remember that by construction, $\|f_{ij}\|$ is a symmetric matrix, which is therefore diagonalisable.
[2] Note that the Gaussian curvature has dimension L^{-2}, while the mean curvature has dimension L^{-1}.

The cylinder is intrinsically flat – this is shown by the Euclidean form (6.16) of its metric. However, regarded as a surface embedded in $\mathbf{R}^3$ it has non-zero mean curvature.

We note in passing that the mean curvature depends on the chosen orientation in $\mathbf{R}^3$ for the normal to S: this clearly shows that K_M depends not only on S but also on the embedding of S in $\mathbf{R}^3$. More elaborate definitions can be found in B. Carter (1991).

We have emphasised above that the curvature[3] is related to the surface considered. We verify this here. The metric of $\mathbf{R}^3$,

$$ds^2 = \delta_{ij}dx^i dx^j \qquad (i, j = 1, 2, 3) \tag{6.30}$$

combined with the local equation for S,

$$x^3 \equiv z \sim \frac{1}{2}f_{ij}x^i x^j \qquad (i = 1, 2), \tag{6.31}$$

where we have chosen the x^i as coordinates of the points of S, immediately gives the metric g_{ij} of the surface:

$$g_{ij} \sim \delta_{ij} + f_{ik} \cdot f_{jl}x^k x^\ell. \tag{6.32}$$

The first term δ_{ij} is the Euclidean term; the second represents a necessarily non-Euclidean part related to the curvature, given its dependence on f_{ij}. The expression (6.32) for the metric also shows that the curvature – and thus f_{ij} – can only be obtained from g_{ij} by eliminating the dependence on the x^ℓ, and thus by *differentiating twice*: the curvature involves the second derivatives of the metric tensor.

6.3   The meaning of intrinsic curvature

To see the physical meaning of Gaussian curvature we use an example of a two-dimensional manifold given by Poincaré (1968). Consider a perfectly plane surface – to fix ideas, consider a perfectly polished marble table, checked by a laser, etc. – and assume that two-dimensional beings living there have some knowledge of physics. In particular, they have graduated rulers, all made of the *same* metal. If the marble table has a uniform temperature (see **Fig. 6.10**) the rulers of all the experimenters will have the same properties and will measure the same lengths for the sides of a triangle ABC wherever it is situated and whatever its orientation (**Fig. 6.10.a**). The beings will verify that Pythagoras's theorem holds, and conclude that the space they inhabit is Euclidean. Now assume that the temperature varies from point to point, as a given function $T(x, y)$. The proper lengths ds of the two-dimensional experimenters' rulers will be given in the neighbourhood of the point (x, y) by

$$ds^2 = \frac{dx^2 + dy^2}{[1 + \chi T(x, y)]^2}, \tag{6.33}$$

---

[3] From now on we shall only be interested in the intrinsic (Gaussian) curvature and we shall simply refer to "the curvature".

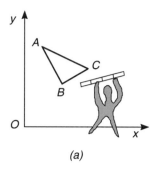 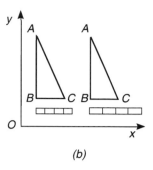

(a)   (b)

Fig. 6.10. Poincaré's table. When the plane xOy has constant temperature (a) the two-dimensional beings regard their space as Euclidean. By contrast, if (b) the temperature is not uniform, rulers (standards of length) vary from point to point, Pythagoras's theorem no longer holds, and space can be regarded as curved.

where χ is the expansion coefficient of the rulers (assumed constant), assuming that the marble table *does not* expand or contract. Under these conditions, an experimenter measuring, for example, the sum of the angles of the right-angled triangle ABC, would find that this differs from π, and conclude that his space was not Euclidean. He would also find that its intrinsic curvature was given by

$$K = \chi \nabla^2 T, \tag{6.34}$$

where we have neglected terms of order χ^2. However, using only these measurements he would have *no* information about any hypothetical Euclidean space $\mathbf{R}^n$ in which his space was embedded; for him this would be a matter of pure theoretical or theological speculation.

What else can we learn from this example?

It is first of all clear that the marble table is only flat for an "external" observer in three dimensions, while for its two-dimensional inhabitants it appears intrinsically curved.

This example shows us the importance of the metric ds^2, which is directly linked to possible measurements of length in this fictitious space and its physical relation to curvature. We can picture this (slightly wrongly, see below) by saying that *the standards of length vary from point to point* in a curved space.

Finally, we can show the connection and the difference between coordinates and distances. Distances, measured using rulers (standards of length) are evaluated using the metric ds^2, while coordinates only *label* the points of the space (**Fig. 6.11**).

6.4   Surfaces in $\mathbf{R}^n$ – Riemann spaces

An ordinary surface in $\mathbf{R}^3$ can be defined in several ways:

$$z = z(x, y) \tag{6.35}$$

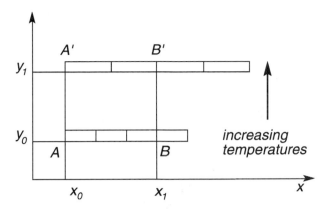

Fig. 6.11. Distances and coordinates. The points A, A' and B, B' have the same abscissas x_0 and x_1 respectively. These points are the corners of a rectangle defined by Cartesian coordinates (x, y). However, the distance AB appears to be larger than the distance $A'B'$ if the temperature increases in the y-direction: the rulers expand on being transported from y_0 to y_1.

$$f(x, y, z) = 0 \tag{6.36}$$

$$\left\{ \begin{array}{l} x = x\left(\xi^1, \xi^2\right) \\ y = y\left(\xi^1, \xi^2\right) \\ z = z\left(\xi^1, \xi^2\right) \end{array} \right. \tag{6.37}$$

and the intersection of two surfaces is in general a manifold of lower dimension, a curve. What happens in $\mathbf{R}^n$? How do our standard ideas generalise?

So as to retain the symmetry between the variables, we consider only the definitions (6.36) and (6.37). A *hypersurface* V_n of $\mathbf{R}^n$ is a manifold[4] (continuum) of $n - 1$ dimensions, which we can define analytically by

$$f\left(x^1, x^2, \ldots, x^n\right) = 0 \tag{6.38}$$

or by a parametric representation (an embedding of V_{n-1} in $\mathbf{R}^n$)

$$\left\{ \begin{array}{l} x^1 = x^1\left(\xi^1, \xi^2, \ldots, \xi^{n-1}\right) \\ x^2 = x^2\left(\xi^1, \xi^2, \ldots, \xi^{n-1}\right) \\ \quad \cdots \qquad \cdots\cdots\cdots \\ x^n = x^n\left(\xi^1, \xi^2, \ldots, \xi^{n-1}\right) . \end{array} \right. \tag{6.39}$$

The metric defined on such a hypersurface is *induced* by that of $\mathbf{R}^n$,

$$ds^2 = \delta_{ij} dx^i dx^j, \tag{6.40}$$

and is given by

$$ds^2 = \delta_{ij} \frac{\partial x^i \left(\xi^1, \xi^2, \ldots, \xi^{n-1}\right)}{\partial \xi^k} \frac{\partial x^j \left(\xi^1, \xi^2, \ldots, \xi^{n-1}\right)}{\partial \xi^\ell} d\xi^k d\xi^\ell \tag{6.41}$$

$$\equiv g_{k\ell} d\xi^k d\xi^\ell, \tag{6.42}$$

[4] See *Appendix D*.

with

$$g_{k\ell} = \frac{\partial x^i}{\partial \xi^k} \cdot \frac{\partial x^j}{\partial \xi^\ell} \delta_{ij} \qquad (6.43)$$

$$(k, \ell = 1, 2, \ldots, n-1; \ i, j = 1, 2, \ldots, n).$$

Of course, the representation (6.38) also allows us to find a form of the metric tensor $g_{k\ell}$ on the hypersurface considered, as one of the coordinates can be expressed as a function of the other $n-1$.

Similarly, a manifold S_p of p dimensions, embedded in $\mathbf{R}^n$, will be given by $n-p$ relations

$$\begin{cases} f_1\left(x^1, x^2, \ldots, x^n\right) = 0, \\ f_2\left(x^1, x^2, \ldots, x^n\right) = 0, \\ \quad \cdots\cdots\cdots \\ f_{n-p}\left(x^1, x^2, \ldots, x^n\right) = 0, \end{cases} \qquad (6.44)$$

between the x^i, i.e. parametrically by

$$\begin{cases} x^1 = x^1\left(\xi^1, \xi^2, \ldots, \xi^p\right) \\ x^2 = x^2\left(\xi^1, \xi^2, \ldots, \xi^p\right) \\ \cdots \quad \cdots\cdots\cdots \\ x^n = x^n\left(\xi^1, \xi^2, \ldots, \xi^p\right). \end{cases} \qquad (6.45)$$

The number of *independent* parameters (with independent *coordinates*), here p, defining the points of S_p, is the *dimension* of S_p. Let S_p and S_q be two manifolds of dimension p and q, respectively. If their intersection is not empty, we have in general

$$\mathrm{dim}.\left[S_p \cap S_q\right] = p + q - n. \qquad (6.46)$$

As before, the metric induced by $\mathbf{R}^n$ on S_p is found in the analogous manner, and we again have the form (6.43) with however $(k, \ell) = 1, 2, \ldots, p$ (rather than $n-1$).

We are mainly interested in n-dimensional manifolds V_n (see *Appendix D*), without reference to any Euclidean space they may be embedded in, which have a metric $g_{ij}[(i, j) = 1, 2, \ldots, n]$ and submanifolds S_p embedded in them. The foregoing remains essentially valid except for (6.43), which becomes

$$g_{k\ell}[S_p] = g_{ij}[V_n]\frac{\partial x^i}{\partial \xi^k} \cdot \frac{\partial x^j}{\partial \xi^\ell}, \qquad (6.47)$$

where the $\{x^i\}$ now stand for the coordinates in V_n and where the indices (k, ℓ) run from 1 to p.

A manifold (a "continuum" in Poincaré's expression) of n dimensions, equipped with a notion of distance – a metric – defined by a quadratic form is called a *Riemann space*, and plays a central role in relativistic theories of gravity. We note again that a manifold can always be approximated *locally* by its ("flat") tangent space. We shall meet this property again in *Chapter 7*, where it will allow us to express the *equivalence principle*.

6.5   Intrinsic curvature of a manifold

We have seen above that the Gaussian curvature of a two-dimensional surface S of $\mathbf{R}^3$ is defined as an invariant property related to the curvature of curves drawn in S. We have also seen that the curvature of S is directly proportional to the second derivatives of the metric of S (induced by that of $\mathbf{R}^3$).

The first remark suggests that the curvature of a manifold V_n of n dimensions could be constructed and defined using the curvatures of two-dimensional surfaces embedded in it. However, as the *direction* of these surfaces (at a point) is arbitrary, the curvature will have a certain arbitrariness which will reveal itself in the existence of a tensor expressing all possible orientations of two-dimensional surfaces.

To find this tensor, consider at a point x of V_n, two arbitrary n-vectors a^μ and b^μ. We form the surface S generated by the geodesics through x tangent to the plane defined by the two n-vectors a^μ and b^μ. The orientation of S is fixed by these two vectors and is thus arbitrary. The metric of S induced by that of V_n is then given by the relation (6.47), with $(k, \ell) = 1, 2$ and $(i, j) = 1, 2, \ldots, n$. The Gaussian curvature of this surface is then $K = \mathrm{Det} \, \| f_{k\ell} \|$, where $f_{k\ell}$ is found using the expression (6.47) for the metric $g_{k\ell}$ of S, using the relation (6.32) between these two quantities. A careful calculation then gives

$$K = \frac{R_{ijmn} a^i b^j a^m b^n}{\left(g_{in} g_{jm} - g_{im} g_{jn} \right) a^i b^j a^m b^n},\tag{6.48}$$

which clearly shows the dependence of K on the vectors a^μ and b^μ, and thus on S. In the latter relation, the tensor R_{ijmn}, the *Riemann tensor*, is given by

$$R_{ijmn} = \frac{1}{2} \{ \partial_{nj} g_{im} - \partial_{ni} g_{jm} - \partial_{jm} g_{in} + \partial_{im} g_{jn} \}$$
$$+ g_{rs} \{ \Gamma^r_{mi} \Gamma^s_{jn} - \Gamma^r_{ni} \Gamma^s_{mj} \},\tag{6.49}$$

where all indices run from 1 to n. In (6.48) R_{ijmn} is the only part which is simultaneously independent of S *and* depends linearly on the second derivatives of the metric of V_n. It must therefore represent the curvature of V_n: the Riemann tensor *is* the curvature tensor.

At this point of the study of the curvature of an n-dimensional manifold, it is worth remarking that we have spoken of vectors, tensors, etc., in V_n without these being really defined; we refer to *Chapter 3* for a discussion of curvilinear coordinates, most of which is valid for curved spaces. We shall return to this question later in this chapter.

We return to the relation (6.49) defining the curvature tensor R_{ijmn}. This tensor also appears when we try to *parallel transport an n-vector* around a closed path. If A^i is such a vector, it is easy to show that its change δA^i after one circuit is

$$\delta A^i = \frac{1}{2} R^i{}_{jmn} A^j \oint x^n dx^m,\tag{6.50}$$

an expression which is used to introduce the curvature in most books on relativity. Thus a necessary and sufficient condition for $\delta A^i = 0$, whatever closed circuit is considered, is that the Riemann tensor should vanish. Put another way, $\delta A^i = 0$ if and only if space is locally flat.

The tensor R_{ijmn} also appears in many other properties involving the curvature, and these can also be used to define it. Thus the *geodesic deviation* equation (see *Chapter 7*),

$$\frac{d^2 \delta x^i}{ds^2} + \partial_j \Gamma^i_{k\ell} \frac{dx^k}{ds} \cdot \frac{dx^\ell}{ds} \delta x^j = R^i{}_{jkl} \delta x^k \frac{dx^j}{ds} \frac{dx^\ell}{ds}, \tag{6.51}$$

(where δx^i is the deviation between two geodesics) also defines this tensor. This is *a priori* rather surprising as we might expect that very different manifestations of curvature might bring in different tensors constructed from g_{ij} and its first derivatives. The fact that R_{ijkl} *always* appears suggests that this tensor is *unique*, which brings us to the second remark made at the beginning of this chapter.

If we had tried to define the curvature using among other things the second partial derivatives $\partial_{ij} g_{k\ell}$ of the metric tensor of V_n, we would have run into the problem that they do not constitute a tensor. However, if we had constructed the most general tensor linear in the second derivatives of the metric, we would inevitably have ended up with the Riemann tensor:

Theorem : *The most general fourth-order tensor linear in second derivatives of the metric tensor is the Riemann tensor.*

[Proof: see for example S. Weinberg (1972).]

Given this, there is nothing surprising in the appearance of R_{ijkl} everywhere that curvature is involved.

6.6   Properties of the curvature tensor

The Riemann tensor (6.49) has a number of algebraic properties which follow from its explicit form. By raising the first index, it is easy to prove the relation

$$R^i{}_{jk\ell} = \partial_k \Gamma^i_{j\ell} - \partial_\ell \Gamma^i_{jk} + \Gamma^r_{j\ell} \Gamma^i_{rk} - \Gamma^r_{jk} \Gamma^i_{r\ell}. \tag{6.52}$$

Similarly, we immediately see that $R_{ijk\ell}$ obeys two symmetry relations:

$$R_{ijk\ell} = R_{k\ell ij} = R_{ji\ell k}, \tag{6.53}$$

$$R_{ijk\ell} = -R_{jik\ell} = -R_{ij\ell k} \tag{6.54}$$

and the cyclic relation

$$R_{ijk\ell} + R_{i\ell jk} + R_{ik\ell j} = 0 \tag{6.55}$$

These relations allow us to find the number of independent components of the curvature tensor. The indices (i, j, k, ℓ) run from 1 to n, so any pair of indices – either the first two or the second two – of $R_{ijk\ell}$ can, using the property (6.54), take $A = \frac{1}{2}n(n-1)$ values. As the pairs (i, j) and (k, ℓ) are symmetrical, [Eq. (6.53)] they can take $\frac{1}{2}A(A+1)$ values, i.e.

$$
\begin{aligned}
\frac{1}{2}A(A+1) &= \frac{1}{2}\frac{n(n-1)}{2} \cdot \left[\frac{1}{2}n(n-1)+1\right] \\
&= \frac{1}{8}n(n-1)(n^2-n+2) \equiv B_n.
\end{aligned}
\tag{6.56}
$$

Finally, the cyclic relation (6.55) imposes C_n^4 constraints, so that the number of independent components is r_n,

$$
r_n = B_n - C_n^4 = \frac{1}{12}n^2(n^2-1).
\tag{6.57}
$$

Thus, for $n=2$, $R_{ijk\ell}$ has only *one* independent component: a surface of $\mathbf{R}^3$ has only one intrinsic curvature. This is no longer true for $n=3$, which has six independent components, or for $n=4$, the case we shall confine ourselves to henceforth[5], which has 20 independent components. For $n=1$ we get the *a priori* slightly surprising result that a curve is not ... curved! However, we should remember that the Riemann tensor characterises only its intrinsic curvature: while a curve is intrinsically flat it is by contrast extrinsically curved, i.e. when regarded as a manifold embedded in $\mathbf{R}^n$ ($n>1$).

To the properties (6.53)–(6.55) of the Riemann tensor we should add the identity

$$
\nabla_\lambda R_{\mu\nu\alpha\beta} + \nabla_\beta R_{\mu\nu\lambda\alpha} + \nabla_\alpha R_{\mu\nu\beta\lambda} \equiv 0,
\tag{6.58}
$$

the *Bianchi identity*.

We can form several useful tensors from the Riemann tensor, such as the *Ricci tensor*

$$
R_{\mu\nu} \equiv R^\lambda{}_{\mu\lambda\nu}
\tag{6.59}
$$

which is symmetric by virtue of (6.53). Using the antisymmetry property (6.54) of $R_{\mu\nu\alpha\beta}$ we can easily show that the Ricci tensor is the only second order tensor that we can form by contraction of the Riemann tensor: other contractions give either 0 or $\pm R_{\mu\nu}$. Further, the scalar

$$
R \equiv R^\mu{}_\mu = R^{\alpha\beta}{}_{\alpha\beta},
\tag{6.60}
$$

the *curvature scalar*, is the only nonzero scalar that we can form[6] from $R_{\mu\nu\alpha\beta}$.

Finally, the Bianchi identities (6.58) contracted over the first and last and the second and third indices give

$$
\nabla_\lambda R - 2\nabla_\alpha R^\alpha{}_\lambda = 0,
\tag{6.61}
$$

where we have used the definitions (6.59) and (6.60) as well as the properties (6.53)–(6.55) of the curvature tensor. The latter relation can be written

$$
\nabla_\lambda \left[R^{\lambda\mu} - \frac{1}{2}g^{\lambda\mu}R\right] = 0.
\tag{6.62}
$$

---

[5] Greek indices will run, as above, from 0 to 3. Thus $R_{\mu\nu\alpha\beta}$ is the curvature tensor of a four-dimensional manifold.

[6] The relation (6.55) shows that the pseudo-scalar $\frac{1}{\sqrt{|g|}}\varepsilon^{\mu\nu\alpha\beta}R_{\mu\nu\alpha\beta}$ vanishes (g is the determinant of $g_{\mu\nu}$).

The tensor

$$G_{\mu\nu} \equiv R_{\mu\nu} - \frac{1}{2}g_{\mu\nu}R \tag{6.63}$$

is the *Einstein tensor*, which will play a central role in the following (see *Chapter 8*).

The definitions used for space–time – whose metric is pseudo-Euclidean – are closely copied from the case of manifolds V_n with a positive–definite metric. Space–time is thus (locally) pseudo-Euclidean if and only if $R_{\mu\nu\alpha\beta} = 0$. In this connection we note that the conditions $R = 0$ or $R_{\mu\nu} = 0$, or $G_{\mu\nu} = 0$, are necessary but not sufficient for space–time to be (locally) flat.

6.7   **Space–time as a Riemannian manifold**

When we perform experiments in the laboratory, space–time appears to us to be Minkowski space, i.e. flat, at least on our lengthscale. Put another way, even if space–time is curved, it appears to us to be locally flat, provided that we only consider closely neighbouring events. This means that the (curved) space–time manifold can (locally) be replaced by its tangent 4-plane. This kind of manifold, locally (pseudo-)Euclidean, equipped with a metric which reduces in suitable coordinates to $\eta_{\mu\nu}$, constitutes by definition a *Riemannian manifold* (see *Appendix D*). We shall see in *Chapter 7* that the *equivalence principle* is related to this fact. For the moment we attempt to establish a purely algebraic relation with the foregoing. In particular, we shall verify that one can always find locally a coordinate system (i) in which the metric tensor $g_{\mu\nu}$ reduces to the (pseudo-)Euclidean form $\eta_{\mu\nu}$ and (ii) where the first derivatives of the metric vanish.

Thus consider a change of variables – coordinates – of the form

$$\{x^\alpha\} \longleftrightarrow \{x^{\alpha'}\} \quad \text{or} \quad x^\alpha = x^\alpha(x^{\lambda'}) \tag{6.64}$$

with

$$A^\alpha{}_{\mu'} \equiv \frac{\partial x^\alpha(x^{\lambda'})}{\partial x^{\mu'}}. \tag{6.65}$$

The metric tensor thus transforms as (see *Chapter 4*)

$$g_{\mu'\nu'}(x') = A^\mu{}_{\mu'}A^\nu{}_{\nu'}g_{\mu\nu}\left[x(x')\right]. \tag{6.66}$$

Consider a point M with coordinates x_0^λ and another point with coordinates $x^{\lambda'}$ *close to* M, so that we can expand the $A^\mu{}_{\mu'}$ and the $g_{\mu\nu}$ in a Taylor series

$$A^\mu{}_{\mu'}(x') = A^\mu{}_{\mu'} + \left(x^{\lambda'} - x_0^{\lambda'}\right)\frac{\partial A^\mu{}_{\mu'}}{\partial x^{\lambda'}}$$
$$+ \frac{1}{2}\left(x^{\lambda'} - x_0^{\lambda'}\right)\left(x^{\sigma'} - x_0^{\sigma'}\right)\frac{\partial^2 A^\mu{}_{\mu'}}{\partial x^{\lambda'}\partial x^{\sigma'}} + \cdots \tag{6.67}$$
$$= A^\mu{}_{\mu'} + \left(x^{\lambda'} - x_0^{\lambda'}\right)\frac{\partial^2 x^\mu}{\partial x^{\mu'}\partial x^{\lambda'}}$$
$$+ \frac{1}{2}\left(x^{\lambda'} - x_0^{\lambda'}\right)\left(x^{\sigma'} - x_0^{\sigma'}\right)\frac{\partial^3 x^\mu}{\partial x^{\mu'}\partial x^{\lambda'}\partial x^{\sigma'}} + \cdots \tag{6.68}$$

$$g_{\mu\nu}(x') = g_{\mu\nu} + \left(x^{\lambda'} - x_0^{\lambda'}\right) \frac{\partial g_{\mu\nu}}{\partial x^{\lambda'}}$$

$$+ \frac{1}{2}\left(x^{\lambda'} - x_0^{\lambda'}\right)\left(x^{\sigma'} - x_0^{\sigma'}\right) \frac{\partial^2 g_{\mu\nu}}{\partial x^{\lambda'}\partial x^{\sigma'}} + \cdots \qquad (6.69)$$

where $A^{\mu}{}_{\mu'}, g_{\mu\nu}$ and their derivatives are evaluated at M on the right hand side of these last two equations. Using (6.67) and (6.68) in the transformation law (6.66) for the metric tensor, we find

$$g_{\mu'\nu'}(x') = A^{\mu}{}_{\mu'}A^{\nu}{}_{\nu'}g_{\mu\nu}$$

$$+ \left(x^{\lambda'} - x_0^{\lambda'}\right)\left\{A^{\mu}{}_{\mu'}A^{\nu}{}_{\nu'}\partial_\lambda g^{\nu}_\mu + A^{\mu}{}_{\mu'}g_{\mu\nu}\frac{\partial^2 x^{\nu}}{\partial x^{\nu'}\partial x^{\lambda'}} + A^{\nu}{}_{\nu'}g_{\mu\nu}\frac{\partial^2 x^{\mu}}{\partial x^{\mu'}\partial x^{\lambda'}}\right\}$$

$$+ \frac{1}{2}\left(x^{\lambda'} - x_0^{\lambda'}\right)\left(x^{\sigma'} - x_0^{\sigma'}\right)[\cdots]_{\mu'\nu'\lambda'\sigma'} + \cdots \qquad (6.70)$$

Now consider the possible choices of coordinate changes (6.64)–(6.65). The matrix $A^{\mu}{}_{\mu'}$ has 16 completely arbitrary elements, while the derivatives $\partial^2 x^{\nu}/\partial x^{\nu'} \cdot \partial x^{\lambda'}$ have $4 \times \left[\frac{1}{2}4(4+1)\right] = 40$ coefficients, which are also arbitrary. In the expression $[\cdots]_{\mu'\nu'\lambda'\sigma'}$, (which we do not write out as it is complex and there is no need) there are derivatives of the third order, of the form $\partial^3 x^{\mu}/\partial x^{\mu'} \cdot \partial x^{\lambda'} \cdot \partial x^{\sigma'}$. Use of the symmetries between the indices $(\mu', \lambda', \sigma')$ shows that there are $80(= 4 \times 20)$ of these. By a suitable choice of the 16 $A^{\mu}{}_{\mu'}$ we can certainly choose the ten components of $g_{\mu'\nu'}$ so that $g_{\mu'\nu'} = \eta_{\mu'\nu'}$: the six degrees of freedom remaining then correspond to the six degrees of freedom for the Lorentz transformations in the tangent space. (Three for the velocity $\vec{v}$ and three for a spatial rotation.) Similarly, the 40 derivatives $\partial_{\mu'}g_{\alpha'\beta'}$ can be made to vanish identically because of the freedom to choose *exactly* the 40 coefficients $\partial x^{\mu}/\partial x^{\lambda'} \cdot \partial x^{\sigma'}$. By contrast, we cannot in general make the 100 second derivatives $\partial_{\mu'\nu'}g_{\alpha'\beta'}$ vanish as we only have 80 degrees of freedom $\partial x^{\mu}/\partial x^{\lambda'} \cdot \partial x^{\sigma'} \cdot \partial x^{\gamma'}$. There thus remain 20 degrees of freedom for the second derivatives of the metric tensor: these correspond precisely to the 20 independent components of the curvature tensor $R_{\mu\nu\alpha\beta}$. These properties are sometimes called the *"theorem of local flatness"*, which can be expressed analytically by the existence of local coordinates such that

$$g_{\mu'\nu'} = \eta_{\mu'\nu'} + O\left(x'^2\right). \qquad (6.71)$$

In summary, one can always find a *local* coordinate system in which the metric tensor takes the (pseudo-)Euclidean form (6.7) and the connection symbols $\Gamma^{\mu}_{\alpha\beta}$ vanish at a point. When the curvature tensor vanishes identically, which is true in Minkowski space, one can find global coordinates, rather than simply at a point, for which $g_{\mu\nu} = \eta_{\mu\nu}$ and $\Gamma^{\mu}_{\alpha\beta} \equiv 0$. We shall discuss the physical significance of these properties in *Chapter 7*.

6.8 Some properties of tensors in curved space

Most of the definitions and properties of tensors in Minkowski space in curvilinear coordinates remain valid, with a few small modifications which we give here.

Consider a 4-vector A^μ of the space–time manifold, defined as in *Chapter 3*. Its covariant derivative is[7]

$$\nabla_\alpha A^\mu \equiv \partial_\alpha A^\mu + \Gamma^\mu_{\alpha\nu} A^\nu, \tag{6.72}$$

and its second covariant derivatives are

$$\nabla_\alpha \nabla_\beta A^\mu = \partial_{\alpha\beta} A^\mu + \partial_\alpha \Gamma^\mu_{\nu\beta} A^\nu + \cdots \tag{6.73}$$

Exchanging the indices α and β and subtracting gives

$$\left(\nabla_\alpha \nabla_\beta - \nabla_\beta \nabla_\alpha\right) A^\mu = R^\mu_{\ \nu\alpha\beta} A^\nu; \tag{6.74}$$

or, *in curved space covariant derivatives do not commute*. In Minkowski space we have $[\nabla_\alpha, \nabla_\beta] = 0$ in every curvilinear coordinate system. Curvature introduces some important changes in tensor algebra. We note that the relation (6.74) could also have been used to define *the* curvature tensor.

We now give without proof (see the exercises) several useful relations which hold in curved as well as flat space. Thus

$$\Gamma^\mu_{\mu\nu} = \partial_\nu \ln \sqrt{|g|} \tag{6.75}$$

$$\nabla_\mu A^\mu = \frac{1}{\sqrt{|g|}} \partial_\mu \left(A^\mu \sqrt{|g|}\right) \tag{6.76}$$

$$\int_V \sqrt{|g|} d^4x \nabla_\mu A^\mu = \int_{\partial V} d\Sigma_\mu A^\mu \tag{6.77}$$
$$(Gauss's\ theorem)$$

$$d\Sigma_\mu = \frac{1}{3!} \sqrt{|g|} \varepsilon_{\mu\nu\alpha\beta} \, dx^\nu \wedge dx^\alpha \wedge dx^\beta \tag{6.78}$$

$$\nabla_\mu A^{\mu\nu} = \frac{1}{\sqrt{|g|}} \partial_\mu \left(\sqrt{|g|} A^{\mu\nu}\right) + \Gamma^\nu_{\mu\lambda} A^{\mu\lambda} \tag{6.79}$$

$$\nabla_\mu A^{\mu\nu} = \frac{1}{\sqrt{|g|}} \partial_\mu \left(\sqrt{|g|} A^{\mu\nu}\right) \tag{6.80}$$
$$(for\ antisymmetric\ A^{\mu\nu}).$$

6.9   Three arguments for curved space–time

Although the real question ought to be "why should space–time be flat?", our intuition rests on the simplicity of Euclidean concepts. Thus we have to make the opposite argument. In fact, there are simple reasons to believe that our space–time is curved,

---

[7] The relation (6.72) and the properties of the previous section show that there always exists a local coordinate system such that $\nabla_\alpha A^\mu = \partial_\alpha A^\mu$.

even if none of these arguments is truly compelling: experiment can never establish a theoretical concept, even though it may strongly suggest it.

(1) The relation $E = mc^2$, associating a mass with *every* form of energy (which has to be verified experimentally; see *Chapter 7*), together with the experimental equality of inertial and gravitating mass (weak equivalence principle) implies that everything has weight. There is no gravitationally neutral force in Nature. This means that even without interaction forces all bodies are constantly accelerated and the very notion of inertial motion loses meaning, except possibly for short distances and time intervals, i.e. locally.

Thus, space–time in the presence of gravity becomes locally (pseudo-)Euclidean, locally Minkowskian. It no longer possesses a system of timelike straight lines associated with global systems of inertia via inertial motions, which are themselves global. Motion which is as "straight" as possible i.e. "geodesic" motion, then depends on the distribution and motion of the energy sources in space–time through the law of universal attraction, or, more generally, a covariant law for the gravitational interaction. This motion can never be reduced to uniform straight–line motion *in the whole of space–time* by a suitable choice of coordinates: space–time is curved.

(2) A different type of argument, by A. Schild (1967), uses the redshift of a a photon "climbing" in a gravitational field. This redshift was measured by R.V. Pound and G.A. Rebka (1960) and R.V. Pound and J.L. Snider (1965), and observed for Sirius B by W.S. Adams (1925). Similar observations have since been performed for other white dwarfs by J.L. Greenstein, J.B. Oke and H.L. Shipman (1971).

We recall (see *Chapter 4*) that if an observer situated at altitude z_0 sends a signal (photons) with period T_0, an observer at altitude z (see **Fig. 6.12**) receives these signals with period $T > T_0$, even though the observers are at rest with respect to each other. In Minkowski space all light rays emitted by the observer at z_0 are "parallel" to each other and must therefore arrive at z spaced at $T = T_0$. This contradiction shows that *Minkowski space is inadequate* for describing gravitational phenomena and only a curved space can remove the contradiction. One sometimes describes this effect by saying that in a curved space the standards of length (and time) vary from point to point because of gravity. This figurative way of speaking is not really correct. In fact, a given atomic clock at altitude z_0 has exactly the same period as an identical clock at altitude z_1. It is only when one compares the readings of the two clocks, for example using electromagnetic signals, that one is aware of a difference. This difference – certainly caused by space–time curvature – is clearly not caused by a real change of the standard of time in the presence of a gravitational field, [i.e. a change of the behaviour of clocks (excuding gravitational clocks with pendulums)] but by the connection established between two distinct points of space–time. This will become clearer in *Chapter 7* (redshift). In the Pound–Rebka experiment, the presence of gravitation means that the comparison of time standards gives different results at z_0 and at z.

It is interesting to note that in Schild's discussion the nature of the gravitational

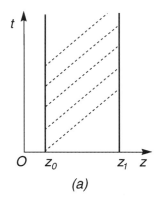
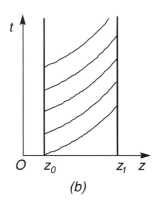

(a)   (b)

Fig. 6.12. Space–time diagram (in Minkowski space) for the Pound–Rebka experiment (1960). In (a) signals emitted by an observer at altitude z_0 above sea level propagate at the speed of light, i.e. along the generators of the light–cone, with period T_0, and are received by an observer at altitude z with the same period. This is contrary to the Pound–Rebka experiment, which shows that $T > T_0$. Minkowski space is incompatible with experiment. At (b), one takes account of the curvature of light rays in the terrestrial gravitational field: the argument still holds. To explain the Pound–Rebka experiment we require that the time standard at z_0 should differ from that at altitude z, so that space–time is curved.

field itself plays no role: we could be discussing a scalar, spinor, vector field, etc. The gravitational field induces space–time curvature without specifying the nature of the curvature.

(3) The curvature of light rays, observed near the Sun and soon near Jupiter, verified by gravitational lensing (see *Chapter 7*), also requires a curved space–time. The structure of Minkowski space in essentially determined[8] by the *light cone*, which is the same everywhere. The light cone induces the *metric structure* of Minkowski space, defined by the tensor $\eta_{\mu\nu}$. The curvature of light rays near the Sun shows that the light cone is deformed by gravitation, in a way that depends on how close one is to the surface. As the light cone gives the space–time metric, we have to conclude that the latter varies spatially. By contrast, Schild's argument given above shows that the time standard (a "clock") varies with gravity. These two experimental facts allow one to show that gravitation makes the standards of both time and distance vary: space–time is thus curved.

We add that since light rays constitute our standard of "straightness", the latter loses all meaning if gravity is present. The straight lines of Minkowski space, including null rays, have no general physical meaning, and are not intrinsic geometric objects of space–time.

---

[8] This is why we emphasized the close connection between light cone properties and Minkowski space in *Chapter 2*.

Exercises

6.1   Consider the two-dimensional space of "Poincaré's table" which is heated. Heat propagation is determined by the relation

$$\nabla^2 T = S,$$

where S is the heat source. The metric is

$$ds^2 = \frac{dx^2 + dy^2}{(1 + \chi T)^2}.$$

Assume that S and T are axisymmetric.

(i) Calculate $T(r)$ if $S = \text{const}$.

(ii) Calculate the curvature K of this space and show that the metric can be written in the form

$$ds^2 = \frac{d\bar{r} + \bar{r}^2 d\theta^2}{\left(1 + \frac{K}{4}\bar{r}^2\right)^2},$$

where $\bar{r}$ and θ can be expressed (give this precisely) as functions of x and y.

(iii) When K is negative, show that the metric can be written as

$$ds^2 = |K|^{1/2} \left[d\chi^2 + \text{sh}^2\chi d\theta^2 \right].$$

[Reference: H.P. Robertson and T.W. Noonan (1968).]

6.2   Consider the surface S_3 in $\mathbf{R}^4$ defined by

$$\begin{cases} w = a\cos\chi \\ z = a\sin\chi\cos\theta \\ y = a\sin\chi\sin\theta\sin\varphi \\ x = a\sin\chi\sin\theta\cos\varphi. \end{cases}$$

(i) Find the metric of S_3 in the coordinate system $\{\chi, \theta, \varphi\}$.

(ii) Find the Christoffel symbols and deduce the components of the curvature tensor.

6.3   Let S_p and S_q be two surfaces of p and q dimensions, respectively, embedded in $\mathbf{R}^n$. Show that, in general, $\dim[S_p \cap S_q] = p + q - n$.

6.4   Consider a two-dimensional Riemannian manifold with the (pseudo-)metric

$$ds^2 = a^2 \left[1 - \text{ch}^2\xi^0 \cdot \sin^2\xi^1\right] d\xi^{02} - a^2\text{sh}^2\xi^0 d\xi^{12}$$

where (ξ^0, ξ^1) are curvilinear coordinates in the manifold and a is a constant with the dimensions of length.

(i) Give the components g_{ij} of the metric tensor and calculate *directly* the Christoffel symbols Γ_{ij}^k.

(ii) Write down the geodesic equation *starting from the definition*

$$\delta \int ds = 0,$$

and deduce the Christoffel symbols by identification with the geodesic equation (3.142).

(iii) Calculate the curvature tensor. Is the manifold curved?

(iv) Make the coordinate change $\{\xi\} \to \{\xi'\}$ defined by

$$\begin{cases} \xi'^0 = a\,\mathrm{ch}\,\xi^0 \\ \xi'^1 = a\,\mathrm{sh}\,\xi^0 \cos \xi^1 \end{cases}$$

    (a) Calculate the direct and inverse transformation coefficients, i.e. the tensors $\partial \xi^i / \partial \xi'^j$ and $\partial \xi'^i / \partial \xi^j$.

    (b) Calculate the new form of the metric tensor in two different ways.

    (c) Calculate the new Christoffel symbols in three different ways.

(v) Calculate the volume element $\sqrt{|g|}\,d\xi^1 d\xi^2$.

6.5   Show that

(i) $\Gamma^\mu_{\mu\nu} = \frac{1}{2}\partial_\nu \ln |g|$.

(ii) $g^{\alpha\beta}\Gamma^\rho_{\alpha\beta} = -\dfrac{1}{\sqrt{|g|}}\partial_\lambda \left[g^{\lambda\rho}\sqrt{|g|}\right]$.

(iii) $\nabla_\mu F^{\mu\nu} = \dfrac{1}{\sqrt{|g|}}\partial_\mu \left[\sqrt{|g|}F^{\mu\nu}\right]$.

(iv) $\nabla_\mu A^\mu = \dfrac{1}{\sqrt{|g|}}\partial_\mu \left[\sqrt{|g|}A^\mu\right]$.

6.6   Verify that $\left(\nabla_\mu \nabla_\nu - \nabla_\nu \nabla_\mu\right) A_\alpha = -R^\lambda_{\ \mu\nu\alpha}A_\lambda$.

6.7   Show that if $u^\mu \cdot u_\mu = +1$, then $u^\mu \nabla_\alpha u_\mu = 0$.

6.8   Let $T^{\mu\nu}$ be a tensor. Show that

$$\left(\nabla_\mu \nabla_\nu - \nabla_\nu \nabla_\mu\right) T^{\alpha\beta} = R^\alpha_{\ \lambda\mu\nu}T^{\lambda\beta} + R^\beta_{\ \lambda\mu\nu}T^{\alpha\lambda}.$$

6.9   Let ϕ be a scalar field. Calculate $\left(\nabla_\alpha \nabla_\beta - \nabla_\beta \nabla_\alpha\right)\phi$.

6.10   Consider a four-dimensional space–time with metric

$$ds^2 = A(r)dt^2 - B(r)dr^2 - r^2 d\theta^2 - r^2 \sin^2 \theta d\varphi^2.$$

Find the geodesic equation and the Christoffel symbols.

6.11   Let S^2 be the usual sphere in $\mathbf{R}^3$.

(i) Write down its metric tensor g_{ij} and connections Γ^k_{ij}.

(ii) Find the curvature and Ricci tensors. Find the curvature scalar.

6.12   Same questions for the sphere S^3 embedded in $\mathbf{R}^4$.

6.13   Same questions for the hyperboloid H^3 embedded in $\mathcal{M}^4$ with the equation $x^{0^2} - \mathbf{x}^2 = a^2$ ($a = $ const.).

7 The Principle of Equivalence

The thought experiment ("*Gedankenexperiment*") constituted by Einstein's lift studied in *Chapter 1* shows that in a uniform and homogeneous gravitational field it is always possible to remove the latter. It is also clear that for a gravitational field which is neither homogeneous nor uniform there is always a time and distance scale such that space–time has *approximately* these properties. This means that *locally* in time and space (i.e. for very short intervals and distances) it is possible to remove the mechanical effects of gravity. We recall (see *Chapter 1*) that this possibility was related to the assumed equality between the inertial and gravitational mass of a body:

$$m_{\text{inertial}} = m_{\text{gravitational}} , \tag{7.1}$$

a theoretical equality suggested by an experimental equality and verified to great precision (see below). The equality (7.1), the *postulate* (7.1), is called the *weak equivalence principle*.

This principle is then extended not just to mechanical effects caused by gravitation, but to *all* physical laws, from electromagnetism to quantum phenomena. This extension, called the *strong equivalence principle* requires, for example, that in a freely falling reference frame, light propagates in straight lines, that the equations of electromagnetism are are just Maxwell's equations, etc. A further extension, the *ultra-strong equivalence principle*, asserts the validity of the principle for gravity itself.

The equivalence principle[1] requires a fundamental revision of our concepts of time and space. Saying that *locally* the laws of physics are the usual (special-relativistic) laws is the same as saying that locally space–time *is* Minkowski space. It also requires that at each point of space–time we can always find a system of Lorentzian coordinates, thus corresponding to a "freely falling" reference system. For minimal coherence we must be able to match all these local coordinate systems. All this is in fact characteristic of a pseudo-Riemannian manifold $[V^4, g_{\mu\nu}(x)]$ of four dimensions, with a (pseudo-) metric $g_{\mu\nu}$ (see *Appendix D*), which describes how gravitation acts on time and length standards, and thus on all physical phenomena.

However, if the equivalence principle tells us that space–time *is* a pseudo-Riemannian manifold, and thus *a priori* a curved space–time, it tells us nothing about how it is curved, i.e. about the metric tensor $g_{\mu\nu}(x)$. This is why the principle allows a large range

---

[1] Unless explicitly stated to the contrary, we use the expression "equivalence principle" for the ultra-strong version.

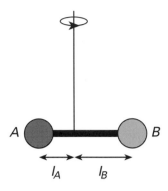

Fig. 7.1. The principle of the Eötvös–Dicke experiments.

of relativistic theories of gravity. We shall study only one of these, general relativity, in the next chapter.

In this chapter we shall expore further the consequences of the equivalence principle and its experimental foundations.

7.1   The weak equivalence principle and the Eötvös–Dicke experiments

We have already explained that Galileo's experiments with falling bodies gave the first indication of the equality of gravitational and inertial masses. Newton well understood the existence of these two types of mass, and using a simple pendulum (see *Chapter 1*) he showed that they were equal to about one part in a thousand. However, at the end of the last century the need for greater precision was apparent, and was possible because of technical progress.

The principle of this measurement (i.e. of the equality $m_{in} = m_{gr}$) is simple; it consists of putting oneself in a physical situation where the two types of mass appear differently in the equilibrium of the apparatus. Thus inertial mass is related to a centrifugal force (Earth's rotation) while the gravitational mass appears through weight forces.

(1) The Eötvös experiment (1890) is of this type; it was repeated several times with various improvements by R. Eötvös, D. Pekár and E. Fekete (1922), and greatly impressed Einstein. The principle is the following. Consider (**Fig. 7.1**) a system consisting of a torsional pendulum whose moving part is two masses A and B of different materials, at distances ℓ_A and ℓ_B from the torsion wire. Assume that the two bodies A and B react differently to the local gravitational field (the Earth's, in this case). This would mean that the ratio of gravitating to inertial mass of the two bodies was not a universal constant (which could always be set equal to unity by a suitable choice of units). In this case, the equilibrium position of the pendulum would be slightly different from that corresponding to masses A and B which reacted identically.

We consider this in detail by examining the forces to which the pendulum is subjected (**Fig. 7.2**). A given mass is subject to its weight $P = mg$ acting vertically, and a horizontal

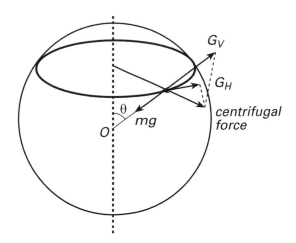

Fig. 7.2. The forces acting on a mass on the Earth's surface.

force $F_H = m\omega^2 R \sin \varphi$, where g is the acceleration of gravity, ω the angular velocity, R the radius of the Earth and φ the latitude of the experiment (originally Budapest). The horizontal forces (**Fig. 7.3**) acting on the pendulum are balanced by the torsion of the wire, and we have

$$\mathscr{C} = \ell_A F_{HA} - \ell_B F_{HB} \tag{7.2}$$

where $\mathscr{C}$ is the couple exerted by the forces. In equilibrium the vertical forces obey

$$\ell_A \cdot (m_{gA}g - m_{iA}G_v) = \ell_B \cdot (m_{gB}g - m_{iB}G_v) \tag{7.3}$$

where G_v is the vertical component of the centrifugal acceleration

$$G_v = \omega^2 R \cos \varphi. \tag{7.4}$$

The relations (7.2) and (7.3) give

$$\mathscr{C} = \ell_A m_{iA} G_H \left\{ 1 - \left[\frac{m_{gA}}{m_{iA}}g - G_v\right]\left[\frac{m_{gB}}{m_{iB}}g - G_v\right]^{-1} \right\}, \tag{7.5}$$

which, using the fact that $G_v \ll g$, can be written as

$$\mathscr{C} \simeq \ell_A m_{gA} G_H \left\{ \frac{m_{iA}}{m_{gA}} - \frac{m_{iB}}{m_{gB}} \right\}, \tag{7.6}$$

where the horizontal component G_H of the centrifugal acceleration is given by

$$G_H = \omega^2 R \sin \varphi. \tag{7.7}$$

The relation (7.6) shows clearly that if the ratio of inertial to gravitating mass is not a universal constant (i.e. if $m_i \neq m_g$, in suitable units, for every choice of materials) the pendulum will rotate, ($\mathscr{C} \neq 0$), and this can be observed.

In reality this experiment is performed twice, exchanging the two masses A and B.

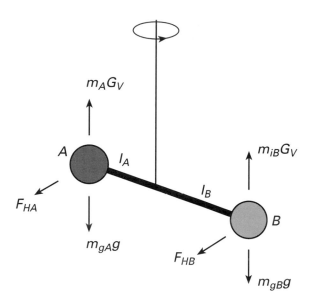

Fig. 7.3. The forces on the moving part of the pendulum.

This gives a reference position for the moving part of the pendulum and allows the possible measurement of a larger deviation (double) of the apparatus.

Eötvös's apparatus is represented in **Fig 7.4**. The unusual form of the instrument arose [see R.H. Dicke (1961)] because of planned applications in geophysics[2]. The different heights of the two masses was required in order to measure vertical gradients of the Earth's gravitational field. Unfortunately this is a source of almost unquantifiable errors. It makes the apparatus sensitive to gradients in the gravitational field and therefore to the positions of nearby massive bodies, such as that of the experimenter himself. The experiment was performed first in 1889 by Eötvös, and then in collaboration with D. Pekár and E. Fekete in 1922, using various materials (copper, glass, etc.), and no deviation was detected. The final result (1922) was

$$\left| \frac{m_i - m_g}{m_g} \right| \lesssim 3 \times 10^{-9}.$$

R.H. Dicke (1961, 1965) made the following criticisms of the experiment [P. Worden, F. Everett (1974), see also F. Everett, W.W. Hansen (1975)]: (i) its sensitivity to gradients in the gravitational field; (ii) the possibility of convection currents arising from (un-measurable) temperature gradients within the apparatus; (iii) magnetic contamination of the weights: a few times 10^{-6}g of a magnetic material interacting with the Earth's magnetic field would produce a deviation about 1000 times the probable error claimed by Eötvös; (iv) the diffraction of the telescope used to read off the deviation of the

---

[2] To which Eötvös made significant contributions.

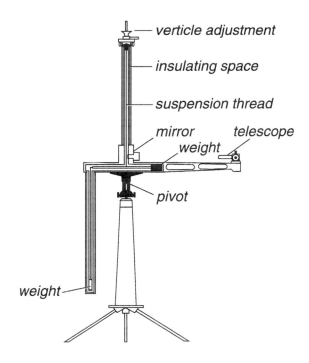

verticle adjustment

insulating space

suspension thread

mirror   telescope
     weight

pivot

weight

Fig. 7.4. Eötvös's apparatus.

mirror fixed to the moving part of the pendulum is a source of error (the claimed probable error corresponded to 5×10^{-3} times the smallest division, while the diffraction
limit of the telescope itself was about 40); (v) the presence of an operator, needed to
read off the measurement, produced a gravitational perturbation about 200 times the
probable error.

(2) For these reasons (and several others) R.H. Dicke, R. Krotkov and P.H. Roll
decided to repeat the experiment. Dicke's version is rather different in using the Sun's
gravity, i.e. the Earth's acceleration towards it.

Although the acceleration towards the Sun is about half ($0.62\mathrm{cm/s^2}$) that resulting
from the centrifugal force caused by the Earth's rotation ($1.4\mathrm{cm/s^2}$), such an experiment
has several advantages. First, the apparatus is fixed, avoiding the need to turn it
(exchange the masses A and B): this rotation occurs naturally every 24 hours. By
eliminating an operation this immediately removes a source of error. Further, a positive
result of the experiment would give a motion of the pendulum with a period of 24
hours (**Fig. 7.5**). To see this, assume that one of the weights (e.g. A) of the Eötvös–Dicke
experiment falls towards the Sun slightly more rapidly than the other (B). When the
experiment began at (say) 0 hours, the apparatus was by definition in equilibrium. Six
hours later the Earth has made a quarter turn (**Fig. 7.5**) and because of the forces on
A and B (which differ, by hypothesis), the equilibrium position of the pendulum has
changed. Six hours later, i.e. twelve hours after the start of the experiment, the position

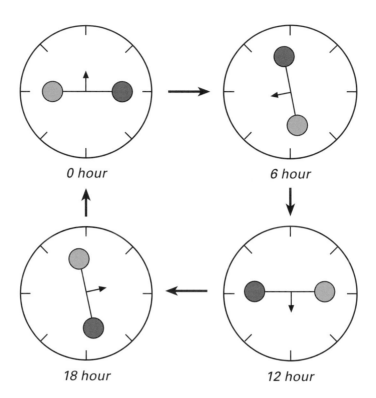

Fig. 7.5. Periodic variation of the equilibrium position of the Eötvös–Dicke pendulum. If the two masses A and B feel slightly different horizontal forces (i.e. if $m_{iA}/m_{gA} \neq m_{iB}/m_{gB}$), the equilibrium position varies with a period of 24 hours. The figure shows (exaggerated) the equilibrium positions at 6-hour intervals, corresponding to a rotation of the Earth through 90°. Remember that the position of the pendulum with respect to the Earth is fixed. Note that A (black) and B (grey) reverse every 12 hours.

of the pendulum is the opposite of the initial one, etc. This would give a variation of period 24 hours in the position of the pendulum.

However, besides these advantages, there are several difficulties, such as the weakness of the acceleration towards the Sun, diurnal perturbations (tides, temperature effects caused by the day–night cycle, etc.), which have to be dealt with carefully.

We now briefly indicate how Dicke and his collaborators (1964) overcame the problems with the Eötvös experiment and improved its accuracy.

First, the whole apparatus (**Fig. 7.6**) was enclosed in a pit, carefully thermally insulated, it being possible to measure the temperature and one of its gradients remotely. The pendulum was designed so that the effects of gradients in the gravitational field were minimal, unlike in the Eötvös experiment: the arrangement of the weights at the vertices of an equilateral triangle reduced the quadrupole moment of the pendulum. The latter consisted of two copper weights and one gold weight (or lead chloride,

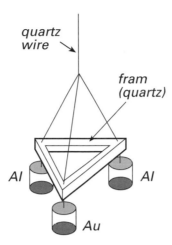

quartz wire

fram (quartz)

Al

Al

Au

Fig. 7.6. Dicke's pendulum [after P. Roll *et al.* (1964); R.H. Dicke (1961)].

copper, platinum, aluminium, etc.) suspended from a frame made of quartz (which has many stability properties: insensitivity to the Earth's magnetic field, resistance to electrical charging, rigidity, etc.). Finally, the whole torsional pendulum apparatus was placed in a good vacuum enclosure (10^{-8} mm of mercury), significantly reducing the effects of temperature gradients or pressure fluctuations on the moving parts of the apparatus.

To prevent the excitation of nonlinear oscillations of the pendulum with periods of 24 hours, the apparatus was electronically damped. Such oscillations could be excited by various phenomena such as seismic effects. Besides the torsional modes, oscillations of the entire apparatus or of the quartz fibre (acting like a spring) are possible. These must be damped as they tend to change the equilibrium position in a random manner.

Finally, the position of the moving part of the pendulum was measured *remotely*, without human interference with the apparatus. A light ray was reflected from one of the quartz bars of the pendulum and its deviation transmitted by means of a photoelectric cell. This signal was then used to return the pendulum to the zero position, the force used being itself recorded. In a period of about 10 s this system could detect an angular deviation of the order of 10^{-7} degrees.

This experiment found that inertial and gravitating masses were equal to an accuracy [P.G. Roll *et al.* (1964)]

$$\left| \frac{m_i - m_g}{m_g} \right| < 1.5 \times 10^{-11}$$

(within one standard deviation).

(3) As the main limitations of this experiment were caused by seismic perturbations, thermal effects and gravitational field gradients [see the discussion by F. Everitt (1975)], the experiment was repeated with a pendulum having a smaller quadrupole moment

and a better seismic environment. V.B. Braginsky and V.I. Panov (1971) gave

$$\left| \frac{m_i - m_g}{m_g} \right| < 0.45 \times 10^{-12}$$

(within one standard deviation); although this experiment has been criticised several times.

(4) Recently E.G. Adelberger *et al.* (1990) have greatly improved the Eötvös experiment and also tested the weak equivalence principle in the Earth's gravitational field. As mentioned in *Chapter 1*, they intended to test various models for a possible fifth force. However, they slightly improved the result of Roll *et al.* (1964) and, still within one standard deviation, found

$$\frac{\Delta m}{m} = \begin{cases} 0.2 \pm 1.0) \times 10^{-11} & \text{for a Cu/Be couple} \\ 0.5 \pm 1.3) \times 10^{-11} & \text{for an Al/Be couple.} \end{cases}$$

(5) Before closing this section we mention a space experiment. This experiment, now called STEP (Satellite Test of the Equivalence Principle) has since been accepted by ESA (European Space Agency), and in April 1991 was in phase A, studied by P.W. Worden and F. Everitt (1974) [see also F. Everitt (1975)] which should determine the ratio m_i/m_g with an accuracy of order 10^{-17}, or even 10^{-20} in more elaborate versions. The main advantage of a space experiment is the elimination of random seismic vibrations, which are the limiting factor for laboratory experiments. Moreover, the acceleration (which introduces the inertial mass) possible in such an experiment is far greater than in laboratory experiments. However, an Eötvös experiment on board a satellite could not give accuracy greater than 10^{-14} to 10^{-15}: the motion of the satellite allows gradients of the gravitational field, which increase with the eccentricity of the satellite orbit. This effect cannot be separated from that which one wants to measure except in an extremely long experiment. Another type of measurement is required.

The idea is as follows (**Fig. 7.7**). Two coaxial cylinders made of different materials are placed in orbit, with their common axis maintaining a fixed direction in space. Both masses feel the attraction of the Earth and the centrifugal acceleration of the orbit. Any difference in the fall of one of the bodies towards the Earth would give a periodic relative displacement of the two cylinders. The displacement is clearly very small: to measure the ratio m_i/m_g to an accuracy of 10^{-17} one needs to measure a displacement of less than 1Å! This kind of measurement clearly requires special techniques (cryogenics, W.M. Fairbanks (1974)) even without taking into account the preliminary studies of the various perturbing effects: motions of the satellite, "tides" in the liquid helium required to reach low temperatures... Even the position of the various electronic components is not (gravitationally) neutral!

(6) We now examine how far the weak equivalence principle is obeyed by the various interactions encountered in nature, i.e. electromagnetic, strong, weak and gravitational. It is *a priori* clear that as gravitational interactions inside the nucleus are of order $Gm_{\text{proton}}/R_{\text{nucleus}}c^2 \sim 10^{-39}$ and thus negligible, so we limit ourselves to the other forces.

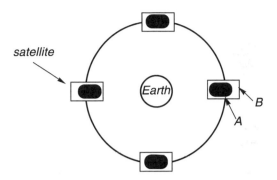

Fig. 7.7. Principle of a space-based measurement of the ratio m_i/m_g. Two coaxial cyclinders A and B, made of different materials, are in orbit about the Earth. If one of the bodies falls more rapidly than the other towards the Earth, this produces a relative motion of A and B with the period of the satellite.

To find the order of magnitude of these effects, assume that an interaction of type K (weakly) violates the weak equivalence principle, so that we can write

$$m_G = m_I + \sum_K \eta_K \frac{E_K}{c^2} \tag{7.8}$$

where η_K is a "violation coefficient" which depends on the nature K of the interaction considered, and E_K is the contribution to the internal energy of the body studied (in an Eötvös-type experiment) for the same type of interaction. For two bodies of different composition A and B in the same gravitational field g, the acceleration is

$$\gamma(A \text{ or } B) = \left\{ 1 + \sum_K \eta_K \frac{E_k(A \text{ or } B)}{m_I(A \text{ or } B)c^2} \right\} g, \tag{7.9}$$

so that the ratio

$$\eta \equiv 2\frac{\gamma(A) - \gamma(B)}{\gamma(A) + \gamma(B)} = \sum_K \eta_K \left[\frac{E_K(A)}{m_I(A)c^2} - \frac{E_K(B)}{m_I(B)c^2} \right] \tag{7.10}$$

gives a measure of the violation of the weak equivalence principle. We have seen above that experiments give $|\eta| < 10^{-11}$ or 10^{-12}. Thus, attributing *all* the possible violation of the equality $m_I = m_G$ to the interaction of type K, we get an *upper bound* for η_K.

As an example, consider a nucleus composed of A nucleons, of which Z are protons. Its electrostatic energy is

$$E_{es} = \frac{3}{5}\frac{Z^2 e^2}{R_0 A^{1/3}}, \tag{7.11}$$

where we have used the semi-empirical formula $R = R_0 A^{1/3}$ ($R_0 = 1.24 \times 10^{-13}$cm) for the radius of a uniformly charged nucleus. For bodies as different as platinum ($Z = 78$, $A = 195$) and aluminium ($Z = 13$, $A = 27$) used in these experiments –

which might give a measurable effect – we have

$$\left| \frac{E_{es}(\text{Pt})}{m_I(\text{Pt})} - \frac{E_{es}(\text{Al})}{m_I(\text{Al})} \right| \sim 10^{-3},$$

and finally $\eta \lesssim 10^{-8}$–10^{-9}. This extremely simple calculation shows not only that electrostatic interactions effectively do not violate the weak equivalence principle, but also the need for *models* describing the materials and the interaction between their various components. Here the nucleus has been treated as a uniformly charged sphere, *and* we have used the semi-empirical formula for its radius.

It is usual to employ a well-established *semi-empirical*[3] formula for nuclear interactions, which gives the binding energy of a nucleus $[A, Z]$,

$$E_{\text{nucl}}(\text{MeV}) = 15.68A - 18.56A^{2/3} - 0.717\frac{Z^2}{A^{1/3}} - 28.1\frac{(A - 2Z)^2}{A} + E_P, \qquad (7.12)$$

where, using the relation $R = R_0 A^{1/3}$, we can recognise the binding energy per nucleon, its electrostatic energy, and where E_P, the pairing energy, is given by

$$E_P = \frac{1}{2}(-1)^Z[1 + (-1)^4] \times \frac{12}{A^{1/2}}. \qquad (7.13)$$

Electromagnetic interactions other than electrostatic ones inside the nucleus are much more complicated, as they require a detailed model of the nuclear structure. These interactions give rise to magnetostatic forces caused by proton currents within the nucleus, and forces coupling the spins of the nucleons and the magnetic fields they induce.

Similarly, the binding energy of atomic electrons can be estimated [M.G. Bowler (1976)] using the Thomas–Fermi model of the atom, giving[4]

$$E_{\text{at}} = 15.73Z^{7/3}, \qquad (7.14)$$

and thus a value for η_{at}.

Weak interactions similarly require a precise model of the nucleus in which the nucleons interact according to the Weinberg–Salam theory. This has been calculated by M.P. Haugan and C.M. Will (1976), who give $\eta_{\text{weak}} \lesssim 10^{-2}$, while an order-of-magnitude estimate based on the Fermi model [M.G. Bowler (1976)] gives $\eta_{\text{weak}} \lesssim 10^{-1}$–$10^{-2}$ if we take $\delta m/m \lesssim 10^{-12}$, and $\eta_{\text{weak}} \lesssim 10^{-1}$ if we limit ourselves to Dicke *et al.*'s results, i.e. $\delta m/m \lesssim 10^{-11}$. It is thus not completely proven that the weak interaction obeys the equivalence principle [see also J.P. Hsu (1978)].

Table 7.1 summarizes these results.

---

[3] See L. Valentin (1975).
[4] L.D. Landau, E.M. Lifschitz (1965).

Table 7.1. *Upper bounds for violations of the weak equivalence principle for various interactions, deduced from Eötvös-type experiments [from C.M. Will (1981)]*

Interaction	$\eta \equiv$ violation of $m_g = m_I$
electrostatic interactions (nucleus)	$\eta_{es} < 4 \times 10^{-10}$
magnetostatic interactions (nucleus)	$\eta_{ms} < 6 \times 10^{-6}$
coupling of nucleon magnetic moments with the magnetic field of the protons	$\eta_{HF} < 2 \times 10^{-7}$
electrostatic interactions of atomic electrons	$\eta_{TF} < 5 \times 10^{-7}$
nuclear interactions	$\eta_{nucl} < 5 \times 10^{-10}$
gravitational forces between nucleons	$\eta_{grav} < 10^{+27}$!
weak interactions	$\eta_{weak} < 10^{-2}$

7.2 The equivalence principle and minimal coupling

We have seen that the equivalence principle is well verified by experiment. We shall now state the principle in its full generality and discuss its main consequences, in particular, how far this too is experimentally justified. We follow the formulation of C.M. Will (1981), which, besides being precise, also clearly separates the various assumptions.

Einstein's equivalence principle: (i) the inertial mass of a body is equal to its (passive) gravitating mass, (ii) the result of any local non-gravitational experiment is independent of the velocity of the apparatus in a freely falling reference system, (iii) the result of any local non-gravitational experiment[5] is independent of the time and place at which it is performed [C.M. Will (1981)].

(1) This statement prompts a number of remarks. We have already discussed (i) – weak equivalence – and shown that it is very well justified experimentally. However, there may be situations in which it does not hold. For example, do particles and antiparticles fall in the same way? Experiment suggests that they do, to a good approximation, at least in the cases of electrons and positrons [see the discussion by M.G. Bowler (1976)], the $K_0 \bar{K}_0$ system[6] [M.L. Good (1961)], neutrinos and anti-neutrinos[7] [S. Pakvasa, W.A. Simmons, T.J. Weiler (1989)]. Similarly, the universality of free–fall is measured (albeit not conclusively) for electrons [F.C. Witteborn, W.M. Fairbank (1968)] and neutrons [L. Koester (1976); V.F. Sears (1982)].

---

[5] ...of negligible mass, so that its gravitational field can be ignored in a freely falling reference system. The experiment thus appears as a "test particle".

[6] It seems that Good's argument is incorrect [T. Damour, private communication]; it used the classical gravitational field ϕ, while only $\nabla\phi$ has meaning (and only $\nabla\nabla\phi$ has meaning in general relativity).

[7] If these are Majorana neutrinos (two-component neutrinos which are identical to their antiparticles), the experiments test the equality of free fall for the left and right neutrinos.

Quite generally, different behaviour of particles and antiparticles in a gravitational field would imply the violation of charge conjugation invariance. This is experimentally verified to an accuracy better than 5×10^{-6}; it is therefore reasonable to assert that particles and antiparticles obey the weak equivalence principle to better than about 10^{-4}.

(2) An important case where weak equivalence cannot be checked is that of gravity itself. Eötvös-type experiments cannot be used, because of the weakness of the inter-action. Such experiments formally give $\eta_{\text{grav}} < 10^{+27}$! One has to use very massive objects, such as the Earth and the Moon in the gravitational field of the Sun, or Mars and Jupiter in the same field. If η_{grav} stands for the "violation coefficient" of the weak equivalence principle by gravity, i.e. if we set

$$m_{\text{grav}} = m_{\text{in}} + \eta_{\text{grav}} E_{\text{grav}}, \tag{7.15}$$

the equations of motion of the Earth and the Moon in the Sun's gravitational field become

$$\begin{cases} \vec{\gamma}_E = -\dfrac{m_{GE}}{m_{IE}} G \dfrac{\vec{x}_E M_\odot}{|\vec{x}_E - \vec{x}_S|^3} - \dfrac{m_{GE}}{m_{IE}} G \dfrac{(\vec{x}_E - \vec{x}_M)}{|\vec{x}_E - \vec{x}_M|^3} m_{GM} \\[2mm] \vec{\gamma}_M = -\dfrac{m_{GM}}{m_{IM}} G \dfrac{\vec{x}_M M_\odot}{|\vec{x}_M - \vec{x}_S|^3} - \dfrac{m_{GM}}{m_{IM}} G \dfrac{(\vec{x}_M - \vec{x}_E)}{|\vec{x}_E - \vec{x}_M|^3} m_{GE} \end{cases} \tag{7.16}$$

where the indices E, M, S refer to the Earth, Moon and Sun and $M_\odot$ is the Sun's *active* gravitating mass. The relative acceleration of the Earth and Moon caused by any violation of $m_I = m_G$ is then given by

$$\vec{\gamma} \equiv \vec{\gamma}_M - \vec{\gamma}_E \quad \text{to order } \eta_{\text{grav}} \tag{7.17}$$

$$= \eta_{\text{grav}} G \left[\frac{E_{\text{grav}}(\text{Earth})}{m_{\text{in}}(\text{Earth})c^2} - \frac{E_{\text{grav}}(\text{Moon})}{m_{\text{in}}(\text{Moon})c^2} \right] M_\odot \frac{(\vec{x}_S - \vec{x}_E)}{|\vec{x}_S - \vec{x}_E|^3} - \frac{(\vec{x}_M - \vec{x}_E)}{|\vec{x}_M - \vec{x}_E|^3}$$

$$\times \eta_{\text{grav}} G \left[\frac{m_{\text{grav}}(\text{Earth}) E_{\text{grav}}(\text{Moon})}{m_{\text{in}}(\text{Moon})c^2} + \frac{m_{\text{grav}}(\text{Moon}) E_{\text{grav}}(\text{Earth})}{m_{\text{in}}(\text{Earth})c^2} \right]. \tag{7.18}$$

While the second term is a purely quantitative correction to the relative motion of the Earth and the Moon, which is circular to a good approximation, the first term in (7.18) is a *qualitative* modification giving what is known as the *Nordtvedt effect* [K. Nordtvedt (1968a, b, c)]. A simple calculation [see C.M. Will (1981)] shows that the Moon's trajectory is slightly polarised towards the Sun (see **Fig. 7.8**). If we assume that the Earth and Moon are uniform spheres, their gravitational energy

$$E_{\text{grav}} = -\frac{3}{5} \frac{G m_{\text{grav}}^2}{R}, \tag{7.19}$$

divided by $m_{\text{in}} c^2$ is -4.6×10^{-10} for the Earth and -2×10^{-11} for the Moon, giving a deviation (in centimetres)

$$\delta r(t) \simeq 920 \eta_{\text{grav}} \cos \left[(\omega_0 - \omega_s) t \right], \tag{7.20}$$

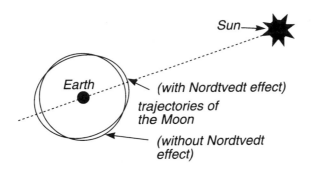

Fig. 7.8. The Nordtvedt Effect. If the Earth and Moon do not "fall" with the same velocity towards the Sun, the Moon's trajectory is slightly deformed towards the Sun (greatly exaggerated in the figure).

where ω_0 is the angular velocity of the Moon about the Earth and ω_s is the angular velocity of the Sun about the Earth.

The results of this experiment give $\eta_{\text{grav}} = 0.00 \pm 0.03$ [J.G. Willams *et al.* (1976)] and $\eta_{\text{grav}} = 0.001 \pm 0.015$ [I.I. Shapiro, C.C. Counselman, R.W. King (1976)], according to two independent analyses of the same data. Thus the values found for η_{grav} are consistent with zero. This corresponds to equality between the Moon's inertial and gravitating masses to order a few times 10^{-11}.

We stress that these measurements constitute *gravitational experiments* which go *beyond* the strong equivalence principle stated above, and are relevant to the ultra-strong principle (see below).

We note also the recent suggestion by T. Damour and G. Schäfer (1991) of using data from the binary pulsars PSR 1913+16, PSR 1953+29 and PSR 1855+09 to detect any violation of the equivalence principle to a higher order, i.e. $[E_{\text{grav}}/mc^2]^2$.

(3) Point (ii) of the equivalence principle asserts that in a freely falling (thus locally inertial) reference frame, non-gravitational physical experiments are independent of the *uniform* velocity of the apparatus. Thus Lorentz invariance must be satisfied. In *Chapter 3* we discussed experimental tests of Special Relativity and their theoretical framework. Almost all of these tests are local.

Point (iii) requires that the laws of physics should be the same throughout the Universe, and at all times. Spatial invariance is tested by experiments on the gravitational redshift [R.V. Pound, G.A. Rebka (1960); R.V. Pound, J.L. Snider (1965)] or comparisons of clocks carried in aeroplanes or rockets [see C.M. Will (1981)]. Here again the equivalence principle is well verified (see below for the gravitational redshift).

Tests of the validity of physical laws at all epochs have to assume that their form is fixed and find upper limits on the variation of the constants appearing in them, such as the fine structure constant $e^2/\hbar c$, the strong interaction constant, etc. In fact, if these fundamental constants varied even slightly many new effects would be observable. Atomic and nuclear spectra would be changed, as would the stability properties of

Table 7.2. *Orders of magnitude for the upper limits on annual variations of the coupling constants of known interactions, assuming that in each case only one constant varies. Shlyakhter's numbers are based on analysis of data from the Oklo natural nuclear reactor [M. Maurette (1976); J.M. Irvine (1983); Y.V. Petrov (1977)]*

Quantity	Upper limit on annual variation	Authors
G	10^{-11}	Damour *et al.* (1988)
e^2	10^{-14}	Davies (1972)
e^2	10^{-17}	Shlyakhter (1976)
g_{nucl}	10^{-18}	Shlyakhter (1976)
g_{nucl}	10^{-14}	Davies (1972)
g_{weak}	10^{-12}	Shlyakhter (1976)
g_{weak}	10^{-10}	Davies (1972)

nuclei, the radii of planets and their orbital evolution, nucleosynthesis of light elements (cosmology) and heavy elements (stars), etc. In short, all kinds of physical, geophysical and astrophysical phenomena would reveal such an effect. Many of these effects are reviewed and analysed briefly by P. Sisterna and H. Vucetich (1990). Table 7.2 gives the order of magnitude of the variations of the coupling constants of the four fundamental interactions[8] [see also C.M. Will (1981), J.M. Irvine (1983)]. We give a few examples to show how these numbers arise in simple cases. Note however that J. Bekenstein (1982) has criticised these results, arguing the necessity for dynamical models for the evolution of physical constants. He has given a model of this kind for the electromagnetic interaction, based on natural postulates, and concluded that these results imply only marginal constancy of $e^2/\hbar c$. He has, however, shown that Eötvös-type experiments would allow one to demonstrate the spatial and temporal constancy of $e^2/\hbar c$.

We consider first a possible variation of the gravitational constant G: suppose that a mass m is in a circular orbit (at least for the purposes of finding an order of magnitude) about another object of mass M. Let R be the radius of the orbit. Then, equating the gravitational attraction to the centrifugal force, one gets

$$\frac{GMm}{R^2} = \frac{mv^2}{R},$$

(7.21)

where v is constant. The period of the "planet" of mass m is thus

$$T = 2\pi \left(\frac{R^3}{GM} \right)^{1/2}.$$

Differentiating, we get the rate of change of T and R if G varies in time

$$\frac{1}{T}\frac{dT}{dt} = \frac{3}{2}\frac{1}{R}\frac{dR}{dt} - \frac{1}{2}\frac{1}{G}\frac{dG}{dt}.$$

(7.22)

---

[8] We also have to consider possible time variations of other quantities such as particle masses, etc.

Thus, the orbits of the planets would have been different in the past, the dates of eclipses would have changed, etc. [V. Canuto (1990)]. Measurements of the solar system thus lead to a maximum relative variation $|\dot{G}/G| < 10^{-11}$ per year, assuming of course that M (and the interactions contributing to it) do not vary in time.

We now consider the fine structure constant $\alpha \equiv e^2/\hbar c$. This enters squared in the wavelengths of spectral lines emitted by atoms. Observing very distant objects such as quasars[9] [J.N. Bahcall, W. Sargent, M. Schmidt (1967)] or radio galaxies [J.N. Bahcall, M. Schmidt (1967)] samples light emitted several billion years ago. Thus, comparison of the wavelengths of lines in quasars (or distant radio galaxies) with those observed in the laboratory allows one to place limits on the variation of the fine structure constant α over time. As the frequency of a line varies as α^2, we have $\dot{v}/v = 2\dot{\alpha}/\alpha$ and find $\dot{\alpha}/\alpha < 10^{-12}$ per year.

There is a second argument placing a limit on the variation of the fine structure constant [see: P.J. Peebles and R.H. Dicke (1962); F.J. Dyson (1967); P.C.W. Davies (1972)]. This uses the fact that two isobars[10] have masses which roughly speaking differ essentially by the Coulomb term in the mass formula [cf. Eq. (7.12)]

$$E_{\text{coul}} = .717\frac{Z^2}{A^{1/3}} \qquad \text{(in MeV)}, \tag{7.23}$$

which is implicitly proportional to e^2. Thus, in a beta-decay (or beta-capture) between two isobars

$$\begin{cases} {}^A_Z X \rightarrow {}^A_{Z+1} X + e + \bar{v}_e \\ {}^A_Z X \rightarrow {}^A_{Z-1} X + e^+ + v_e, \end{cases} \tag{7.24}$$

the energy carried off by the leptons[11] depends on the difference ΔE_{coul},

$$\Delta M = \Delta E_{\text{coul}} = 0.717\frac{2Z-1}{A^{1/3}}(\text{MeV}). \tag{7.25}$$

If the lifetime of the element ${}^A_Z X$ is long enough that e^2 could have varied appreciably, we would have

$$\frac{1}{\Delta M}\frac{\delta\Delta M}{\delta t} = \frac{1}{\alpha}\frac{\delta\alpha}{\delta t}, \tag{7.26}$$

which would affect the lifetime of ${}^A_Z X$ and thus the relative abundances of the two isobars. Such pairs of isobars do exist, notably Rhenium/Osmium ($A = 187$), whose importance P.J. Peebles and R.H. Dicke (1962) pointed out: the lifetime of Rhenium is of order 4×10^{11} years, but is unfortunately very difficult to measure accurately in the laboratory.

The most accurate estimates of possible variations in the various coupling constants [A.I. Shlyakhter (1976)] are given by the isotopic abundances in the "natural reactor" at Oklo (Gabon). This is a fortuitous concentration of Uranium 235, large enough that

---

[9] Taking quasar redshifts as cosmological in origin, as is generally assumed.
[10] Nuclei with the same number A of nucleons and which thus differ in their numbers Z of protons.
[11] Light particles: $e, v_e, e^+, \bar{v}_e$, etc...

two billion years ago it spontaneously started to react. This can be deduced from the much lower percentage of Uranium 235 in the ore than the usual terrestrial abundance (0.72%). The abundance of the two isotopes of Samarium (147 and 149) in Oklo is lower than it should be, showing the effect irradiation by the neutrons produced in the fission of Uranium about 1.8×10^9 years ago. This gives an extremely precise value for the resonant neutron capture cross-section of Samarium 149 1.8×10^9 years ago, which can be compared with modern laboratory values. This leads to the values quoted by A.I. Shlyakhter (1976). The article by J.M. Irvine (1983) gives a clear and quite detailed account.

Of course, the numerical values given above assume implicitly that other physical constants do not vary. Thus one assumes Planck's constant, the electron mass etc. are all fixed [B.E.J. Pagel (1977) gives limits on the variation of the proton mass with respect to that of the electron, using quasar absorption lines.] A brief discussion of possible mass variations can be found in H.C. Ohanian (1976).

We conclude that there is currently no evidence for any change in the laws of physics over time, and to some extent in space. The equivalence principle and its consequences are thus verified in all their forms.

(4) The importance of the equivalence principle arises not only from its various experimental bases, but also from the fact that it essentially implies the existence of a (locally) curved four-dimensional space–time equipped with a metric $g_{\mu\nu}(x)$ revealing the existence of the gravitational field. We stress, however, that the principle says nothing about the laws that the metric tensor must obey: many theories, including Einstein's General Relativity, satisfy the equivalence principle, but differ in the equations obeyed by the metric tensor and/or the additional classical fields assumed to describe gravity. In all cases matter is always coupled to the metric tensor in such a way that in a *locally inertial* (freely falling) system, the non-gravitational laws of physics are the usual ones of special relativity.

We can thus from now on find the form of laws in curved space–time from the usual relativistic forms: all we have to do in general is (mentally) replace the Lorentzian indices by tensor indices pertaining to general coordinate changes in the space–time manifold V^4, and replace partial derivatives by covariant derivatives

$$(a)\, \partial_\mu \rightarrow (b)\, \nabla_\mu. \tag{7.27}$$

For example, energy–momentum conservation becomes

$$(a)\, \partial_\mu T^{\mu\nu} = 0 \rightarrow (b)\, \nabla_\mu T^{\mu\nu} = 0, \tag{7.28}$$

the form on the left being manifestly covariant only under Lorentz transformations, while the equation on the right is covariant under general changes of coordinates in V^4. As a result we note that although the space–time indices of the two relation (7.28) are very different (they relate to different transformations), they conserve the *signature* of the metric. Similarly, using (7.27), Maxwell's equations in curved space can be obtained from those in Minkowski space

$$(a) \begin{cases} \partial_\mu F^{\mu\nu} = 4\pi J^\nu \\ \partial_\mu {}^*F^{\mu\nu} = 0 \end{cases} \rightarrow (b) \begin{cases} \nabla_\mu F^{\mu\nu} = 4\pi J^\nu \\ \nabla_\mu {}^*F^{\mu\nu} = 0 \end{cases} \tag{7.29}$$

where the Levi–Civita tensor relating $F^{\mu\nu}$ and $^*F^{\mu\nu}$ becomes

$$(a)\, \varepsilon^{\mu\nu\alpha\beta} \rightarrow (b)\, \frac{1}{\sqrt{|g|}} \varepsilon^{\mu\nu\alpha\beta}. \tag{7.30}$$

We should note an important difference between equations (7.27), (7.28) and (7.29) on the one hand, and equations (3.135) and (3.133), on the other hand. In the latter case it is always possible to find a coordinate system in which equations (*a*) reduce *everywhere* to equations (*b*), while for equations (7.28) and (7.29) this is only possible *locally*, in a freely falling reference frame.

Finally, we should examine the rule (7.27) very closely. If the Minkowskian equations we wish to write down in the presence of gravity have second or higher derivatives, ambiguities can arise because covariant derivatives are *non-commutative*

$$\left[\nabla_\mu, \nabla_\nu\right] X^{\cdots}_{\cdots} \neq 0 \tag{7.31}$$

In practice it is usually possible to remove this ambiguity [see C.W. Misner, K.S. Thorne, J.A. Wheeler (1973) for a more detailed discussion], rather in the same way as when quantising a classical system.

The rule (7.27), with these restrictions, is known as *minimal coupling*: It is the minimum required (i) to arrive at equations which are covariant under arbitrary coordinate changes in V^4, and (ii) to couple matter and gravity via the geometry of V^4. We could find other generalisations of the Minkowskian equations by adding curvature-dependent terms to ∇_μ. For example, the equation

$$\nabla_\mu F^{\mu\nu} + \lambda J^\mu R_\mu{}^\nu{}_{\alpha\beta} F^{\alpha\beta} = 4\pi J^\nu, \tag{7.32}$$

would constitute such a non-minimal generalisation. In the absence of gravity the curvature tensor vanishes identically, and we would recover the usual Maxwell equations if we chose Lorentzian coordinates.

We thus see that the equivalence principle does not completely fix the coupling of matter and gravity.

(5) We conclude this section by briefly mentioning the ultra-strong equivalence principle. We have already noted above that the absence of the Nordtvedt effect offers support for this version of the principle as applied to the equality of inertial and gravitating masses. Gravitational energy does not differ from other forms of energy as far as its contribution to the gravitating mass is concerned.

Violation of the ultra-strong equivalence principle would be revealed, for example, by the dependence of a *local* gravitational experiment on the cosmological context, or possible long-range fields. If, on the other hand, the principle is satisfied, this would mean that the various fields present in nature would always be coupled in the same way, independent of position and time, and with *constant coupling*, independently of any cosmological structure. In this context the Hughes–Drever experiment allows one to eliminate such supplementary fields to a very good approximation [see C.M. Will (1981)].

An example of the violation of this principle [see V. Canuto and I. Goldman (1983); V. Canuto (1990)] could be found in a possible discrepancy between the usual clocks (atomic, nuclear, etc.) and gravitational clocks. This is compatible with the fact that gravity is the only important interaction on cosmological scales. This inequivalence would be revealed by a time-dependent scale factor between time intervals ds_{grav} indicated by a gravitational clock and those ds_{at} indicated by an atomic clock

$$ds_{grav} = \beta(t)ds_{at} \qquad (7.33)$$

with $\beta(t_0) = 1$. Observation (laser reflections from the Moon or radar reflections from the inner planets) and geophysical measurements give $\dot{\beta}(t) < 10^{-11} \text{yr}^{-1}$.

We should explain the idea of gravitational and atomic clocks used above. We limit ourselves here to the simplest cases. An atomic clock involves only electromagnetic forces, and thus obeys the strong equivalence. A rough example is constituted by a particle of charge q and inertial mass m_{in} in a circular orbit around a charge Zq, so that

$$m_{in}\frac{V^2}{R} = \frac{Zq^2}{R^2},$$

with the period

$$T_{at} = 2\pi\sqrt{\frac{R^3 m_{in}}{Zq^2}}.$$

A gravitational clock uses gravity, through the intermediary of its gravitating mass m_{grav},

$$m_{grav} = m_{in} + \eta\frac{E_{grav}}{c^2}$$

(η: coefficient of violation of the equivalence principle; E_{grav}: gravitational energy involved). A gravitational clock, like the atomic one, has a period

$$T_{grav} = 2\pi\sqrt{\frac{m_{in}}{m_{grav}}\frac{R^3}{GM}}.$$

We note in passing that two planets with exactly the same (m_{in}, R), orbiting the same central star of mass M, but with different chemical compositions – and thus different passive gravitational masses m_g – would have different orbital periods... We compare now the behaviour at the same point of an atomic and a gravitational clock, freely falling in a constant gravitational field g. The atomic clock has a free-fall velocity

$$v_{at} = gt,$$

while the gravitational clock will have

$$v_{grav} = \frac{m_{grav}}{m_{in}}gt = \left(1 + \eta\frac{E}{m_{in}}\right)gt.$$

The two clocks thus have relative velocity

$$v \equiv \frac{v_{grav} - v_{at}}{1 - v_{grav} \cdot v_{at}} = \eta\frac{E}{m_{in}}gt + O(\eta^2),$$

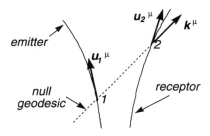

Fig. 7.9. The gravitational redshift.

so that if the clocks originally had the same period the longitudinal Doppler effect would produce changes

$$\Delta T_{\text{grav}} = v\Delta T_{\text{at}} + O(v^2)$$
$$= \eta \frac{E}{m_{\text{in}}} gt\Delta T_{\text{at}} + O(\eta^2).$$

Of course the coefficient η might itself change in time (on a larger scale than that of the two clocks, e.g. cosmological). Thus a relation like (7.33) holds generally.

7.3 The gravitational redshift

We have seen that in the presence of gravity space–time can be correctly represented as a four-dimensional metric manifold $\{V^4, g_{\mu\nu}(x)\}$. We now examine some consequences, the first being a specifically gravitational *time dilatation*, which produces a photon *redshift* (see *Chapter 4* for a more intuitive argument).

We consider an atomic transition at the space–time point 1 $[1 \equiv (x_1^0, x_1^1, x_1^2, x_1^3)]$ (see **Fig. 7.9**) where the gravitationl field is $g_{\mu\nu}(1)$. A photon emitted from an atom of 4-velocity $u^\mu(1)$ has the wave 4-vector $k^\mu(1)$, and is observed at point 2 by an observer with 4-velocity $u^\mu(2)$. At 2 the gravitation field is $g_{\mu\nu}(2)$. The energy (or frequency) of the photon emitted at 1 is

$$E(1) = k^\mu u_\mu|_1 \equiv v(1) \tag{7.34}$$

while that of the photon observed at 2 is then

$$E(2) = k^\mu u_\mu|_2 \equiv v(2). \tag{7.35}$$

In the rest frames of the emitter and the observer $E(1)$ and $E(2)$ reduce to the frequencies $v(1)$ and $v(2)$. The redshift is

$$z = \frac{\lambda_{\text{obs}} - \lambda_{\text{em}}}{\lambda_{\text{em}}}, \tag{7.36}$$

and we have also

$$1 + z = \frac{[u^\mu k_\mu]_{\text{em}}}{[u^\mu k_\mu]_{\text{obs}}}. \tag{7.37}$$

This relation is completely general, and applies in many physically important cases[12]. Here we restrict ourselves to the case of a *static* gravitational field, i.e. there exists a coordinate system in which $\partial_t g_{\mu\nu}(\vec{x}, t) = 0$; in this case we can always choose coordinates such that $g_{0i}(\vec{x}) = 0$.

Equation (7.37) can then be rewritten as

$$1 + z = \frac{\left[u^0 k_0\right]_{\text{obs}}}{\left[u^0 k_0\right]_{\text{em}}} \tag{7.38}$$

$$= \left(\frac{g_{00}(\text{em})}{g_{00}(\text{obs})}\right)^{1/2} \frac{k_{0\,\text{obs}}}{k_{0\text{em}}}, \tag{7.39}$$

recalling that $g_{\mu\nu} u^\mu u^\nu = 1$, and thus that

$$u^0 = \frac{1}{\sqrt{g_{00}}}, \tag{7.40}$$

in a comoving reference frame. As we also have $\partial_t g_{\mu\nu} = 0$, the zero component of the wave 4-vector k_μ is constant[13] along the null geodesic from event 1 to event 2. We show this briefly. The geodesic equation (see below) is

$$\frac{\nabla k_\mu}{d\xi} = 0 = \frac{dk_\mu}{d\xi} + \Gamma^\alpha_{\mu\beta} k_\alpha k^\beta = 0; \tag{7.41}$$

$$(\xi: \textit{affine parameter})$$

or, evaluating the Christoffel symbols for the zero component of k_μ,

$$\frac{dk_0}{d\xi} - \frac{1}{2}\left[\partial_\beta g_{0\alpha} + \partial_0 g_{\beta\alpha} - \partial_\alpha g_{0\beta}\right] = 0. \tag{7.42}$$

Since $g_{\mu\nu}$ is diagonal and $\partial_0 g_{\beta\alpha} = 0$, we find $dk_0 = 0$ and thus $k_{0\text{obs}} = k_{0\text{em}}$. Hence the redshift is given by

$$1 + z = \left(\frac{g_{00}(\text{em})}{g_{00}(\text{obs})}\right)^{1/2}. \tag{7.43}$$

We shall see in the next section that for weak fields the relation above reduces to that found heuristically in *Chapter 4*. The spectral shift z is quite generally proportional to the gravitational parameter GM/Rc^2. The values of this parameter[14] for systems known up to the 1960s showed that only white dwarfs could give a measurable effect: typically $z \sim 2 \times 10^{-4}$. Observations [see W.S. Adams (1925); J.L. Greenstein and V.L. Trimble (1967); G. Gatewood, J. Russel (1974); J.L. Greenstein *et al.* (1977); G. Gatewood, C. Gatewood (1978); J. Hershey (1978); H.L. Shipman (1979); G. Wegner (1980)] gave

---

[12] It can be used to find the cosmological redshift or the usual formula for the Doppler effect [see G.F.R. Ellis (1971)].

[13] This property is a special case of the following theorem [see K.S. Thorne (1971)]: "*If the metric tensor $g_{\mu\nu}$ is independent of a coordinate x^β, the numerical value of the covariant β-component of the 4-momentum of a massive particle or a photon is constant along a geodesic.*" The proof is identical to that given above.

[14] See Table 4.1.

results in rough agreement with the relativistic predictions. However, these results have large errors since the spectral lines of white dwarfs are often quite broad, so that measuring a displacement of order 10^{-4} is difficult.

For the Sun there is much less of a problem as the spectral lines are much narrower: the effect to be measured is only of order 2×10^{-6}... However, the main problem is not in measuring the displacement of the lines, but the fact that the Sun's atmosphere is subject to random convective motions which introduce unknown Doppler effects. As the convection currents are roughly vertical, and the line of sight is orthogonal to these motions at the Sun's limb, this effect is largely eliminated for lines observed there. Repeated measurements [J.E. Blamont, F. Roddier (1961); J. Brault (1963); J.L. Snider (1972) (1974)] confirm the relativistic prediction to an accuracy of order 5%.

However, the first accurate measurement (about 1%) of the gravitational redshift was made in the laboratory, i.e. on the Earth [R.V. Pound, G.A. Rebka (1960); R.V. Pound, J.L. Snider (1965)]. Although the spectral shift at two levels [differing by of order 22 m in the experiment by R.V. Pound and G.A. Rebka (1960)] is very small ($\sim 10^{-15}$), it is possible to produce very narrow lines using the Mössbauer effect[15]. These lines are close to their natural widths[16].

Later, considerable improvements in clock stability [10^{-15}–10^{-16} over times of 10 to 100 seconds] allowed comparisons at different altitudes: for example, comparison of a hydrogen maser clock[17] on the Earth with a similar clock in a rocket confirmed the relativistic prediction to an accuracy better than 2×10^{-4} [R.F.C. Vessot, M.W. Levine (1979); R.F.C. Vessot (1984)]. Other experiments all give results with the same trend [see C.M. Will (1981)].

We have already pointed out the theoretical importance of the gravitational red-shift: it represents one of the strongest arguments in favour of metric theories of relativistic gravity (including General Relativity). But it also constitutes an experimental verification of point (iii) of the strong equivalence principle, i.e. the independence of non-gravitational experiments of the place and time where they are performed. Any such dependence would imply a spectral shift of the form [see C.M. Will (1981)]:

$$z = (1 + \alpha)\Delta U/c^2, \tag{7.44}$$

where ΔU is the potential difference and α is a "violation coefficient": the experiment by R.F.C. Vessot and M.W. Levine (1979) gives $|\alpha| < 2 \times 10^{-4}$.

7.4   Geodesic motion

Consider a particle of mass m sufficiently small that we may neglect its gravitational field, moving in a Riemannian space–time $\{V^4, g_{\mu\nu}\}$: this is the motion of a *test particle*

---

[15] This is the *recoilless* emission (or absorption) of a photon by a crystalline solid at low temperature. See e.g. A. Abragam (1964).

[16] For example, the width of the gamma-ray line produced in the transition ${}^{67}Ga \rightarrow {}^{67}Zn$, at 93 keV is $\Delta\nu/\nu \sim 5.2 \times 10^{-16}$.

[17] This type of clock is described by R.F.C. Vessot (1974).

and only gravitational interactions are considered at this stage. We must consider this simple case first if we wish to find a relation between the metric tensor $g_{\mu\nu}$ and the usual gravitational potential ϕ, at least in the limit of low velocities and weak fields.

In a freely falling reference frame the effect of gravity can be cancelled by a suitable choice of local coordinates, and the test particle has (locally) a uniform straight-line motion. The equivalence principle requires that this motion should be locally the same as that in Minkowski space. It thus describes the longest[18] space–time path between two points, i.e. the arc of a geodesic:

$$\delta \int ds = 0 \tag{7.45}$$

or

$$\delta \int \left[g_{\mu\nu} \frac{dx^\mu}{d\xi} \frac{dx^\nu}{d\xi} \right]^{1/2} d\xi = 0. \tag{7.46}$$

The latter equation gives exactly the same type of Euler–Lagrange equations for the geodesics as in Special Relativity, written in curvilinear coordinates [see *Chapter 3*], i.e.

$$\frac{d^2 x^\mu}{ds^2} + \Gamma^\mu_{\alpha\beta} \frac{dx^\alpha}{ds} \frac{dx^\beta}{ds} = 0, \tag{7.47}$$

where we have chosen $d\xi = ds$. However, we should remember that there is an important different between the two cases. In Special Relativity we can always find a *global* coordinate system such that $g_{\mu\nu}$ reduces to $\eta_{\mu\nu}$ and $\Gamma^\mu_{\alpha\beta} \equiv 0$, i.e. so that the geodesic equation (7.47) reduces to the usual equation

$$\frac{d^2 x^\mu}{ds^2} = 0. \tag{7.48}$$

By contrast, in the relativistic gravitational case, this is only possible *locally*. Moreover, while the use of curvilinear coordinates in Special Relativity is in general a useless luxury, it is an absolute necessity here, as space–time is a curved manifold. Further, while the tensor $\eta_{\mu\nu}$ only expresses the structure of the light cone in Minkowski space, $g_{\mu\nu}$ also reveals the effects of gravitation *via* the curvature tensor. We note also that the geodesic equation (7.47) could have been deduced from the locally Minkowskian equation (7.48), which must hold by virtue of the equivalence principle, by passing to an arbitrary coordinate system in the space–time V^4 considered [see the discussion by S. Weinberg (1972)].

We return now to the geodesic equation (7.47). The second term is sometimes interpreted as the gravitational force, and thus the metric tensor as a gravitational potential, as the $\Gamma^\mu_{\alpha\beta}$ are expressed in terms of derivatives of the $g_{\mu\nu}$. This purely formal interpretation is solely based on the analogy between the equation (7.47) and that of relativistic dynamics (3.65), and misses the "kinematic" (geometric) character of gravity. In fixing the metric properties of the space–time it also determines the geodesics.

---

[18] In the sense of the metric $\eta_{\mu\nu}$. It is easy to show that path is not the shortest: if A and B are separated by a timelike interval it is always possible to pass from A to B by means of two null straight lines.

Moreover, to determine a force, we need to evaluate the deviations of motions *relative to free motions*, while the free motion is precisely represented by the geodesics, which are purely geometric (kinematic) properties. In reality it is the combination $\{g_{\mu\nu}, \Gamma^{\mu}_{\alpha\beta}\}$ which plays the role of a gravitational potential. We shall see in the next section that "gravitational forces" involve the first derivatives of this combination through the curvature tensor.

The physical interpretation of the geodesic equation (7.47) becomes clearer in the *Newtonian approximation*: the velocities are assumed small compared with the velocity of light,

$$\left|\frac{d\mathbf{x}}{ds}\right| \ll c\frac{dt}{ds};$$ (7.49)

and the gravitational field is weak enough that we can set

$$g_{\alpha\beta} = \eta_{\alpha\beta} + h_{\alpha\beta},$$ (7.50)

where $h_{\alpha\beta}$ is a first order correction

$$|h_{\alpha\beta}| \ll 1 \qquad \text{and} \qquad |h_{\alpha\beta}|^2 \ll |h_{\alpha\beta}|.$$ (7.51)

The first condition, Eq. (7.49), required the geodesic equation to reduce to

$$\frac{d^2x^{\mu}}{ds^2} + \Gamma^{\mu}_{00}\left(\frac{dx^0}{ds}\right)^2 \sim 0,$$ (7.52)

while the second – linearisation of the terms involving the metric tensor – gives

$$\Gamma^{\mu}_{00} \sim \frac{1}{2}\eta^{\mu\nu}\partial_{\nu}h_{00},$$ (7.53)

where we have considered a *stationary* gravitational field, i.e. such that $\partial_0^{(n)}h_{\alpha\beta} \equiv 0$ for all n. Under these conditions, the geodesic equation reduces to

$$\begin{cases} \dfrac{d^2t}{ds^2} = 0 \\ \dfrac{d^2\mathbf{x}}{ds^2} = -\dfrac{1}{2}\nabla h_{00}. \end{cases}$$ (7.54)

By identification with the Newtonian equation

$$\frac{d^2\mathbf{x}}{dt^2} = -\nabla\phi,$$ (7.55)

where ϕ is the *gravitational potential*, we deduce

$$h_{00} = 2\phi,$$ (7.56)

and finally

$$g_{00} = 1 + 2\phi/c^2.$$ (7.57)

Thus, for a mass M at the origin of coordinates we have

$$g_{00} = 1 - 2\frac{GM}{rc^2}.$$ (7.58)

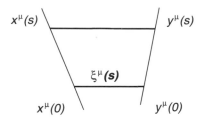

Fig. 7.10. Geodesic deviation in Minkowski space. The deviation, ξ^μ, is a linear function of the proper time. Any deviation from linearity shows the existence of a force.

It is interesting to note that in the Newtonian approximation only g_{00} appears. This explains how one can *a priori* consider a scalar field in order to describe relativistic gravitation [G. Nordström (1913)]. However, this kind of relativistic generalisation of gravitation predicts, for example, a *recession* of the perihelion of Mercury rather than an advance...

We note finally that the gravitational redshift equation (7.43) gives

$$z = \left(\frac{g_{00}(\text{em})}{g_{00}(\text{obs})} \right)^{1/2} - 1 = \frac{\phi_{\text{em}} - \phi_{\text{obs}}}{c^2}; \tag{7.59}$$

and we thus recover the intuitive results of *Chapter 4*.

7.5   Geodesic deviation

Like inertial motion in Minkowski space, geodesic motion in the curved space–time $\{V^4, g_{\mu\nu}\}$ of gravity is not discernable by an intrinsically attached observer. Thus, a geodesic observer cannot establish the existence of a gravitational field except by comparison with a nearby motion, for example another geodesic. We make this more precise. In Minkowksi space, two geodesic motions (see **Fig. 7.10**) are defined by the relations

$$\begin{cases} x^\mu(s) = u^\mu s + x^\mu(0) \\ y^\mu(s') = w^\mu s' + y^\mu(0) \end{cases} \tag{7.60}$$

so that the deviation $\xi^\mu(s) \equiv x^\mu(s) - y^\mu(s)$ is a *linear* function of the proper time s, or more generally of any affine parameter $\alpha \equiv as + b$. Thus any deviation from this linearity reveals the existence of a force, and hence the non-geodesic character of the motion of one or other (or both) motions.

Consider now a Riemannian space–time $\{V^4, g_{\mu\nu}\}$, and two neighboring geodesics described by $x^\mu(s)$ and $x^\mu(s) + \delta x^\mu(s)$, $\delta x^\mu(s)$ being the *geodesic deviation* (s is the proper time or any other affine parameter). Then we have (see **Fig. 7.11**)

$$\frac{d^2 x^\mu}{ds^2} + \Gamma^\mu_{\alpha\beta} \frac{dx^\alpha}{ds} \frac{dx^\beta}{ds} = 0 \tag{7.61}$$

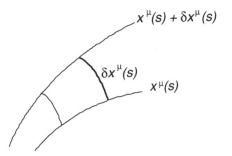

Fig. 7.11. Geodesic deviation in a Riemannian space–time.

$$\frac{d^2\,(x^\mu + \delta x^\mu)}{ds^2} + \Gamma^\mu_{\alpha\beta}(x + \delta x)\frac{d\,(x^\alpha + \delta x^\alpha)}{ds}\cdot\frac{d\,\left(x^\beta + \delta x^\beta\right)}{ds} = 0. \qquad (7.62)$$

We restrict ourselves to the case of small deviations, which allows us to neglect second-order and higher terms $(\delta x)^2$, etc. Using the exact form of the Christoffel symbols $\Gamma^\mu_{\alpha\beta}$, we restrict ourselves to the first order in δx^α and subtract (7.61) from (7.62), giving

$$\frac{d^2}{ds^2}\delta x^\mu + \partial_\rho\Gamma^\mu_{\alpha\beta}\frac{dx^\alpha}{ds}\cdot\frac{dx^\beta}{ds}\delta x^\rho = R^\mu_{\nu\alpha\beta}\frac{dx^\nu}{ds}\frac{dx^\alpha}{ds}\delta x^\beta \qquad (7.63)$$

which is the equation of *geodesic deviation*. This equation expresses the *relative* deviation of two freely falling particles in a gravitational field. The presence of the Riemann–Christoffel tensor clearly shows the influence of gravitation, and that in principle "gravitational forces" (i.e. $R_{\mu\nu\alpha\beta}$) are measurable from the geodesic deviation. The fact that the tensor $R_{\mu\nu\alpha\beta}$ appears in the expression for the "gravitational force" – i.e. on the right hand side of Eq. (7.63) – shows that the combination $\{g_{\mu\nu}, \Gamma^\lambda_{\alpha\beta}\}$ constitutes the "gravitational potential": $R_{\mu\nu\alpha\beta}$ involves derivatives of these quantities. In Minkowski space, $R_{\mu\nu\alpha\beta} \equiv 0$, and Eq. (7.63) reduces to the geodesic equation; moreover, we can always find a coordinate system in which $\Gamma^\mu_{\alpha\beta} \equiv 0$; δx^μ is then a *linear* function of the affine parameter s: any deviation from linearity reveals the presence of a gravitational field *via* $R_{\mu\nu\alpha\beta}$, of course in the absence of other forces which could also produce such a deviation.

We add finally that the geodesic deviation equation will be used extensively in calculations of experiments on gravitational waves (see *Chapter 8*).

7.6 The metric tensor in spherical symmetry

In many practical problems – the solar system, compact stars, cosmology, etc. – we have to deal with spherically symmetrical systems. This invariance of the system under *spatial rotations* about a point (usually taken as the origin of coordinates) allows us to write the metric tensor in a particularly simple form in a suitably chosen coordinate system. In such a system the most general (rotationally invariant) *spatial* tensors one can write are δ^{ij} and $x^i x^j$. It follows that the metric must have the form

$$ds^2 = A(r,t)dt^2 - B(r,t)dt\frac{\mathbf{x} \cdot d\mathbf{x}}{r} - C(r,t)d\mathbf{x}^2 - D(r,t)\frac{(\mathbf{x} \cdot d\mathbf{x})^2}{r^2}, \qquad (7.64)$$

where r is the distance from the origin, where we have put $t \equiv x^0$ and where A, B, C and D are *arbitrary* dimensionless functions of r and t. We have thus reduced from ten to four the number of unknown functions defining the metric $g_{\mu\nu}$. We can, moreover, eliminate two of the four functions A, B, C, D by exploiting the invariance of ds^2 under coordinate transformations of the type

$$\begin{cases} x^i = f_1\left(t', r'\right) x'^i \\ t = f_2\left(t', r'\right) \end{cases} \qquad (7.65)$$

where f_1 and f_2 are arbitrary functions. We note that the transformations (7.65) preserve invariance under rotations, i.e. the form (7.64) of the metric. We finally get (the proof is given below)

$$g_{\alpha\beta}(r,t) = \begin{array}{c} \\ t \\ r \\ \theta \\ \varphi \end{array} \begin{array}{c} t \quad\quad r \quad\quad \theta \quad\quad \varphi \\ \begin{pmatrix} e^{\nu(r,t)} & 0 & 0 & 0 \\ 0 & -e^{\lambda(r,t)} & 0 & 0 \\ 0 & 0 & -r^2 & 0 \\ 0 & 0 & 0 & -r^2\sin^2\theta \end{pmatrix} \end{array}, \qquad (7.66)$$

where ν and λ are arbitrary functions of t and r, and we have used polar coordinates and made the change of variables $t' \to t$ and $r' \to r$.

Proof:[19] We begin by showing that the non-diagonal term g_{0i} in the metric can be reduced to zero. We look for a transformation of the type (7.65) which will effect this. As this should only affect t, we seek this in the form

$$\begin{cases} x^i = x'^i \\ t = h(t', r'). \end{cases} \qquad (7.67)$$

Using the transformation laws for the tensor $g_{\mu\nu}$ we have

$$g'_{0i} = \frac{\partial h}{\partial t'} \cdot \left[\frac{\partial h}{\partial x'^i}g_{00} + g_{0i}\right], \qquad (7.68)$$

so that if we choose h such that

$$\frac{\partial h}{\partial x'^i} = -\frac{g_{0i}}{g_{00}}, \qquad (7.69)$$

we have $g'_{0i} \equiv 0$. The latter relation gives

$$\frac{\partial h}{\partial r'} = -\frac{B}{A}; \qquad (7.70)$$

we may thus always choose $B \equiv 0$.

We show now that we can find coordinates such that $C \equiv 1$. A transformation of the type

$$\begin{cases} t' = t \\ x'^i = k(r', t')x^i \end{cases} \qquad (7.71)$$

---

[19] See J.L. Anderson (1967); see also S. Mavridès (1973).

is suitable. Inserting this in the expression (7.64) for the metric, the spatial part becomes

$$g'_{ij} = C(t,r) \cdot k^2(r',t')\delta_{ij} + \text{term in } x'^i \cdot x'^j. \tag{7.72}$$

Now *choosing* $k = C^{-1/2}$, we get the desired result. We can easily show, as is required, that the new transformation does not reintroduce a term g_{0i}.

In the usual polar coordinates, the form (7.66) for the metric is immediate, and is called the *standard form*, the coordinates being *standard coordinates* or *Schwarzschild coordinates*. There are many other forms, for example the *isotropic* form

$$ds^2 = E(r,t)dt^2 - G(r,t) \cdot \left[dr^2 + r^2(d\theta^2 + \sin^2\theta d\varphi^2)\right]. \tag{7.73}$$

These are the two main forms we shall use in the following. For a metric which is also *static* (i.e. the metric does not depend on t) we can pass from the standard to the isotropic form *via* the transformation [see e.g. S. Weinberg (1972)]

$$r' = \exp \int \left\{ 1 + \frac{r^2}{C} \left[D + \frac{B^2}{4r^2 A} \right] \right\}^{1/2} \frac{dr}{r}. \tag{7.74}$$

7.7   Overview of the PPN formalism

In this chapter we have given arguments in favour of the idea of a curved space–time, stated the equivalence principle and discussed some of the experimental evidence for it as well as its theoretical implications. However, we have still to decide what *field equations* the gravitational potentials $g_{\mu\nu}$ obey – this is the aim of the next chapter. Clearly, these equations constitute a further postulate, and in principle are largely arbitrary, despite the various physical constraints [see *Chapter 8*] we can impose.

A procedure which allows us to use experimental data to select *classes* of acceptable theories from those possible is the use of the PPN *"parametrized post-Newtonian"* formalism, which we shall present in a simplified version given by H.P. Robertson (1962) and R. Eddington (1922); the general formalism is explained in the books by C.M. Will (1981) and C.W. Misner, K.S. Thorne and J.A. Wheeler (1973).

What is the PPN formalism and what are its limitations? In the limit of slow motions and weak gravitational fields we must recover the usual Newtonian equations. However, non-trivial effects of space–time curvature must appear to the next order (e.g. v^2/c^2), which is called the *post-Newtonian approximation*. We wish to *parametrize* this and possibly higher-order approximations to distinguish among the infinity of possible equations which the metric components $g_{\mu\nu}$ may obey. Clearly, such PPN effects can be measured in gravitating systems such as the solar system (e.g. deflection of light rays by the Sun, advances of planetary perihelia, etc.), for which the gravitational parameter GM/Rc^2 is weak ($< 10^{-6}$), but not in the binary pulsar PSR 1913+16, which emits gravitational waves (for which $v = c$).

We illustrate this formalism in the simple case of a spherically symmetrical system such as the Sun. We write the metric in the isotropic form (7.73). If we identify[20] the radial coordinate r with the usual radial distance, we can easily check that the only dimensionless quantity that we can form from $G, M_\odot, c$ and r is $GM_\odot/rc^2 \equiv \chi$. As r varies from 7×10^{10}cm [the Sun's radius] to infinity, the variable χ is always smaller than 10^{-6}. It is therefore legitimate[21] to expand the gravitational potentials g_{00} and g_{ii} [or equivalently the functions $E(r)$ and $G(r)$ of the metric (7.73), assumed static] in powers of χ (the $g_{\mu\nu}$ are dimensionless):

$$ds^2 = c^2 \left[1 - 2\alpha \frac{GM_\odot}{rc^2} + 2\beta \left(\frac{GM_\odot}{rc^2} \right)^2 + \cdots \right] dt^2$$
$$- \left[1 - 2\gamma \frac{GM_\odot}{rc^2} + \cdots \right] d\mathbf{x}^2, \tag{7.75}$$

where we have reinstated the factors c. We have implicitly assumed that space–time is flat when we are far from the mass $M_\odot$ which curves it, i.e. that space–time is asymptotically Minkowskian. This is why the expansion (7.75) starts with 1. Note also that the parameter χ can be interpreted as v^2/c^2, where v^2 is a characteristic velocity (e.g. of a planet in orbit); this is clearly a relativistic expansion. Thus, different theories of relativistic gravity will give differing values of the parameters $\alpha, \beta, \gamma, \ldots$

We return to the expansion (7.75) and assume we can truncate it at the first order in χ in g_{ii}:

$$g_{ii} \sim 1 - 2\gamma \frac{v^2}{c^2}. \tag{7.76}$$

For the two terms in the metric to be of the same order, we have to expand g_{00} *to second order* in χ. Thus, since

$$c^2 dt^2 = c^2 \frac{dt^2}{d\mathbf{x}^2} d\mathbf{x}^2 = \frac{c^2}{v^2} d\mathbf{x}^2, \tag{7.77}$$

we must have

$$g_{00} \sim 1 - 2\alpha \frac{v^2}{c^2} + 2\beta \frac{v^4}{c^4}, \tag{7.78}$$

so that

$$c^2 \frac{v^4}{c^4} dt^2 \sim \frac{v^2}{c^2} d\mathbf{x}^2, \tag{7.79}$$

to order of magnitude.

We see that for most practical purposes the predictions of different theories only depend on the parameters (α, β, γ) and not on the precise forms of the field equations. Observational data give limits on the possible values of these parameters and thus

---

[20] This identification is not obvious (and indeed is incorrect) and requires a careful discussion which we omit.
[21] This is physically a weak hypothesis but mathematically strong.

eliminate many theories. Einstein's general relativity (see *Chapter 8*) leads to the Schwarzschild metric:

$$ds^2 = \frac{\left(1 - \frac{GM_\odot}{2rc^2}\right)^2}{\left(1 + \frac{GM_\odot}{2rc^2}\right)^2} c^2 dt^2 - \left(1 + \frac{GM_\odot}{2rc^2}\right)^4 \cdot \left[dr^2 + r^2 d\theta^2 + r^2 \sin^2\theta d\varphi^2\right]. \qquad (7.80)$$

and thus to $\alpha = \beta = \gamma = 1$, values which are compatible with all current measurements.

We remark finally that if we assume that gravity is correctly described by a metric theory, α does not appear as a true parameter, as it can always be absorbed into the definition of mass. The Newtonian approximation of (7.80) gives

$$\Gamma^i_{00} = -\alpha \frac{M_\odot G}{r^2}, \qquad (7.81)$$

which is simply the Newtonian gravitational force. We can thus always choose $\alpha = 1$ (definition of the mass $M_\odot$).

In this connection, the gravitational redshift measured, for example, in the Pound–Rebka experiment gives no information about the parameters β and α: it simply confirms the metric character of relativistic theories of gravity.

7.8 The classical tests

There are four classical tests of general relativity: the gravitational redshift, curvature of light rays near large masses, advance of planetary perihelia and time delay of radar signals in the solar system [I.I. Shapiro (1964, 1968); I.I. Shapiro *et al.* (1971)]. In the solar system, the variable $GM_\odot/rc^2$ is always very small, so that the PPN formalism is entirely sufficient for estimating these effects. We shall therefore evaluate them in this formalism: this is legitimate for all metric theories of gravity, and will give estimates for the PPN parameters. We have already discussed the gravitational redshift, so we consider only the other three tests.

Further, we shall only calculate the deflection of light rays by large masses, relegating to an exercise the calculation of the radar echo delay, and referring the reader to one of the texts quoted for the advance of planetary perihelia [e.g. S. Weinberg (1972)]. We shall, however, discuss the results obtained as functions of the PPN parameters. Below, we take $\alpha = 1$, following the remarks at the end of the last section.

(1) We begin with the planetary perihelion advance. We can either follow Einstein in regarding the relativistic terms as small perturbations of the classical equations, or solve the geodesic equations corresponding to the metric (7.75). We thus find

$$\delta\Psi = \frac{2\pi GM_\odot}{(1-e^2)a}(2 - \beta + 2\gamma), \qquad (7.82)$$

where $\delta\psi$ is the perihelion advance expressed in radians per revolution, with e the

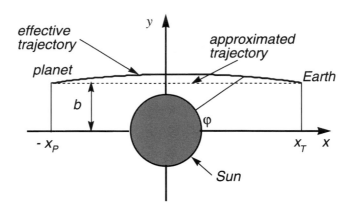

Fig. 7.12. Geometry of the Earth–Sun–planet radar beam system in the ecliptic plane, when the planet is at superior conjunction. The radar beam would follow the dotted path if space were Euclidean. The curvature of the radar beam is greatly exaggerated here: in reality it can be approximated by two straight lines.

eccentricity of the planetary orbit and a the semi-major axis. Applied to Mercury[22], General Relativity ($\alpha = \beta = \gamma = 1$) predicts an advance of 43.03 seconds of arc per century, very close to the 43.11 ± 0.45 observed. However, any oblateness of the Sun (producing a quadrupole moment) would also give a perihelion advance for Mercury (see Exercise 1.7). Thus, if the measurement of R.H. Dicke and H.M. Goldenberg (1967) (1974) [R.H. Dicke (1974)] which gives a polar diameter for the Sun about $(5.0 \pm .7) \times 10^{-5}$ smaller than its equatorial diameter is correct, and if this oblateness reflects the interior mass distribution, the resulting quadrupole moment[23] would cause a perihelion advance for Mercury of about 3.4 seconds of arc per century. This would represent a deviation of about 8% from the General Relativistic prediction. Long baseline interferometry allows detection of the bending of radio signals emitted by various astronomical objects, and gives $\gamma = 1.0002 \pm 0.002$ [D.S. Robertson *et al.* (1991)].

(2) We turn to radar echos from planets. If we send a radar signal from the Earth to a given planet, this does not travel in a straight line – as in Euclidean space – but in a curved space, corresponding to the spatial sections of the space–time (see **Fig. 7.12**) with metric

$$d\sigma^2 \equiv -ds^2 = \left[1 + 2\gamma \frac{GM_\odot}{r}\right] \cdot \left[dr^2 + r^2 d\theta^2 + r^2 \sin^2\theta d\varphi^2\right]. \tag{7.83}$$

In the PPN coordinate system used here the interval of *time coordinate* Δt between emission of the signal and its reception at Earth after reflection at the planet is given

---

[22] Mercury makes 415 revolutions per century, has an eccentricity $e = 0.2056$ and a semi-major axis $a = 57.91 \times 10^{11}$cm.

[23] See the discussions by S. Weinberg (1972) and C.M. Will (1981).

by

$$\frac{1}{2}\Delta t \sim x_T + x_P + (1+\gamma)GM_\odot \ln\left(\frac{4x_T x_P}{b^2}\right), \tag{7.84}$$

where x_T is the Earth–Sun distance, x_P is the planet–Sun distance and b is the impact parameter (see **Fig. 7.12**) of the radar beam, so that $b \approx R_\odot$ if we consider superior conjunction of the planet, which is the case giving the largest effect. We note that $b/x_T \ll 1$ and $b/x_P \ll 1$. Actually it is not Δt which is observable but the corresponding interval of *proper time* $\Delta\tau$ for the terrestrial observer ($\Delta\tau^2 \equiv ds^2 = g_{00}dt^2 - 0$), i.e.

$$\Delta\tau = \left(\sqrt{g_{00}}\right)_{\text{Earth}} \Delta t = \left(1 - 2\frac{GM_\odot}{\sqrt{x_T^2 + b^2}}\right)\Delta t. \tag{7.85}$$

Relation (7.84) above corresponds only to the case of **Fig. 7.12**; the general case can be found in Weinberg (1972). The first term in this relation is the usual term giving the outward and return journey of the radar signal at the speed of light $c = 1$, while the second term gives the real relativistic effect caused by the curvature of space near the Sun.

Relation (7.84) can be obtained[24] by noting that the radar photons propagate along a null geodesic $ds^2 = 0$, i.e. using the PPN metric (7.75),

$$\left(1 - \frac{2GM_\odot}{r}\right)dt^2 \simeq \left(1 + 2\gamma\frac{GM_\odot}{r}\right) \cdot \left[dr^2 + r^2 d\varphi^2\right]. \tag{7.86}$$

To find Δt, we have only to integrate dt: dr and $d\varphi$ become integrable by noting that

$$r = \sqrt{x^2 + y^2}, \qquad \cos\varphi = \frac{y}{r}. \tag{7.87}$$

Moreover, as the beam is bent very little, $y \approx b$. With

$$\frac{d\varphi}{dx} = \frac{y}{r} \qquad \text{and} \qquad \frac{dr}{dx} = \frac{x}{r}, \tag{7.88}$$

we thus get

$$dt \sim \left(1 + (1+\gamma)\frac{GM_\odot}{\sqrt{x^2 + b^2}}\right)dx. \tag{7.89}$$

Integrating this expression from x_P to x_T and keeping only the lowest-order terms in b/x_T and b/x_P gives (7.84) immediately.

For a radar reflection from Mercury, which takes about 20 minutes, the delay expected from General Relativity ($\gamma = 1$) is of order 240 μs, and thus eminently measureable. The experiment is however not only difficult to perform (particularly because of the weakness of the echo) but also extremely complex[25] to analyse as the theoretical corrections and possible sources of error are numerous and difficult to evaluate. We note that the Earth's motion must be taken into account: as the Earth's orbit is approximately circular we have $v^2 \sim GM_\odot/r$, and the time dilatation factor

---

[24] See M.G. Bowler (1976) or N. Straumann (1984).
[25] An overview is given in Weinberg (1972). See also the theoretical discussion by I.I. Shapiro (1966), D.K. Ross and L.I. Schiff (1966).

$\sqrt{1-v^2}$ then gives a factor $3/2$ in place of the 2 in the correction (7.85). Also, in the relations (7.84) and (7.85) the quantities x_T, x_P and b are not sufficiently accurately known (to about 1.5 km for an accuracy of 10 μs in $\Delta\tau$). Further, the reflection of the radar beam at Mercury is not from a point but from a significant area[26] which causes a dispersion in the arrival times of the radar echo. Another difficulty results from the fact that the solar corona is a dispersive medium whose properties vary markedly in time. We see that it is quite difficult to extract physically significant results from this type of experiment.

The experiment has been performed for Mercury and Venus [I.I. Shapiro (1968); I.I. Shapiro *et al.* (1971), see **Fig. 7.13**] and also on the artificial satellites Mariner VI and VII, which have the advantage of avoiding the problems of planetary surface topography, but have the disadvantage of being very sensitive to non-gravitational perturbations such as the solar wind and radiation pressure [J.D. Anderson *et al.* (1975)]. The Viking mission placed two transponders on Mars (a signal received at one frequency is reflected at another, allowing one to eliminate the effects of reflection from a large area of the planet) which substantially improved the accuracy of the experiment [R.D. Reasenberg *et al.* (1979)]. Analysis of the results [R. Hellings *et al.* (1983)] gives

$$\gamma = 1 - (0.7 \pm 1.7) \times 10^{-3}, \tag{7.90}$$

as well as a value[27] for the parameter β

$$\beta = 1 - (2.9 \pm 3.1) \times 10^{-3}, \tag{7.91}$$

confirming the general relativistic values for these quantities (*Chapter 8*). However it seems that these figures may be optimistic. We shall limit ourselves to the values given by R.D. Reasenberg *et al.* (1979), i.e. $\gamma = 1.0002 \pm 0.002$.

(3) The final classical test (historically the first) involves the deflection of light rays passing near the Sun. This effect was measured for the first time in the eclipse of 1919, and assured the triumph of General Relativity. In astrophysics this effect is particularly important as it gives rise to *gravitational lensing* (see the next section), which is well observed today, and can be used to give information about galaxies and cosmology.

The trajectory of a light ray in a gravitational field can be found from the geodesic equation or from the fact that $ds^2 = 0$ along it. With the metric (7.75) the latter relation gives

$$\frac{dt}{dr} = 1 + (1+\gamma)\frac{GM_\odot}{r} + O\left(\left[\frac{GM_\odot}{r}\right]^2\right). \tag{7.92}$$

This relation shows that light behaves *as if* its velocity were not constant in a

---

[26] The reflective properties of Mercury's surface are calibrated experimentally at inferior conjunction, eliminating the delay caused by relativistic gravity.

[27] Only γ appears in the simple relation (7.84). However, several other parameters appear in a fuller analysis of the results, particularly β.

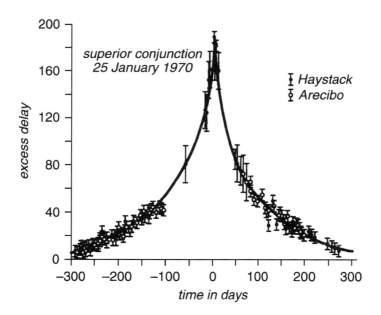

Fig. 7.13. Results of radar reflection experiments on Venus [from I.I. Shapiro *et al.* (1971)].

gravitational field. This is equivalent to a refractive index

$$n(r) = \frac{c}{v(r)} = \frac{1}{v(r)} = \text{Eq. (7.92).} \tag{7.93}$$

Thus *Fermat's Principle*

$$\delta \int n(r)d\xi = 0 \tag{7.94}$$

($d\xi$ is the Euclidean element of length along the light ray) gives the optical path by means of the Lagrange equations (see *Appendix C*) corresponding to (7.94), i.e.

$$\frac{d}{d\xi}\left(n(r)\frac{d\mathbf{x}}{d\xi}\right) = \nabla n(r). \tag{7.95}$$

[see M. Born and E. Wolf (1975)]. The desired angular variation is simply the difference between the directions of the asymptotes, i.e.

$$\delta \equiv \left\{\frac{d\mathbf{x}}{d\xi}\bigg|_{+\infty} - \frac{d\mathbf{x}}{d\xi}\bigg|_{-\infty}\right\} \cdot \mathbf{e}_2 \tag{7.96}$$

($\mathbf{e}_2$ is a unit vector along $0y$).

Integrating equation (7.95) along the unperturbed trajectory $y = b$ makes only a second-order error; further using the fact that at infinity (where the metric (7.75) is Minkowskian) $n(r) = 1$, we then find (**Fig. 7.14**) successively

$$\delta \equiv \mathbf{e}_2 \cdot \left\{\frac{d\mathbf{x}}{d\xi}\bigg|_{+\infty} - \frac{d\mathbf{x}}{d\xi}\bigg|_{-\infty}\right\}$$

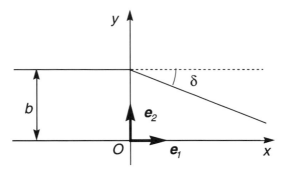

Fig. 7.14. Deviation of light rays.

$$= \int_{-\infty}^{+\infty} \mathbf{e}_2 \cdot \nabla n(r) d\xi \tag{7.97}$$

$$\sim \int_{-\infty}^{+\infty} \frac{d}{dr} n(r) \frac{\mathbf{x} \cdot \mathbf{e}_2}{r} \, dx \tag{7.98}$$

$$= \int_{-\infty}^{+\infty} n'[(x^2 + y^2)^{1/2}] \frac{b}{(x^2 + y^2)^{1/2}} dx. \tag{7.99}$$

Now inserting the expression (7.93) for $n(r)$ in the latter equation gives

$$\delta = (1+\gamma)2 \frac{GM_\odot}{b}. \tag{7.100}$$

For $b = R_\odot$, the deviation is

$$\delta = \frac{(1+\gamma)}{2} 1.75''.$$

General relativity thus predicts a deviation of 1.75″ of light rays close to the Sun's limb.

The derivation above used Fermat's Principle in General Relativity. It is not obvious that this principle holds. In fact it can easily be shown [see e.g. M.A. Tonnelat (1964); N. Straumann (1984); etc.] to hold in the form

$$\delta \int dt = 0, \tag{7.101}$$

but for static fields only.

A proof of the relation (7.100), invariant under changes of PPN coordinates, can be found in C.M. Will (1981) [see also the remarks in Weinberg (1972)].

> There is a final point to clarify: why does the approximate calculation of *Chapter 4* give one-half of the General Relativistic value? We quote C.M. Will's (1981) very clear explanation of this "anomaly": *"these calculations are correct as far as they go. But the result of these calculations is the deflection of light relative to local straight lines, as defined by rigid rods; however, because of space curvature around the Sun, determined by the PPN parameter γ, local straight lines are bent relative to asymptotic straight lines from the Sun by just enough to yield the remaining factor 'γ/2'. the first factor '1/2' holds in any*

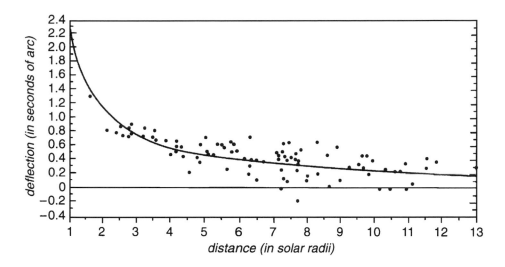

Fig. 7.15. Results of measurements of light deflection by the Sun [from H. von Klüber (1960)].

metric theory, the second 'γ/2' varies from theory to theory. Thus, calculations that purport to derive the full deflection using the equivalence principle alone are incorrect [see Schiff (1960), and the critique by Rindler (1968)]."

The effect itself was first observed in 1919 by F.W. Dyson, A.S. Eddington and C. Davidson, and many times since then. The principle of the measurement is the comparison of photographs of stars near the Sun during a total eclipse with photographs of the same region of the sky at another epoch, for example six months earlier. After a large number of careful corrections it is possible to measure changes in the positions of the stars. The most recent measurements were made in Mauretania and Texas in 1973. **Figure 7.15** summarizes these and other measurements[28].

The Sun's atmosphere is an important source of systematic error, and there have been suggestions of using light rays passing close to Jupiter, which eliminates this problem. For $\gamma = 1$ (general relativity) the deviation is about 0.02″, too small to be measured from the Earth. By contrast the Hipparcos satellite should be able to perform this measurement, as in principle it can determine stellar positions to an accuracy of 2.0×10^{-3} seconds of arc.

Progress in the techniques of long baseline interferometry in the radio should allow one to reach an accuracy of a few times 10^{-4} seconds of arc. This experiment has been carried out since 1969 [D.O. Muhleman *et al.* (1970)] and currently gives the most accurate value for the parameter γ from light-deflection experiments[29]: each year the quasars 3C273, 3C279 and 3C48 (which are very powerful radio sources) pass very close to the Sun, as seen from Earth.

[28] A list of the various measurements is given by L. Bertotti *et al.* (1962).
[29] See the review by E.B. Fomalont and R.A. Sramek (1977).

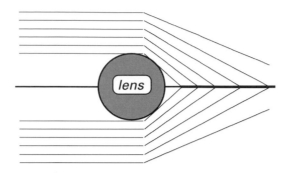

Fig. 7.16. A gravitational lens. In contrast to a normal optical lens, for which light rays from infinity are concentrated at the focus, a gravitational lens has no focal length. Parallel light rays from infinity are focussed on a semi-infinite straight line. A gravitational lens is completely achromatic.

However, even more than visible light, radio waves are dispersed by the solar corona (and by the Earth's ionosphere) causing a deflection which has to be estimated.

7.9   Gravitational lenses

Gravitational lenses[30], of which the first was observed by D. Walsh, R. Carswell and R. Weymann in 1979, have a long history which began with the idea of light deflection by large masses in General Relativity. It appears that the idea of a "gravitational lens" (see **Fig. 7.16**) is due to O. Lodge (in 1919) and the first calculation was made by A. Einstein (1936) on the basis of a suggestion by R.W. Mandl (1935). Since then there have been a large number of calculations and discussions.

The distinctive properties of gravitational lenses result from their ability to form *multiple images* of a point object. This is not the only effect. The images show an *amplification* of the source's apparent luminosity, as well as a *time delay* which is characteristic of the source, the lensing object, and the relative distances of these objects and the observer. The main interest of gravitational lenses, apart from confirming Einstein's General Relativity, is that they allow in principle the determination of the cosmological parameters specifying the expansion of the Universe (Hubble constant H_0) and its spatial curvature (deceleration parameter q_0) [see S. Refsdal (1964a, b, 1966)]. We shall study these properties for the simple case of a point source S and a lensing object (deflector) D, also assumed pointlike (see **Fig. 7.17**). There is no difficulty in principle in generalising these results to extended objects. Note that we could develop a "gravitational geometric optics" using Fermat's principle as enunciated above[31].

---

[30] A historical account is given by J.M. Barnothy (1989). See also S. Liebes (1964) C. Vanderriest (1984) (1986), P. Schneider (1992) and F. Link (1969).

[31] The geometrical optics approximation clearly holds here, as the distances involved (including the sizes of the source and lensing object) are always far greater than light wavelengths.

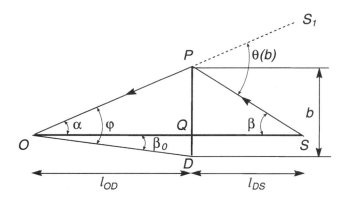

Fig. 7.17. Geometry of the deflection of light emitted by a source S by a lensing object D observed at O. b is the impact parameter, and ℓ_{OD} and ℓ_{DS} the observer/deflector and deflector/source distances respectively. The observable quantities are in principle ℓ_{OD}, ℓ_{DS} and $\varphi \equiv \beta_0 + \alpha$. All the angles shown are in practice very small. $\theta(b)$ is the deviation of the light rays; S_1 is the image of S seen by O. Note that S is not directly observable.

(1) We begin by fixing the conditions under which the observer O can see multiple images of S. Consider a light ray reaching O after deflection by D, which has mass M. The geometry of the triangle OPS gives[32]

$$\theta(b) = \alpha + \beta = \frac{4GM}{b}. \tag{7.102}$$

All the angles here are very small (see **Fig. 7.17**)[33] so

$$\frac{PQ}{\ell_{OD}} = \alpha, \qquad \frac{PQ}{\ell_{DS}} = \beta, \qquad PQ = PD - QD, \qquad QD = \ell_{OD}\beta_0; \tag{7.103}$$

and (7.102) gives

$$\frac{4GM}{b} = (b - \beta_0\ell_{OD}) \cdot \left(\frac{1}{\ell_{OD}} + \frac{1}{\ell_{DS}} \right), \tag{7.104}$$

which is shown diagrammatically in **Fig. 7.18**.

Figure 7.18 shows that for a pointlike source and deflector there are always two images of the source, on either side of the source/observer axis [Eq. (7.104) has one solution of either sign], one close and the other distant. If the source, deflector and observer lie on a line, the two solutions are equal and opposite. This symmetry implies that the image is a ring, called the Einstein ring [O. Chwolson (in 1924), A. Einstein (in 1936)].

In general, the source is not directly observable, rather than its image. The other observables are the distance and position of the deflector, the distance of the source

---

[32] We adopt General Relativity, for which $\gamma = 1$ and thus $2.(1 + \gamma) = 4$.
[33] At most of order a few seconds of arc.

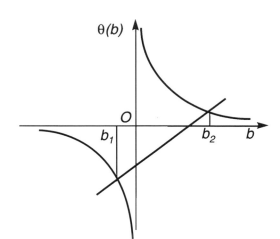

Fig. 7.18. Condition for a light ray from S to be observed. For a pointlike source and deflector there are always two solutions, and thus two images of the source.

and the angular separations $\varphi_i \equiv \alpha_i + \beta_0$ (i = 1,2) of the deflector and the images S_i. Using

$$b_i = (\alpha_i + \beta_0)\ell_{OD} = \varphi_i \ell_{OD} \qquad (7.105)$$

(7.104) gives

$$\varphi_1 \cdot \varphi_2 = \frac{4GM}{\ell_{OD} \ell_{OS}} \cdot \ell_{DS}, \qquad (7.106)$$

which in theory allows us to find the mass of the deflector in terms of observables. Similarly, we get a relation for β_0, i.e. the angular position of the source.

In practice, the only distant pointlike objects we know are quasars, and the only lensing objects giving observable effects (apart from so-called microlensing effects) are galaxies [F. Zwicky (1937a, b)]. Both objects are at cosmological distances, so their distances are given by the Hubble law as

$$\ell = H_0^{-1} z, \qquad (7.107)$$

where H_0 is a constant between 50 and 100 kilometres per second per megaparsec[34] and z is the redshift. Hence, in terms of observables (7.106) becomes

$$\varphi_1 \cdot \varphi_2 = \frac{4GM}{z_D \cdot z_S} H_0(z_S - z_D), \qquad (7.108)$$

where z_S and z_D are the redshifts of the source and the deflector respectively. If the mass M is known, this relation gives a value for the Hubble constant H_0. In fact z_S is accessible only because the images S_1 and S_2 have the *same* redshift:

$$z_{S_1} = z_{S_2} = z_S. \qquad (7.109)$$

---

[34]   1 Pc = 3.26 light-years.

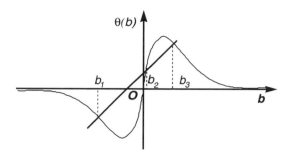

Fig. 7.19. The curve $\theta(b) - b$ **for an extended deflector with mass density** $\sim b^{1/4}$. There are now either one or three images, which may be degenerate.

This property obviously is very important in searching for candidate lensed objects.

In reality things are rather more complicated, as the Hubble law (7.107) is only valid at small distances, and is modified by a term depending on the spatial curvature of the Universe, introducing a new unknown parameter, the deceleration q_0.

Moreover, the lensing galaxy is not pointlike, and the deflection of light at impact parameter b has the form

$$\theta(b) = \frac{4GM(b)}{b}, \tag{7.110}$$

where $M(b)$ is the mass in a sphere of radius b. Thus condition (7.104) is altered, and in consequence so is **Fig. 7.18**. For an elliptical galaxy the mass density is often well approximated by the function [G. de Vaucouleurs (1948)]

$$\rho(b) = \rho_0 e^{-(b/r_0)^{1/4}}, \tag{7.111}$$

or another function of $b^{1/4}$, etc. In this case, in place of the equilateral hyperbola of **Fig. 7.18** we get a more complicated curve (**Fig. 7.19**), and there is now the possibility of one or three images (which may be degenerate). In general there may be many images, but the total number is always odd [see W.L. Burke (1981)].

What forms can the lensed images take? **Figure 7.20** illustrates a simple case [Liebes (1964)], where we have indicated the various possibilities given by the relative positions of the source and the deflector.

A great variety of images are possible for more complicated lensing galaxies, and the interested reader may consult the literature.

(2) The second feature of gravitational lenses, which can be used to confirm that two or more objects of the same redshift are lensed images of the same object, is the time delay between signals from the various images.

This effect results firstly from the fact that light rays coming from different images have different trajectories, (SP_1O and SP_2O, where 1 and 2 denote the two trajectories; see **Fig. 7.17**) and secondly from the fact that the gravitational fields along these trajectories are not the same.

To evaluate the effect we consider a photon travelling along a null geodesic $ds^2 = 0$,

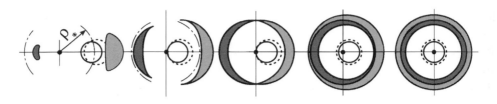

Fig. 7.20. Gravitational lensing in the case of two images S_1 and S_2 for different relative positions of the source and the deflector [from F. Link (1969)]. The black dot is the lensing object. The dotted circle represents the occulted object seen from the occulting object. The full circle shows the occulted object as seen from the Earth in the absence of the lens.

so that

$$[1 - 2G\phi(r)]dt^2 - [1 - 2G\phi(r)]^{-1}dx^2 = 0, \tag{7.112}$$

[$\phi(r)$ is the Newtonian potential of the lensing galaxy], giving

$$dt \sim [1 + 2G\phi(r)]d\ell, \tag{7.113}$$

and thus

$$\Delta T = \int_{\Delta(\text{trajectory})} dt \tag{7.114}$$

$$= \int_{\Delta(\text{trajectory})} d\ell + 2G \int_{\Delta(\text{trajectory})} \phi(r)d\ell. \tag{7.115}$$

The first term on the right hand side is the difference in the optical paths, while the second term comes from the fact that the spatial geometry is not Euclidean. Of course, the potential $\phi(r)$ depends on the mass distribution of the lensing galaxy, and possibly other masses close to the light rays. The first term gives

$$\Delta T_1 = SP_1 + P_1O - SP_2 - P_2O$$
$$\sim \frac{z_S z_D}{H_0(z_S - z_D)}(\varphi_1 + \varphi_2)(\varphi_1 - \varphi_2 - 2\beta_0), \tag{7.116}$$

and is typically of order a year. The second term requires a knowledge of the mass distribution in the deflector; for a point mass the calculation is exactly the same as for the delay of radar echos in the solar system – the associated physics is precisely the same. We thus get[35]

$$\Delta T_2 = 4GM \ln \frac{b_2}{b_1} = 4GM \ln \frac{\varphi_2}{\varphi_1}. \tag{7.117}$$

The two contributions ΔT_1 and ΔT_2 are of the same order of magnitude, as noted by J.H. Cooke and R. Kantowski (1975).

(3) The third effect, the amplification of the brightness of the images, gives more

---

[35] In cosmological applications the term ΔT_2 must be corrected by a factor $(1 + z_D)$ and other terms.

information about the source–deflector system and its cosmological environment. This follows from the conservation of photon number, or of specific intensity. One can show [see e.g. R.K. Sachs (1961)] that given a bundle of null geodesics issuing from a point, with $d\Sigma$ a[36] 2-surface orthogonal to the wave 3-vector, the quantity[37] $Id\Sigma$ (where I is the intensity at a point within the bundle) obeys

$$Id\Sigma|_1 = Id\Sigma|_2 = \text{const.} \qquad (7.118)$$

and is independent of the observer's velocity. It follows that the ratio of the two luminosities is given by

$$\frac{L_1}{L_2} = \frac{d\Sigma_2}{d\Sigma_1}. \qquad (7.119)$$

To find the amplification for a gravitational lens we use the method of S. Refsdal (1964a, b) in the form given by E.E. Clark (1972). Consider a light beam emitted by Σ into the solid angle $d\Omega$; this intercepts an area $d\Sigma_D$ in the plane Π_D orthogonal to the source–observer axis at distance ℓ_{OD} (see **Fig. 7.21**) and therefore contains the deflector. If the latter was absent, the light beam would intercept a surface element $d\Sigma_0$ in the image plane Π_0, and we would have

$$\frac{d\Sigma_0}{d\Sigma_D} = \left(\frac{\ell_{OD} + \ell_{DS}}{\ell_{SD}}\right)^2 \equiv n^2. \qquad (7.120)$$

In reality the light beam is deflected, and intercepts an elementary area $d\Sigma_1$ in the image plane Π_0, corresponding to the image S_1 (not shown on **Fig. 7.21**) of the source S. Of course, there is a similar configuration for the other image of the source. Note that by symmetry $d\chi_0 = d\chi_D \equiv d\chi$. We also have

$$d\Sigma_D = b\,db\,d\chi \qquad (7.121)$$

$$d\Sigma_1 = r\,dr\,d\chi \qquad (7.122)$$

with r defined by $r \equiv \beta_0 n\ell_{OD}$; r is then given by the lens equation (7.104), which becomes

$$r = nb - \ell_{OD}\theta(b). \qquad (7.123)$$

Under these conditions,

$$\frac{d\Sigma_1}{d\Sigma_D} = \frac{r}{b} \cdot \frac{dr}{db}, \qquad (7.124)$$

and using (7.120) we get the amplification of the image L_1/L_0:

$$\frac{L_1}{L_0} = \frac{d\Sigma_0}{d\Sigma_1} = n^2 \frac{b}{r} \left(\frac{dr}{db}\right)^{-1}, \qquad (7.125)$$

---

[36] This depends on the choice of time axis.
[37] $I = A^2$, where A is the electromagnetic amplitude.

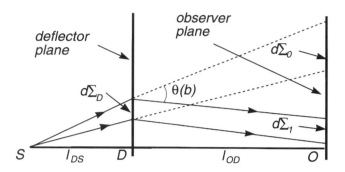

Fig. 7.21. Geometry of image amplification [after E.E. Clark (1972)].

which by differentiation of Eq. (7.123) gives

$$\frac{L_1}{L_0} = n^2 \frac{b}{r} \left. \frac{1}{n - \ell_{\text{OD}} \frac{d\theta(b)}{db}} \right|_{\substack{b=b_1 \\ r=r_1}} .\tag{7.126}$$

Of course, there is a similar relation for the image S_2, and from the geometry of **Fig. 7.17**, we have [for details see S. Refsdal (1964a, b)]

$$\begin{cases} L_1 = \frac{1}{4} \left(2 + \frac{\varphi}{\beta_0} + \frac{\beta_0}{\varphi} \right) L_0 \\[2mm] L_2 = \frac{1}{4} \left(-2 + \frac{\varphi}{\beta_0} + \frac{\beta_0}{\varphi} \right) L_0, \end{cases}\tag{7.127}$$

(where φ is the angular separation of the two images) so that the total luminosity $L_{\text{tot}} \equiv L_1 + L_2$ is given by

$$L_{\text{tot}} = \frac{1}{2} \left(\frac{\varphi}{\beta_0} + \frac{\beta_0}{\varphi} \right) L_0,\tag{7.128}$$

with the difference of the luminosities being L_0. Similarly, the ratio of luminosities of the two images is given by

$$\frac{L_1}{L_2} = \frac{\varphi_1^2}{\varphi_2^2}.\tag{7.129}$$

If the source, deflector and observer are well aligned, i.e. if $\beta_0 \sim 0$, L_{tot} can be large.

(4) The foregoing is elementary, and real situations are much more complicated: extended sources and deflectors produce multiple images with complex forms. We can also consider deflectors which are partially transparent or have no symmetries. We note also that these considerations apply also to *gravitational waves* (*Chapter 8*). Moreover, many cosmological studies use or take account of gravitational lensing. As a result there is a very large current literature (several hundred papers) on this subject, with a growing number devoted to observations.

Exercises

7.1   Starting from the geodesic equations (7.61) and (7.62) for two neighbouring motions, derive the geodesic deviation equation (7.63).

7.2   Verify that the most general spherically symmetric metric described by (7.64) can be found from the only available spatial tensors δ^{ij} and $x^i x^j$.

7.3   (i) Show that the PPN metric given above in the isotropic form (7.75) takes the form

$$ds^2 = \left[1 - 2\alpha\frac{GM}{r} + 2(\beta - \alpha\gamma)\frac{G^2M^2}{r^2} + \cdots\right] dt^2$$
$$- \left[1 + 2\gamma\frac{GM}{r} + \cdots\right] dr^2 - r^2 \left[d\theta^2 + \sin^2\theta d\varphi^2\right]$$

in standard coordinates.

(ii) Calculate the corresponding Christoffel symbols and write down the geodesic equations.

(iii) Calculate the Riemann tensor in this approximation and then that of Einstein.

7.4   In special relativity the Lagrangian of a scalar field ψ is

$$L = \frac{1}{2}\{(\partial\psi)^2 - V(\psi)\},$$

where $V(\psi)$ is a given function of ψ.

(i) Write down the equations of motion.

(ii) What do these become in the presence of a gravitational field if ψ is "minimally" coupled to gravity?

(iii) Give a simple example of non-minimal coupling.

7.5   Let the metric

$$ds^2 = (1 + 2\phi)dt^2 - (1 - 2\phi)dx^2,$$

be defined on V^4. ϕ is a function such that $|\phi| \ll 1$. Write down the geodesic equation. Calculate the components of the Riemann tensor $R_{\mu\nu\alpha\beta}$.

7.6   Verify that Maxwell's equation $\nabla_\mu F^{\mu\nu} = 4\pi J^\nu$ implies the conservation of 4-current: $\nabla_\mu J^\mu = 0$.

7.7   Consider the equation satisfied by the 4-potential A^μ in Minkowski space

$$\Box A^\mu - \partial_\lambda{}^\mu A^\lambda = 4\pi J^\mu.$$

Are the following equations

$$\begin{cases} \nabla_\lambda\nabla^\lambda A^\mu - \nabla_\lambda\nabla^\mu A^\lambda = 4\pi J^\mu \\ \nabla_\lambda\nabla^\lambda A^\mu - \nabla_\lambda\nabla^\mu A^\lambda - R^\mu{}_\lambda A^\lambda = 4\pi J^\mu \end{cases}$$

the correct generalisations in $\{V^4, g_{\mu\nu}\}$? Can they be found from the minimal coupling rule? Are they gauge invariant?

7.8   Consider the equation

$$\nabla_\mu J^\mu = \lambda R ,$$

where λ is a constant and R the Ricci scalar. J^μ is a particle 4-current.

(i) What equation does the one above reduce to in the absence of gravity? Is it the minimal coupling generalisation?

(ii) Do physical arguments allow the conclusion that $\lambda = 0$?

7.9   Prove the theorem of footnote 13.

7.10   Starting from the relation (7.82) giving the perihelion advance for a planet (per revolution), find limits for the PPN parameter γ, using the observational data given in *Chapter 1*.

7.11   It is proposed to calculate to the first order of perturbations the perihelion advance of a planet, whose unperturbed trajectory obeys the equation of motion

$$m\frac{dv}{dt} = -\frac{GmM_\odot}{r^2}e_r. \tag{1}$$

Part 1:

Show that

$$\frac{de_\theta}{d\theta} = -e_r \tag{2}$$

and that

$$\frac{dv}{dt} = \frac{GmM_\odot}{J}\frac{de_\theta}{dt} \tag{3}$$

where

$$J = mr^2\frac{d\theta}{dt}. \tag{4}$$

Deduce the first integral

$$v = \frac{GmM_\odot}{J}(e_\theta + e). \tag{5}$$

where e is a constant vector.

(iii) Verify that

$$e = \frac{1}{Gm^2M_\odot}k \wedge \left[p \wedge L - Gm^2M_\odot e_r\right] \tag{6}$$

where k is the unit vector orthogonal to the plane of the motion.

(iv) We now wish to show that $|e|$ is the eccentricity of the ellipse and that its direction is that of the minor axis. We call φ the angle between e_r and r. Considering the scalar product $mr \cdot v$ and using the equation of an ellipse in polar coordinates (r, θ),

$$r = \frac{a(1 - e^2)}{1 + e\cos\theta}, \tag{7}$$

verify the results above and that $\varphi = \theta + \pi/2$.

Part 2:

We now wish to calculate the perihelion advance of the planet, i.e. φ, caused by the perturbing central force,

$$\delta F = \frac{GmM_\odot}{r} g(r)e_r. \tag{8}$$

(i) Show that

$$\frac{de}{dt} = -\frac{d\theta}{dt} g(r)e_r, \tag{9}$$

and deduce that

$$e.\frac{de}{dt} = -\frac{dr}{dt} \frac{J^2 g(r)}{Gm^2 M_\odot r^2}. \tag{10}$$

(ii) Show that the quantity

$$\frac{1}{2}e^2 + \int^r \frac{J^2 g(r')}{Gm^2 M_\odot r'^2} dr' \tag{11}$$

is conserved.

(iii) Show that

$$e \wedge \frac{de}{dt} = e^2 \psi k, \tag{12}$$

and deduce from (9) that

$$\frac{d\psi}{dt} = \left[\frac{J^2}{Gm^2 M_\odot r} - 1 \right] \frac{g(r)}{e} \frac{d\theta}{dt}. \tag{13}$$

(iv) Integrating the latter relation, then considering just the first order perturbation, show that the advance $\Delta\psi$ of a planetary perihelion in one revolution is given by

$$\Delta\psi \simeq \frac{2}{e} \int_0^\pi g[r(\theta)] \cos\theta d\theta. \tag{14}$$

(v) Application: $g(r) = 3J^2/m^2c^2r^2$.

[Ref.: B. Davies (1983).]

7.12   Calculate the delay Δt of a radar echo from a planet by using the PPN metric in the standard form. Compare with the result found from the isotropic form.

7.13   Why does comparison of radar echo measurements of an Earth–Sun–planet system at inferior and superior conjunction allow one to eliminate the uncertainties relating to the reflection properties of the planetary surface?

7.14   Using Fermat's principle, written in the form

$$\delta \int n(r)d\xi = 0,$$

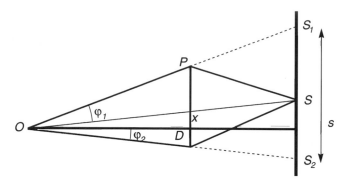

Fig. 7.22. For exercise 19.

(i) show that the trajectories of light rays obey the following equation:

$$\frac{d}{d\xi}\left(n(r)\frac{d\mathbf{x}}{d\xi}\right) = \nabla n(r)$$

(ii) and deduce the deflection of light rays if $n(r)$ is given by the relations (7.93), (7.100).

7.15   In a plasma – such as the solar corona – a (transverse) electromagnetic wave propagates according to the dispersion relation

$$\omega^2 - \mathbf{k}^2 = \omega_p^2,$$

where the plasma frequency ω_p is given by

$$\omega_p^2 = \frac{4\pi\rho e^2}{m}$$

(ρ = electron density).

(i) Calculate the plasma refractive index (ratio of the speed of light to its phase velocity) as a function of ω and ρ.

(ii) Using Fermat's principle (Exercise 13) show that if ρ is a function of r, a light ray is deviated by the plasma. Specify the conditions on $\rho(r)$.

(iii) Calculate the deviation of radio waves by the Sun, given $r > 2.5R_{\odot}$

$$\rho(r) = \frac{10^8}{(r/R_{\odot})^6} + \frac{10^6}{(r/R_{\odot})^2}$$

(ρ is in electrons per cm^3).

7.16   Given a point gravitational lens and a point source find the angular position β_0 of the source as a function of observable quantities.

7.17   Why do the source and its images have the same redshift in a gravitational lens?

7.18   Prove relations (7.127)–(7.129) characterising the light amplification in a gravitational lens.

7.19 In this problem we wish to show that two images of a source situated behind a deflector in uniform straight line motion can have apparently superluminal velocities (we use the same notation as in the chapter). See **Fig. 7.22**. We further assume that the source moves orthogonally to the axis OD, with speed v. We call x the distance of the lens from OD, and s the distance between the two images S_1 and S_2.

(i) Show that $s = (l_{OD} + l_{DS})(\varphi_1 + \varphi_2)$, and express this quantity as a function of x, l_{OD}, l_{OS} and the gravitational radius of the lens. Show also that

$$x = \frac{l_{OD} + l_{DS}}{l_{OD}} vt.$$

(ii) Calculate the relative velocity $\frac{ds}{dt}$ of separation of the images S_1 and S_2 and deduce that if $l_{DS} \gg l_{OD}$, then $\left|\frac{ds}{dt}\right|$ can be arbitrarily large.
[Ref.: L. Kh. Ingel' (1975).]

7.20 Starting from the most general spherically symmetric metric (7.66), i.e.

$$ds^2 = e^\nu dt^2 - e^\lambda dr^2 - r^2 d\Omega^2,$$

(i) show that the nonzero components of the Riemann tensor are

$$
\begin{cases}
R^0{}_{101} = -\tfrac{1}{2}\nu'' + \tfrac{1}{4}\nu'\lambda' - \tfrac{1}{4}\nu'^2 \\
R^0{}_{202} = -\tfrac{1}{2}r\nu'e^{-\lambda} \\
R^0{}_{303} = -\tfrac{1}{2}r\nu'e^{-\lambda}\lambda' \sin^2\theta
\end{cases}
\qquad
\begin{cases}
R^1{}_{212} = \tfrac{1}{2}r\nu'e^{-\lambda} \\
R^1{}_{313} = \tfrac{1}{2}r\lambda'e^{-\lambda}\sin^2\theta \\
R^2{}_{323} = (1 - e^{-\lambda})\sin^2\theta
\end{cases}
$$

(plus those obtained by permuting the indices).

(ii) Deduce that the nonzero components of the Ricci tensor $R_{\mu\nu}$ are

$$R_{00} = e^{\nu-\lambda} \cdot \left[-\frac{1}{2}\nu'' + \frac{1}{4}\nu'\lambda' + \frac{1}{4}\nu'^2 - \frac{\lambda'}{r} \right]$$

$$R_{11} = \frac{1}{2}\nu'' - \frac{1}{4}\nu'\lambda' + \frac{1}{4}\nu'^2 - \frac{\lambda'}{r}$$

$$R_{22} = e^{-\lambda} \cdot \left[1 + \frac{1}{2}r(\nu' - \lambda') \right] - 1$$

$$R_{33} = \left\{ e^{-\lambda} \left[1 + \frac{1}{2}r(\nu' - \lambda') \right] - 1 \right\} \sin^2\theta.$$

(iii) Verify that the curvature scalar is given by

$$R = e^{-\lambda} \left[-\nu'' + \frac{1}{2}\nu'\lambda' - \frac{1}{2}\nu'^2 + \frac{2}{r}(\lambda' - \nu') - \frac{2}{r^2} \right] + \frac{2}{r^2}.$$

7.21 Let $\{\mathbf{K}\}$ and $\{\mathbf{K'}\}$, be respectively inertial and uniformly accelerated observers in $\mathcal{M}^2$. Their worldlines are

$$x = a \quad \text{(for } \{\mathbf{K}\}) \quad \text{and} \quad t^2 - x^2 = -a^2 \quad \text{(for } \{\mathbf{K'}\}).$$

(i) Sketch a space–time diagram in the inertial coordinate system linked to $\{\mathbf{K}\}$.

(ii) $\{\mathbf{K}\}$ sends electromagnetic signals to $\{\mathbf{K'}\}$. Verify that after a time t_0, which you should determine, $\{\mathbf{K'}\}$ can no longer receive the signals emitted by $\{\mathbf{K}\}$. Show that before a time t_1, again to be determined, $\{\mathbf{K}\}$ does not know of the existence of $\{\mathbf{K'}\}$. The *horizon* of $\{\mathbf{K}\}$ relative to $\{\mathbf{K'}\}$ is defined by t_0 and t_1.

(iii) Now consider the coordinate system (ξ, χ)

$$t = \xi \operatorname{sh} \chi, \qquad x = \xi \operatorname{sh} \chi$$

(Rindler coordinates). In this system, how do (a) the metric, (b) the equation of $\{\mathbf{K'}\}$, and (c) the equation of $\{\mathbf{K}\}$ appear?

(iv) What is the equation of the light cone in this coordinate system?

(v) Calculate the redshift of electromagnetic signals emitted by $\{\mathbf{K'}\}$ and received by $\{\mathbf{K}\}$.

8 Einstein's Relativistic Gravitation (General Relativity)

We have seen in the preceding chapters that the changes in the theory of gravity introduced by relativity, chiefly as a consequence of the famous relation $E = mc^2$ (*Chapter 4*), amount to the near-necessity of introducing curved space–time, also a consequence of the Equivalence Principle (*Chapters 6 and 7*). However, this principle does not specify the equations determining the ten components of the metric tensor $g_{\mu\nu}$. The relation $E = mc^2$ suggests that these equations must be *non-linear* (*Chapter 4*). In this chapter we shall study the simplest equations compatible with observation, and show how they arise. These are Einstein's equations, and the theory they define is called general relativity. We shall derive some elementary consequences which are astrophysically important.

The curvature of space–time expresses the effects of gravitation on physical phenomena through a metric tensor $g_{\mu\nu}$, which cannot be reduced to $\eta_{\mu\nu}$ everywhere. However, this does not rule out the existence of other long-range fields which might also be important. The theory of C. Brans and R.H. Dicke (1961) is the best-known example; here a scalar field coexists with the metric tensor. The existence of such fields must ultimately be decided by experiment or observation [see C. Will (1981)]. Currently it appears that the metric tensor alone appears capable of ensuring agreement with observation, and that general relativity is the correct relativistic theory of gravity.

We note again that simple arguments rule out relativistic theories of gravity based solely on a scalar or vector field. First, the simplest relativistic generalisation of Poisson's equation for a scalar gravitational field is

$$\Box\Phi = 4\pi G \times (\text{mass density}). \tag{8.1}$$

As in the relativistic case the mass density (or that of energy and momentum) is represented by the energy–momentum tensor $T^{\mu\nu}$, the only scalar with the dimensions of an energy density which one can construct from this tensor is its trace, so that Eq. (8.1) must take the form

$$\Box\Phi = 4\pi G T^{\mu}{}_{\mu}. \tag{8.2}$$

Although this equation generalises Poisson's equation in a relativistically invariant way, it must be rejected as, when used with the appropriate equations of motion, it predicts a *retardation* of the perihelion of Mercury, and indeed an incorrect order of magnitude. Moreover, it does not allow one to couple gravity and electromagnetism as $T^{\mu}{}_{\mu em} = 0$: in such a model we could have neither a gravitational redshift nor the deflection of light by large masses. Of course, there may be rather contrived ways around these arguments. A gravitational theory based on a vector field can be eliminated since two massive particles would repel rather than attracting each other. Of course, a combination of these three kinds of field (tensor, vector and scalar) could agree perfectly with observation, and such models have been studied.

However, we shall confine ourselves here to Einstein's theory, in which gravity curves space–time, and the metric tensor is the only field describing it.

8.1   Einstein's equations

Before finding these equations, it must be clear that they constitute an additional *postulate*, which is largely independent of the equivalence principle, and which we could therefore assume *a priori*. However, although they are very natural, we prefer to indicate the sources and constraints which lead to them.

We first of all require that the equations of gravity should reduce to Poisson's equation in the Newtonian limit of low velocities and weak gravitational fields. This is a minimal constraint which can be accommodated by requiring the equations to be *second order* in derivatives of the metric tensor, and that the second derivatives should appear *linearly*. These two requirements lead to Poisson's equation, but although they are sufficient, they are certainly not necessary. We must consider the simplest hypotheses. We can argue from the fact that we would like to be able to pose a Cauchy problem for the gravitational field in the usual way[1], so that only derivatives up to the second order can appear in the relativistic equations. Linearity of the second derivatives is obviously the simplest way of obtaining the (linear) Poisson equation, remembering that $g_{00} \sim 1 - 2\Phi/c^2$ in the Newtonian approximation.

The second idea which we use is Einstein's fundamental intuition, which suggested the "relativistic relation"

$$\mathcal{CURVATURE} = \mathcal{MATTER}. \tag{8.3}$$

However, curvature is expressed by the fourth-order Riemann–Christoffel tensor $R_{\mu\nu\alpha\beta}$, while the presence of matter is expressed through the second-order energy–momentum tensor $T_{\mu\nu}$. Thus the relation (8.3) can only be realised by contractions of $R_{\mu\nu\alpha\beta}$, and we write formally

$$K_{\mu\nu} = 8\pi G T_{\mu\nu}, \tag{8.4}$$

where $K_{\mu\nu}$ is a tensor (to be determined) constructed from the Riemann–Christoffel and metric tensors. Of course, equations (8.3) or (8.4) satisfy the earlier criteria as we

---

[1] We wish to be able to find the gravitational field $g_{\mu\nu}$ at a point x of the space–time manifold from a knowledge of the derivatives $\partial_\lambda g_{\mu\nu}$ and $g_{\mu\nu}$ at "time t", i.e. on a spacelike 3-surface. This implies that the equations to be solved must be of second order.

recall that $R_{\mu\nu\alpha\beta}$ is the only fourth-order tensor which is linear in the second derivatives of the metric (and their contractions with the latter tensor). The tensor $K_{\mu\nu}$ must be *symmetric* in μ and ν, and its most general form is thus

$$K_{\mu\nu} = aR_{\mu\nu} + bRg_{\mu\nu} + a\lambda g_{\mu\nu}, \tag{8.5}$$

where a, b and λ are constants. To find them we recall that the energy–momentum tensor is conserved,

$$\nabla_\mu T^{\mu\nu} = 0, \tag{8.6}$$

and thus that $K_{\mu\nu}$ must be too for any $T_{\mu\nu}$, i.e.

$$\nabla_\mu K^{\mu\nu} = 0. \tag{8.7}$$

Now using the relation (3.131) and the Bianchi identities (6.58), i.e.

$$\nabla^\mu g_{\mu\nu} = 0 \tag{8.8}$$

and

$$\nabla_\lambda R_{\mu\nu\alpha\beta} + \nabla_\beta R_{\mu\nu\lambda\alpha} \mid \nabla_\alpha R_{\mu\nu\beta\lambda} = 0, \tag{8.9}$$

respectively, the relation (8.7) becomes

$$\nabla_\mu K^\mu{}_\nu = \left(\frac{a}{2} + b\right)\nabla_\nu R = 0, \tag{8.10}$$

implying $b = -a/2$, as $\nabla_\nu R$ cannot vanish everywhere (except in Minkowski space, which is not curved). The tensor $K_{\mu\nu}$ thus has the form

$$K_{\mu\nu} = a\left[G_{\mu\nu} + \lambda g_{\mu\nu}\right] \tag{8.11}$$

with

$$G_{\mu\nu} = a\left[R_{\mu\nu} - \frac{1}{2}Rg_{\mu\nu}\right] \tag{8.12}$$

where $G_{\mu\nu}$ is the *Einstein tensor*. Finally, the *only* equation which one can write, given the preceding constraints, has the form

$$R_{\mu\nu} - \frac{1}{2}g_{\mu\nu}R + \lambda g_{\mu\nu} = \frac{8\pi G}{a}T_{\mu\nu}, \tag{8.13}$$

and constitutes Einstein's equation if $a = 1$. The constant λ is called the *cosmological constant* and was only introduced by Einstein with the aim of obtaining a static universe.

We still have to verify that Eq. (8.13), with $\lambda = 0$ and $a = 1$, reduces to Poisson's equation in the Newtonian limit. The limit of weak gravitational fields implies that on the left hand side of (8.13) we should retain only terms *linear* in the deviation $h_{\mu\nu}$ of the metric tensor $g_{\mu\nu}$ from $\eta_{\mu\nu}$:

$$g_{\mu\nu} \sim \eta_{\mu\nu} + h_{\mu\nu} \text{ and } \|h_{\mu\nu}\| \ll 1. \tag{8.14}$$

with a similar linearisation of the derivatives.

The limit of low velocities affects the right hand side of (8.13) as $T_{\mu\nu}$ describes the state of matter, including its motion. For example, if we consider a perfect fluid, the energy–momentum tensor has the form

$$T_{\mu\nu} = (\rho + P)u_\mu u_\nu - P g_{\mu\nu}, \qquad (8.15)$$

and the ratio T_{ii}/T_{00} is given by

$$\left| \frac{T_{ii}}{T_{00}} \right| = \left| \frac{(\rho + P)u_i u_i - P g_{ii}}{(\rho + P)u_0 u_0 - P g_{00}} \right|. \qquad (8.16)$$

However, as $P/c^2 \ll \rho$, Eq. (8.16) becomes

$$\left| \frac{T_{ii}}{T_{00}} \right| = \left| v_i^2 - \frac{P}{\rho} \frac{g_{ii}}{g_{00}} \right|. \qquad (8.17)$$

using $u_0 = g_{00}u^0$ (assuming $g_{0i} = 0$, as we may for a static metric in a suitable coordinate system) and $v_i \equiv u_i/u_0$. On the other hand, the ratio g_{ii}/g_{00} is of order unity in the weak-field approximation (8.14), and as $P \ll \rho c^2$, the ratio (8.17) is of order v^2/c^2 and is thus negligible. Quite generally, in the Newtonian approximation the spatial components of the energy–momentum tensor are negligible by comparison with the timelike component:

$$|T_{ij}| \ll T_{00} = \rho \, ; \qquad (8.18)$$

in other words the equivalent energy of the tensions is much smaller than the energy density ρ.

Under these conditions, taking the trace of equation (8.13), we get

$$-R + 4\lambda = \frac{8\pi G}{a} T^\mu{}_\mu \qquad (8.19)$$

$$\sim \frac{8\pi G}{a} T^0{}_0, \qquad (8.20)$$

so that in the Newtonian approximation the relation (8.13) for the zero components gives

$$R_{00} \approx +\lambda g_{00} + \frac{4\pi G}{a} T_{00}, \qquad (8.21)$$

where we have used $T_{00} = g_{00} T^0{}_0$ for a static metric, neglecting the T^{ij}. Now using

$$R_{00} \approx +\partial_i \Gamma^i_{00}, \qquad (8.22)$$

(which follows from linearising the expression for R_{00} and assuming a static metric), we get [cf. Eq. (7.53)]

$$\Gamma^i_{00} \approx \frac{1}{2} \delta^{ij} \partial_j h_{00}; $$

i.e.

$$\frac{1}{2} \nabla^2 h_{00} \simeq +\lambda(h_{00} + 1) + \frac{4\pi G}{a} \rho, \qquad (8.23)$$

or [with $h_{00} = +2\phi$ where ϕ is the Newtonian potential; cf. Eq. (7.57)]

$$\nabla^2 \phi = \lambda(1 + 2\phi) + \frac{4\pi G}{a}\rho, \qquad (8.24)$$

which is indeed the Poisson equation with $\lambda = 0$ and $a = 1$.

A useful form of the Einstein equations is

$$R_{\mu\nu} = 8\pi G \left[T_{\mu\nu} - \frac{1}{2} g_{\mu\nu} T^\lambda{}_\lambda \right], \qquad (8.25)$$

found by replacing the curvature scalar R in (8.13) by the expression (8.19) with $\lambda = 0$ and $a = 1$.

At this point we should note that while the Einstein equations (8.13) or (8.25) provide ten equations for the ten unknown components $g_{\mu\nu}$ of the metric tensor, the four relations (the contracted Bianchi identities)

$$\nabla_\mu G^{\mu\nu} = 0, \qquad (8.26)$$

which result from the relations (8.12) and the Bianchi identities are automatically satisfied, thus leaving only six independent equations. There are thus *a priori* four equations fewer than required. In fact, the partial indeterminacy of the $g_{\mu\nu}$ corresponds to the arbitrariness in the choice of coordinate systems: given specified coordinates, and thus imposing four conditions, Einstein's equations do allow us to determine the metric tensor. We still have to specify the initial conditions and any boundary conditions.

In conclusion, the local geometry of space–time is not determined *a priori* but fixed by the distribution and motion of matter. The latter are themselves in turn influenced by the local space–time geometry. In this sense Einstein's relativistic gravity is a complete theory, at least for what concerns the local properties of space–time: this becomes a dynamical element itself, and is no longer a passive arena in which physical events unfold. On the contrary, these change the space–time. We have to add the important remark that boundary conditions (e.g. at infinity) are not at all fixed by general relativity alone: any *global* hypothesis is cosmological in nature.

We close this section by returning to the *cosmological constant* λ, which we took earlier to be zero. We have already stated that Einstein introduced this constant in order to find a *static* cosmological solution (i.e. neither expanding nor contracting, at a time before Hubble's discovery of the expansion of the Universe) which thus agreed with the then current ideas about the notion of space. Actually a cosmological term had already been introduced into Newtonian gravity by H. Seeliger (in 1894, 1895, 1896) and C. Neumann (1896) with the same motivation as Einstein [see e.g. H. Bondi (1960)], and the only constraint on it is that its effect on planetary (or galaxy) motions must be negligible. Current estimates (A.D. Dolgov and Ya. B. Zeldovich (1981)) give $\lambda < 10^{-57}$ cm^{-2}, this extremely small value itself posing a problem [see Ya. B. Zeldovich and I.D. Novikov (1983) and S. Weinberg (1989); see also S.M. Caroll *et al.* (1992)]. We note that if we transfer the cosmological term $\lambda g_{\mu\nu}$ to the right hand side of Einstein's equation (8.13) with $a = 1$, this term can be regarded as the energy–momentum tensor of the vacuum, as it remains and is conserved even in the absence of matter ($T_{\mu\nu} \equiv 0$) [cf. Eq. (8.8)].

8.2   Other derivations of the Einstein equations

There are many other derivations of Einstein's equations which are more or less equivalent to that given above, but which illuminate the theory from varying standpoints. There are also other derivations which are *very* different in their physical bases. It is all the more surprising that one nevertheless gets the same equations quite naturally in almost every case. All experimental and observational evidence currently supports general relativity [C.M. Will (1981)] rather than any competing theory, so it appears that this theory embodies some fundamental property of matter which is still mysterious.

In this section we shall outline two very different derivations of the Einstein equations. The first is by D. Hilbert (1915), and adopts the same assumptions made above. The other, by several authors, is based on the formal analogy between the equations describing zero rest-mass particles of spin 2 and the linearised Einstein equations.

(1) We begin with Hilbert's approach (1915), which is *variational*, but based on the same assumptions[2] as above. It is interesting for several reasons. Consider an arbitrary physical system composed of non-quantum particles and/or fields, characterised by a Lagrangian $\mathscr{L}_{\text{mat}}$ and obeying (in the absence of gravitation) the usual Euler–Lagrange equations (see *Appendix C*). Let $\mathscr{L}_{\text{grav}}$ be the Lagrangian, as yet undetermined, of the gravitational field, i.e. that which gives the equations satisfied by the metric tensor $g_{\mu\nu}$. In passing we note that even if $\mathscr{L}_{\text{grav}}$ is independent *a priori* of the matter state, $\mathscr{L}_{\text{mat}}$ is by contrast a function of local space–time properties via the metric tensor $g_{\mu\nu}$: $\mathscr{L}_{\text{mat}}$ thus contains the coupling between matter and gravitation, so that we can assume the variational principle

$$\delta S = 0 = \delta \int \sqrt{|g|}d^4x \left[\mathscr{L}_{\text{grav}} + \mathscr{L}_{\text{mat}} \right]. \tag{8.27}$$

The factor $\sqrt{|g|}$ is introduced to keep the element of integration invariant under coordinate changes (see *Appendix B*).

$$\sqrt{|g|}d^4x = \sqrt{|g'|}d^4x'. \tag{8.28}$$

We now determine the Lagrangian $\mathscr{L}_{\text{grav}}$ by applying the requirements that we want equations linear in the second derivatives of the metric, which reduce to Poisson's equation in the non-relativistic approximation. The first requirement immediately gives the Lagragian

$$\mathscr{L}_{\text{grav}} = \text{const.} \times [R - 2\Lambda], \tag{8.29}$$

where R is the only scalar which can be constructed from $g_{\mu\nu}$ and $R_{\mu\nu\alpha\beta}$ which is linear

---

[2] For the relations between Hilbert and Einstein and the discovery of the relativistic equations of gravity, see J. Mehra (1974).

in $\partial_{\alpha\beta}g_{\mu\nu}$: Λ is a constant which for the moment is arbitrary, but will appear as the cosmological constant. The multiplicative constant in (8.29) is fixed by the necessity of recovering Poisson's equation in the Newtonian limit.

The equations of relativistic gravity can now be found using the variational principle (8.27) [see e.g. J.L. Anderson (1967); S. Weinberg (1972); C.W. Misner, K.S. Thorne, J.A. Wheeler (1973)]. We should however note that $\mathscr{L}_{\mathrm{grav}}$ is second-order (in the derivatives of the metric) and we cannot use the standard Euler–Lagrange equations given in *Appendix C*. These must be generalised. However, we can use a simpler procedure[3] which varies $g_{\mu\nu}$ and $\Gamma^{\mu}_{\alpha\beta}$ independently (the latter appears in R). The variation with respect to the Γ gives their expression in terms of the $g_{\mu\nu}$ and $\partial_{\alpha}g_{\mu\nu}$. Using the relations

$$\delta\sqrt{|g|} = \frac{1}{2}\sqrt{|g|}g^{\mu\nu}\delta g_{\mu\nu} \tag{8.30}$$

$$\delta g^{\alpha\beta} = -g^{\alpha\mu}g^{\beta\nu}\delta g_{\mu\nu} \tag{8.31}$$

we thus get

$$\mathrm{const.} \times \left[R_{\mu\nu} - \frac{1}{2}Rg_{\mu\nu} + \lambda g_{\mu\nu}\right] - \frac{\delta\mathscr{L}_{\mathrm{mat}}}{\delta g^{\mu\nu}} - \frac{1}{2}g_{\mu\nu}\mathscr{L}_{\mathrm{mat}}, \tag{8.32}$$

which are indeed the Einstein equations if the constant appearing in the left hand side is taken equal to $(8\pi G)^{-1}$ and if the energy–momentum tensor of the matter is identified with

$$T^{\mu\nu}_{\mathrm{mat}} \equiv \frac{\partial\mathscr{L}_{\mathrm{mat}}}{\partial g_{\mu\nu}} - \frac{1}{2}g^{\mu\nu}\mathscr{L}_{\mathrm{mat}}. \tag{8.33}$$

This is clearly only possible if the tensor (8.33) is conserved and differs from the canonical energy–momentum tensor (see *Appendix C*) by terms with zero covariant divergence. This can be proved without difficulty (see the references above). The relation (8.33) has the advantage of giving a *symmetric* energy–momentum tensor, which is not in general the case with the canonical form [see F. Belinfante (1940); L. Rosenfeld (1940)]; the invariance of the Lagrangian $\mathscr{L}_{\mathrm{mat}}$ under infinitesimal coordinate changes

$$x'^{\mu} = x^{\mu} + \varepsilon^{\mu}(x) \tag{8.34}$$

implies the conservation of the energy–momentum tensor

$$\nabla_{\mu}T^{\mu\nu}_{\mathrm{mat}} = 0.$$

(2) Another approach, originally due to S.N. Gupta (1954) (1957) (1962) and later developed [R.H. Kraichnan (1955); W. Thirring (1961); R.P. Feynman (1962); S. Weinberg (1965); S. Deser (1970) (1971)] is radically different[4]. It is based on the fact that in quantum theory a spin 2 field is represented by a symmetric traceless rank two

---

[3] This method is by A. Palatini (1919). Its analogues in mechanics, electromagnetism, etc. are discussed in C.W. Misner, K.S. Thorne, J.A. Wheeler (1973), pages 493ff.

[4] One of the best summaries is without doubt that by R.P. Feynman (1962), a book which is unfortunately difficult to find; see also S. Deser (1971).

tensor. Also, the idea of interpreting the field $g_{\mu\nu}$ as directly linked to a classical (non-quantum) spin 2 field is very natural. However, instead of starting immediately with the equations for a spin 2 field [M. Fierz, W. Pauli (1939)] we follow the development by R.P. Feynman (1962) and S. Deser (1971). We indicate only the main steps.

Consider a symmetric classical tensor field $h_{\mu\nu}$ and write down the most general Lagrangian which it can obey, given that it is composed of quadratic combinations of the first derivatives $\partial_\lambda h_{\alpha\beta}$: we wish to obtain linear second-order equations. Because we are dealing with a massless field, there is no term which is quadratic in $h_{\alpha\beta}$. The Lagrangian can in general then be written in the form[5]

$$L = a\partial_\sigma h_{\mu\nu}.\partial^\sigma h^{\mu\nu} + b\partial_\nu h^{\mu\nu}.\partial_\sigma h_\mu{}^\sigma + c\partial_\nu h^{\mu\nu}.\partial_\mu h_\sigma{}^\sigma + d\partial_\mu h_\sigma{}^\sigma.\partial^\mu h_\sigma{}^\sigma - \chi h_{\mu\nu} T^{\mu\nu}_{\text{mat}}, \quad (8.35)$$

where χ is a constant which is for the moment arbitrary, and (a, b, c, d) are constants to be determined[6]. In this equation $T^{\mu\nu}_{\text{mat}}$ is the matter energy–momentum tensor and the last term is the only possible coupling leading to linear equations; the tensor $T^{\mu\nu}_{\text{mat}}$ must clearly be conserved. The equations of motion then become (see *Appendix C*)

$$2a\Box h_{\alpha\beta} + b\partial_{\sigma(\alpha}h_{\beta)}{}^\sigma + c\{\partial_{\alpha\beta}h_\sigma{}^\sigma + \eta_{\alpha\beta}\partial_{\mu\nu}h^{\mu\nu}\} + 2d\eta_{\alpha\beta}\Box h_\sigma{}^\sigma = \chi T_{\alpha\beta\text{mat}}. \quad (8.36)$$

The coefficients (a, b, c, d) are now determined by requiring that the left hand side of (8.36) should be gauge invariant (see below). A possible choice is then: $a = 1/2$, $b = -1$, $c = 1$, $d = -1/2$. All other choices give a multiplicative constant which can be absorbed in χ. We may thus finally write

$$\Box h_{\alpha\beta} - \partial_{\sigma(\alpha}h_{\beta)}{}^\sigma + \{\partial_{\alpha\beta}h + \eta_{\alpha\beta}\partial_{\mu\nu}h^{\mu\nu}\} - \eta_{\alpha\beta}\Box h = \chi T_{\alpha\beta\text{mat}}, \quad (8.37)$$

where we have set $h \equiv h_\sigma{}^\sigma$. With the substitution

$$h_{\mu\nu} \to \bar{h}_{\mu\nu} \equiv h_{\mu\nu} - \frac{1}{2}\eta_{\mu\nu}h, \quad (8.38)$$

Eq. (8.37) reduces to

$$\Box\bar{h}_{\mu\nu} - \partial^\alpha\partial_{(\mu}\bar{h}_{\nu)\alpha} + \eta_{\mu\nu}\partial^{\alpha\beta}\bar{h}_{\alpha\beta} = \chi T_{\mu\nu\text{mat}}. \quad (8.39)$$

Equations (8.37) and (8.39) are essentially identical to the linearised Einstein equations (see below). They are, however, not physically correct as *they do not contain the contribution of the gravitational field $h_{\mu\nu}$ itself to the energy–momentum.* We must therefore add this contribution, which is necessarily non-linear in $h_{\mu\nu}$ and $\partial_\lambda h_{\mu\nu}$, to the right hand side of (8.37) and (8.38). If we call this contribution $N^{\mu\nu}$ it must obey

$$\partial_\mu\left[T^{\mu\nu}_{\text{mat}} + N^{\mu\nu}\right] = 0, \quad (8.40)$$

and we get

$$\Box\bar{h}_{\mu\nu} - \partial^\alpha\partial_{(\mu}\bar{h}_{\nu)\alpha} + \eta_{\mu\nu}\partial^{\alpha\beta}\bar{h}_{\alpha\beta} = \chi\left\{T_{\mu\nu\text{mat}} + N_{\mu\nu}(\bar{h}_{\rho\sigma})\right\}. \quad (8.41)$$

---

[5] Note that in the following *all* the tensor indices are Lorentzian.

[6] *A priori* we could add to the Lagrangian a term of the form $\partial_\sigma h^{\mu\nu}.\partial_\nu h^\sigma{}_\mu$; however, an integration by parts in the action integral reduces such terms to terms of the form $\partial_\nu h_\mu{}^\nu.\partial_\sigma h^{\mu\sigma}$.

$N_{\mu\nu}$ can be found from the Lagrangian (8.35) and then appears as quadratic in the $\partial_\lambda h_{\mu\nu}$. Equation (8.41) is then bilinear and must follow from a cubic Lagrangian, etc. This iterative process leads to $N_{\mu\nu}$ which, when added to the left hand side of (8.41), leads exactly to the Einstein equations. The process is quite complex in practice, but with a suitable formalism one can obtain the exact result after only a single iteration [see S. Deser (1970) (1971)].

8.3 The Schwarzschild solution

The first exact solution of the Einstein equations was found by K. Schwarzschild (in 1916) and plays a particularly important role[7] in the verification of general relativity in the solar system (the famous "classical tests"). It is the relativistic counterpart of the Newtonian potential $\phi(r) = -GM/r$: this is a static solution with spherical symmetry, Minkowskian at infinity, decribing the gravitational field of a body at the origin[8].

In standard coordinates the Schwarzschild metric is

$$ds^2 = \left(1 - 2\frac{GM}{r}\right) dt^2 - \frac{1}{\left(1 - 2\frac{GM}{r}\right)} dr^2 - r^2 \left[d\theta^2 + \sin^2\theta \, d\varphi^2\right], \qquad (8.42)$$

and we shall derive this in a heuristic manner. We shall also derive the solution from the Einstein equations.

(1) Consider a mass M situated at the origin $r = 0$ and a test particle moving in the gravitational field of the mass M at distance r. The spherical symmetry of the system implies that there exist circular trajectories for the test particle which at least approximately satisfy $v^2 = GM/r$; in general we set

$$v^2 = a\frac{GM}{r}, \qquad (8.43)$$

where a is a constant to be determined ($a = 2$ for Newtonian parabolic trajectories). An observer linked to the mass M and measuring a time interval dt and a distance interval dr in a reference frame linked to the test particle will find a time interval dt' *dilated* by the usual factor

$$dt = \frac{dt'}{\sqrt{1 - v^2}} \qquad \text{(for } dr' = 0) \qquad (8.44)$$

and a length interval dr' *contracted* as

$$dr = dr'\sqrt{1 - v^2} \qquad \text{(for } dt' = 0). \qquad (8.45)$$

In his local coordinates $\{t', r'\}$ the observer linked to the test particle (assumed very

---

[7] The detailed history of the Schwarzschild solution is given by J. Eisenstaedt (1982).
[8] Strictly speaking this is a *vacuum* solution ($T_{\mu\nu} = 0$) almost everywhere, and the space–time manifold of which it is the metric is *chosen* as $V^4 = \mathbf{R} \times \mathbf{R}^+ \times S^2$; $\mathbf{R}$ corresponds to the time, $\mathbf{R}^+$ to the radial coordinate minus the origin $r = 0$, and S^2 is the usual sphere (see below).

distant from the mass M) must see a Lorentzian metric

$$ds^2 = dt'^2 - dr'^2. \tag{8.46}$$

Now using (8.44) and (8.45) and replacing v by the expression (8.43), we get

$$ds^2 = \left(1 - a\frac{GM}{r}\right) dt^2 - \frac{1}{\left(1 - a\frac{GM}{r}\right)} dr^2. \tag{8.47}$$

We must now find the coefficient a, which is easy if we recall that in the Newtonian approximation we must have $g_{00} \simeq 1 + 2\phi$; we thus immediately get $a = 2$ and (8.47) is the Schwarzschild solution with $d\theta = d\varphi = 0$. Clearly this "derivation" is highly approximate. It has the sole advantage of illustrating two phenomena directly linked to the metric (8.42), namely the changes in the scales of time and length.

(2) We return now to the Schwarzschild solution (8.42) and discuss it further. We consider first $t = $ const., i.e. $dt = 0$. This is a spacelike 3-surface, with positive definite metric

$$d\sigma^2 = \frac{dr^2}{1 - 2\frac{GM}{r}} + r^2 \left[d\theta^2 + \sin^2 \theta d\varphi^2\right]. \tag{8.48}$$

These spacelike 3-surfaces are orthogonal to the time coordinate lines and are clearly all identical[9] whatever the value of t. They also possess spherical symmetry. The 2-surfaces $r = $ const. which they contain are normal 2-spheres, as indicated by their metric

$$d\tilde{\sigma}^2 = r^2 \left[d\theta^2 + \sin^2 \theta d\varphi^2\right], \tag{8.49}$$

with area $S = 4\pi r^2$; this is the simplest interpretation of the radial coordinate r. We note, however, that the metric (8.48) defines only the *local structure* of the 3-surfaces $t = $ const.; it says nothing about their *global* properties[10], or their topology. This situation is quite general in relativity: Einstein's equations determine the local properties of the space–time manifold but not the global properties, which must be imposed *a priori*.

To have a more precise idea of the 3-surfaces $t = $ const., we obtain a representation through a 2-surface of $\mathbf{R}^3$ corresponding to the metric (8.48) with $\theta = \pi/2$. This 2-surface has the metric

$$d\sigma^2 = \frac{dr^2}{1 - 2\frac{GM}{r}} + r^2 d\varphi^2, \tag{8.50}$$

and can be regarded as a surface of revolution with axis Ow, generated by a parabola $w = w(r)$ (**Fig. 8.1**):

$$w^2(r) = 8GM \cdot [r - 2GM]. \tag{8.51}$$

This 2-surface is the paraboloid of L. Flamm (1916). We thus have an embedding of this 2-surface in three-dimensional Euclidean space. However, although x and y correspond to Schwarzschild coordinates, w has no special meaning. We could also simply note that

---

[9]   This is general for static space–times.

[10] For example, if we identify two diametrically opposed points of the usual sphere S^2 we get the projective plane P^2, which has the *same* metric but is not orientable.

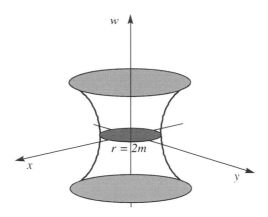

Fig. 8.1. Representation of the 3-surfaces $t = $ const. by Flamm's paraboloid.

only the points of the 2-surface have a meaning, and not the space $\mathbf{R}^3$ in which it is embedded[11].

A second remark concerns the behaviour of the Schwarzschild metric at large distance: when $r \to \infty$, we have

$$\lim_{r \to \infty} ds^2_{\text{Schw}} = ds^2_{\text{Minkowski}}; \tag{8.52}$$

put another way, the Schwarzschild space–time is *asymptotically flat*. However, as r is decreased we reach a critical value, the *Schwarzschild radius* (or gravitational radius)

$$R_G = 2GM/c^2, \tag{8.53}$$

where the metric (8.42) becomes singular. Is this a true space–time singularity, or is it simply a coordinate singularity? A calculation of the curvature tensor shows that this is non-singular (finite) for $r \geq R_G$. This suggests that $r = R_G$ is a coordinate singularity which does not allow us to describe the *whole* space–time but only a part. We have to extend the coordinates to cover the whole space–time. Nevertheless, as Schwarzschild coordinates have some physical meaning, some phenomenon must occur at $r = R_G$. The 3-surface $r = R_G$ divides space–time into two regions, one of them unobservable; this surface is called the *Schwarzschild horizon* and plays an important role in the study of *black holes*.

Considering the coordinates further, at $t = $ const., the size of an object is defined (at $d\theta = d\varphi = 0$) by

$$d\ell^2 = -ds^2 = \frac{dr^2}{1 - 2\frac{GM}{r}}, \tag{8.54}$$

---

[11] The use of Flamm's paraboloid is discussed by W. Rindler (1977).

indicating the connection between the radial coordinate r and the distance ℓ:

$$\ell = \int_0^r \frac{dr}{\left(1 - 2\frac{GM}{r}\right)^{1/2}} = r^{1/2}\left(1 - 2\frac{GM}{r}\right)^{1/2}$$

$$+ 2GM \log\left\{[r - 2GM]^{1/2} + r^{1/2}\right\}. \tag{8.55}$$

For a length characterised by radial coordinates r_1 and r_2 this distance is approximately given by

$$\ell \sim (r_2 - r_1) + GM \log\left(\frac{r_2}{r_1}\right), \tag{8.56}$$

where the second term shows the effect of the space–time curvature. Of course, if r_1 is close to r_2, the measured distance ℓ is close to $r_2 - r_1$.

Similarly, the measured time (at $dr = d\theta = d\varphi = 0$) is given by

$$d\tau^2 = ds^2 = \left(1 - 2\frac{GM}{r}\right) dt^2 \tag{8.57}$$

and reduces at infinity to the usual time of an inertial observer at rest with respect to the mass M.

A final remark about the Schwarzschild metric. In reality the mass M is not a point but distributed over the interior of a certain radius R: for the Sun, for example, $R_\odot = 7 \times 10^{10}$ cm while $R_G = 3 \times 10^5$ cm. In general, as for the Sun, we have $R_G \ll R$ so that the solution (8.42) of the Einstein equations holds only for $r > R$, i.e. outside the star considered; this is an *exterior-vacuum-solution*, which must be *matched* to the interior-solution for the star.

(3) We give a short overview of the derivation of the metric (8.42). Using the metric form (7.66) the only nonzero Christoffel symbols[12] are

$$\Gamma^0_{01} = \frac{1}{2}v' \quad \Gamma^1_{00} = \tfrac{1}{2}v'e^{v-\lambda} \quad \Gamma^2_{33} = -\sin\theta\cos\theta$$

$$\Gamma^1_{11} = \frac{1}{2}\lambda' \quad \Gamma^1_{22} = -re^{-\lambda} \quad \Gamma^1_{33} = -e^{-\lambda}r^2\sin^2\theta \tag{8.58}$$

$$\Gamma^2_{12} = 1/r \quad \Gamma^3_{23} = \cot g\,\theta \quad \Gamma^3_{13} = 1/r$$

so that the only nonzero components of the Ricci tensor $R_{\mu\nu}$ are

$$R_{00} = -e^{(v-\lambda)}.\left[\frac{1}{2}v'' - \frac{1}{4}v'.\lambda' + \frac{1}{4}v'^2 + v'/r\right] \tag{8.59}$$

$$R_{11} = \frac{1}{2}v'' - \frac{1}{4}v'.\lambda' + \frac{1}{4}v'^2 - \lambda'/r \tag{8.60}$$

$$R_{22} = e^{-\lambda}\left[1 + \frac{1}{2}r\,(v' - \lambda')\right] - 1 \tag{8.61}$$

$$R_{33} = R_{22}\sin^2\theta, \tag{8.62}$$

and the Einstein equations are $R_{\mu\nu} = 0$.

___

[12] We also have $\Gamma^2_{12} = \Gamma^3_{13}$ and $\Gamma^\mu_{\alpha\beta} = \Gamma^\mu_{\beta\alpha}$.

The first two equations give $v' = -\lambda'$, which integrates to $v = \lambda + \text{const.}$; we can choose the constant to be zero by a suitable choice of the time coordinate t. Equation (8.61) then gives

$$e^v \cdot \left[1 + rv'\right] = 1. \tag{8.63}$$

Setting $b \equiv \exp v$, this becomes

$$b + rb' = 1, \tag{8.64}$$

and integrates to give

$$b(r) = 1 - \frac{\text{const.}}{r}, \tag{8.65}$$

where the constant of integration is identified by returning to the Newtonian approximation: we thus get const. $= 2GM/c^2$. We have finally to check that the metric (8.42) obeys the Einstein equations $R_{\mu\nu} = 0$, as we have so far only used the *sum* of (8.59) and (8.60). We note finally that we did not need to assume that the space–time was asymptotically flat: this appeared as a consequence of the solution. Moreover, G.D. Birkhoff (in 1916) showed that the Schwarzschild metric is the only spherically symmetric vacuum solution of the Einstein equations (Birkhoff's theorem). Put another way, the local geometry of any vacuum region of a spherically symmetric space–time is necessarily Schwarzschild. This theorem[13] justifies our assertion that the *exterior* geometry of a spherically symmetrical star is that of Schwarzschild.

(4) The *interior* geometry of a spherical star in hydrostatic equilibrium is characterised by the more general metric (7.66), which we can always write in the form

$$ds^2 = [1 + 2\phi(r)]\, dt^2 - \frac{dr^2}{\left(1 - 2\frac{GM(r)}{r}\right)} - r^2 \left[d\theta^2 + \sin^2\theta \cdot d\varphi^2\right], \tag{8.66}$$

where $\phi(r)$ and $M(r)$ are two unknown functions. They are simply a re-writing of the components $g_{00}(r)$ and $g_{rr}(r)$ and at this point do not represent anything in particular. In the Newtonian approximation $\phi(r)$ reduces to the usual Newtonian potential, hence the name *gravitational potential* given to $\phi(r)$. Inserting the form (8.66) into the Einstein equations gives a coupled system – the *Tolman–Oppenheimer–Volkov* equations – (for $M(r)$ and $\phi(r)$) related to the energy density ρ and pressure P of the stellar material. The function $M(r)$ then appears as the mass inside radius r:

$$M(r) = 4\pi \int_0^r dr'\, r'^2 \rho(r'). \tag{8.67}$$

We note again that (8.66) holds for $r \leq R$, where R is the star's radius and that at this point we must have

$$\phi(R) = -GM/R, \quad M \equiv M(R) \tag{8.68}$$

---

[13] The proof is given in many relativity texts and proceeds on lines very similar to the derivation of the Schwarzschild solution given above.

so as to match the metric (8.66)[14] to the Schwarzschild metric exterior to the star. At the origin $r = 0$ the mass plays no role (as in the Newtonian case) and the space–time must be locally Minkowksian, i.e. $\phi(0) = 0$ and $M(0) = 0$.

8.4   The local geometry of Friedman spaces

One of the main applications of relativistic gravitation is cosmology. This has long been "a free space for thinking about General Relativity" [J. Eisenstaedt (1989)], and in its simpler versions is based on the *Cosmological Principle*.

This principle first of all asserts the existence of a *universal time*: the space–time manifold V^4 may be decomposed into a temporal part and a spatial part V^3

$$V^4 = \mathbf{R} \times V^3. \tag{8.69}$$

In a suitably adapted coordinate system one can write the metric in the general form

$$ds^2 = dt^2 - d\eta^2, \tag{8.70}$$

where $d\eta^2$ is the metric of the spacelike manifold V^3. As we can see from the last expression, the coordinate t corresponds to the proper time measured by a "fundamental observer", whose worldline is by definition orthogonal to the spacelike manifolds V^3 (**Fig. 8.2**).

The cosmological principle[15] also asserts that the manifolds V^3 are *homogeneous and isotropic* ("we do not occupy a privileged place in the Universe" and "the Universe looks essentially the same in all directions"). The isotropy of the spatial sections V^3 requires that their metric $d\eta^2$ can be put in the (isotropic) form

$$d\eta^2 = B(t,r)d\mathbf{x}^2 \tag{8.71}$$

where the function $B(t,r)$ must have the form

$$B(t,r) = R^2(t)F^2(r), \tag{8.72}$$

with $R(t)$ and $F(r)$ two unknown functions.

We can give a heuristic proof that the function $B(t,r)$ must have the form (8.72). At time t_0 consider an arbitrary triangle ABC and let $A'B'C'$ be the analogous triangle at time t (resulting from the evolution of the original triangle; thus AA' etc. are worldlines of "fundamental observers"), see **Fig. 8.3**.

By homogeneity and isotropy the two triangles are similar: no point or direction can be picked out. Moreover, the coefficient of similarity must be independent of both the triangle chosen and its position in space. It can thus depend only on time. The g_{ij} must therefore have the same form at different times, up to a conformal time factor.

The function $F(r)$ is then determined by noting that the homogeneity and isotropy

---

[14] Matching two metrics is a delicate problem, and several methods are possible.
[15] A non-technical overview discussing the implications of this principle is given by E.R. Harrison (1981); see also G.F.R. Ellis (1973).

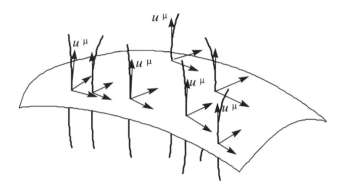

Fig. 8.2. Fundamental observers in a Friedman universe.

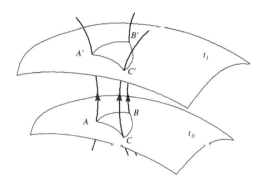

Fig. 8.3. Homogeneity and isotropy of the spatial sections implies the similarity of arbitrary pairs of triangles (ABC) and $(A'B'C')$; the coefficient of similarity depends only on time.

of the spatial sections V^3 implies that they are hypersurfaces of constant curvature, and we get

$$F^2(r) = \frac{1}{\left(1 + \frac{k}{4}r^2\right)^2} \quad \text{[with } k = \pm 1, 0\text{]}, \tag{8.73}$$

so that the cosmological (or Friedman–Robertson–Walker) metric can be written

$$ds^2 = dt^2 - R^2(t)\frac{d\mathbf{x}^2}{\left(1 + \frac{k}{4}r^2\right)^2}, \tag{8.74}$$

in isotropic coordinates, and

$$ds^2 = dt^2 - R^2(t)\left\{\frac{dr^2}{1 - kr^2} + r^2\left[d\theta^2 + \sin^2\theta\, d\varphi^2\right]\right\}, \tag{8.75}$$

in Schwarzschild coordinates.

We note that for $k = 0$, $R^{-2}(t)d\eta^2$ is the usual Euclidean metric; for $k = +1$, $R^{-2}(t)d\eta^2$ is the metric of a 3-sphere S^3, while for $k = -1$, $R^{-2}(t)d\eta^2$ is the metric of a hyperbolic space H^3 [see e.g. N. Efimov (1981) or D. Hilbert and S. Cohn-Vossen (1952)].

The scale factor[16] $R(t)$ is easily found by inserting the form (8.74) of the cosmological metric into the Einstein equations (or the equations of another theory of relativistic gravity). We note finally that the form of the Friedman–Robertson–Walker metric only depends on the assumptions of the cosmological principle (which is essentially an invariance principle) and not on the validity of the Einstein equations or those of any other theory. The latter fix only the scale factor $R(t)$.

If matter is described by a perfect-fluid energy–momentum tensor

$$T_{\mu\nu} = (\rho + P)u_\mu u_\nu - P g_{\mu\nu}, \tag{8.76}$$

with ρ and P the energy density and pressure of the "cosmic fluid", the conservation relation $\nabla_\mu T^{\mu\nu} = 0$ implies

$$P\, dR^3(t) + d[\rho R^3(t)] = 0, \tag{8.77}$$

which can be interpreted as the thermodynamic relation

$$T\, dS = 0 = P\, dV + dU, \tag{8.78}$$

where S is the entropy and U the internal energy of the cosmic fluid.

We now give a less intuitive proof of the forms (8.74) and (8.75).

The first assumption contained in the Cosmological Principle is the existence of a universal time. Although a similar hypothesis is implicit in the whole of Newtonian physics, this is a genuinely independent postulate – studied by H. Weyl from 1923 onwards – in General Relativity. In general there is no foliation of space–time by a family of spacelike hypersurfaces orthogonal to a congruence of timelike lines. Such lines must correspond to irrotational shear-free motions.

Now consider observers attached to galaxies without proper motions with respect to the cosmic fluid, and let u^μ be their 4-velocities, which are thus tangent to their worldlines (**Fig. 8.2**). These observers follow timelike goedesics, so that

$$\frac{d^2 x^\mu}{ds^2} + \Gamma^\mu_{\alpha\beta} u^\alpha u^\beta = 0. \tag{8.79}$$

In a comoving frame, i.e. that of one of these observers, we have

$$\frac{dx^\mu}{ds} = u^\mu = (1, \mathbf{0}), \tag{8.80}$$

and the geodesic equation (7.47) becomes

$$\frac{d^2 x^0}{ds^2} + \Gamma^0_{\alpha\beta} u^\alpha u^\beta = 0 \tag{8.81}$$

$$\frac{d^2 x^i}{ds^2} + \Gamma^i_{\alpha\beta} u^\alpha u^\beta = 0, \tag{8.82}$$

---

[16] Sometimes incorrectly called "the radius of the Universe".

which reduce to

$$\begin{cases} \Gamma^0_{00} = 0 = \frac{1}{2}g^{00}\left[\partial_0 g_{00} + \partial_0 g_{00} - \partial_0 g_{00}\right] \\ \Gamma^i_{00} = 0 = \frac{1}{2}g^{ij}\left[\partial_0 g_{0j} + \partial_0 g_{j0} - \partial_j g_{00}\right], \end{cases} \tag{8.83}$$

showing that the elements of the cosmic fluid are freely falling, and giving immediately

$$\begin{cases} \partial_0 g_{00} = 0 \\ \partial_i g_{00} = 0. \end{cases} \tag{8.84}$$

The most general rotationally invariant metric, written in isotropic coordinates is (see *Chapter* 7)

$$ds^2 = A(t,r)dt^2 - B(t,r)dx^2, \tag{8.85}$$

where $A(t,r)$ and $B(t,r)$ are arbitrary positive functions. Equations (8.84) show that g_{00} depends on neither t nor r, and consequently the metric (8.85) can be rewritten

$$ds^2 = dt^2 - B(t,r)dx^2, \tag{8.86}$$

where we have chosen the units of time t so that $g_{00} = 1$; t is then the proper time of the fundamental observers moving with u^μ (**Fig. 8.2**).

This type of coordinate system is called *comoving*. The surfaces $t = $ const. correspond to $dt = 0$; on these surfaces distances are measured by the family of time-dependent metrics

$$d\sigma^2 \equiv -ds^2 = B(t,r)dx^2. \tag{8.87}$$

These are surfaces of homogeneity, orthogonal to the worldlines of the cosmic fluid, whose existence we have not yet demonstrated.

To do this we decompose the 4-velocity of the cosmic fluid analogously to the Newtonian case [see S. Mavridès (1973)]. Setting [see e.g. J.L. Anderson (1967)]

$$\omega_{\mu\nu} = \left[\nabla_\rho u_\lambda - \nabla_\lambda u_\rho\right]\Delta^\lambda_{\ \nu}\Delta^\rho_{\ \mu} \tag{8.88}$$

$$\sigma_{\mu\nu} = \left[\nabla_\lambda u_\rho + \nabla_\lambda u_\rho - \frac{1}{3}\theta g_{\lambda\rho}\right]\Delta^\lambda_{\ \nu}\Delta^\rho_{\ \mu} \tag{8.89}$$

$$\theta = \nabla_\mu u^\mu, \tag{8.90}$$

where $\Delta_{\mu\nu}$ is the projection tensor orthogonal to u^μ,

$$\Delta_{\mu\nu} = g_{\mu\nu} - u_\mu u_\nu, \tag{8.91}$$

the variation of the 4-velocity u^μ under a variation δx^i of the spatial coordinates *in the* surface $t = $ const. is

$$\delta u^\mu = \delta x^\nu \left[\omega_\nu^{\ \mu} + \sigma_\nu^{\ \mu} + \frac{1}{3}\theta g_{\mu\nu}\right], \tag{8.92}$$

where the variations of u and x are confined to the surface $t = $ const. The various terms of this equation can be interpreted in exactly the same way as in the Newtonian case. For a non-rotating cosmic fluid we have

$$\omega_{\mu\nu} = 0 = \left[\partial_\rho u_\lambda - \partial_\lambda u_\rho\right] \Delta^\lambda{}_\nu \Delta^\rho{}_\mu, \tag{8.93}$$

which can be written

$$\nabla \wedge \mathbf{u} = \mathbf{0} \tag{8.94}$$

or

$$\partial_{[i} u_{j]} = 0.$$

In other words, the 3-vector $\mathbf{u}$ is irrotational and is therefore the gradient of a function $\phi(\mathbf{x}, t)$. This property also holds for u^μ, which is thus orthogonal to a family of spacelike hypersurfaces. The worldlines of this 4-vector field are obviously given by $x^i = \text{const.}$

The cosmological metric was obtained for the first time by A. Friedman (1922, 1924), but H.P. Robertson (1935, 1936) and A.G. Walker (1936) were the first to show that this metric results solely from the Cosmological Principle, and hence is completely independent of the Einstein equations. The mathematical proof can be found in, for example, the book by S. Weinberg (1972). Here we give a derivation which is essentially similar although formally different, consisting of successive simplifications of the general form (8.85) [which we have already reduced to (8.86)] by using the homogeneity and isotropy of the surfaces $t = \text{const.}$ when we need them, rather than systematically as in a mathematical proof [see L.P. Eisenhart (1966)].

We begin with the condition for shear-free motion of the cosmic fluid:

$$\sigma_{\mu\nu} = 0, \tag{8.95}$$

or

$$\left[\nabla_\lambda u_\rho + \nabla_\rho u_\lambda - \frac{1}{3}\theta g_{\lambda\rho}\right] \Delta^\lambda{}_\nu \Delta^\rho{}_\mu = 0. \tag{8.96}$$

In a comoving frame this equation becomes

$$\partial_0 g_{ij} = \frac{1}{3}\theta g_{ij}. \tag{8.97}$$

The homogeneity of the spatial sections implies that the expansion scalar θ can only depend on the time. Two timelike geodesics orthogonal to the surfaces $t = \text{const.}$, considered at two instants (see **Fig. 8.4**) must move apart by the same amount for the same initial distance, i.e. $\theta = \theta(t)$.

Equation (8.97) becomes

$$\frac{\partial}{\partial t} B(t, r) = -\theta(t) B(t, r), \tag{8.98}$$

which has no solution unless $B(t, r)$ has the general form

$$B(t, r) = R^2(t) F^2(r); \tag{8.99}$$

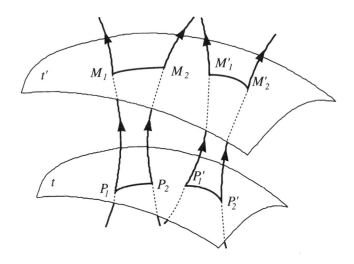

Fig. 8.4. If $d(P_1, P_2) = d(P_1', P_2')$ at time t, then $d(M_1, M_2) = d(M_1', M_2')$ at time t' whatever pairs of points (P_1, P_2) and (P_1', P_2') are considered.

which allows us to write (8.87) in the form

$$d\sigma^2 = R^2(t)F^2(r)dx^2. \tag{8.100}$$

We now have to find the function $F(r)$ explicitly. We use the homogeneity and isotropy of the spatial sections, which are necessarily 3-surfaces of *constant curvature*. Using the only spatial tensor available[17] i.e.

$$\gamma_{ij} = F^2(r)\delta_{ij}, \tag{8.101}$$

the only tensors with the symmetry properties of the curvature tensor which we can form are

$$^{(3)}R_{ijkl} = K \left[g_{ik}g_{jl} - g_{il}g_{jk} \right], \tag{8.102}$$

where K is an arbitrary constant. Now using the explicit expression for $^{(3)}R_{ijkl}$, the Riemann tensor for a 3-manifold

$$^{(3)}R_{ijkl} = \gamma_{in} \left[\partial_k \Gamma^n_{jl} - \partial_l \Gamma^n_{jk} + \Gamma^n_{mk}\Gamma^m_{jl} - \Gamma^n_{ml}\Gamma^m_{jk} \right], \tag{8.103}$$

we get two differential equations

$$-\frac{1}{r}\left[\frac{F'}{rF}\right]' + \left[\frac{F'}{rF}\right]^2 = 0 \tag{8.104}$$

$$2\left[\frac{F'}{F}\right] + \left[\frac{F'}{F}\right]^2 + KF^2 = 0. \tag{8.105}$$

---

[17] The isotropy of space requires that no privileged direction exists in a comoving frame.

The first equation has the solution

$$F(r) = \frac{a}{br^2 + c},$$ (8.106)

with a, b, c arbitrary constants, while the second simply imposes the constraint

$$K = 4\frac{bc}{a}.$$ (8.107)

Absorbing a factor c/a into the scale factor $R(t)$, setting

$$k = \frac{1}{4}\text{sgn}(b/c),$$ (8.108)

and redefining r through $(b/c)r^2 \rightarrow r^2$, we get

$$d\sigma^2 = R^2(t)\frac{d\mathbf{x}^2}{\left(1 + \frac{1}{4}kr^2\right)^2},$$ (8.109)

so that the space–time metric becomes

$$ds^2 = dt^2 - R^2(t)\frac{d\mathbf{x}^2}{\left(1 + \frac{1}{4}kr^2\right)^2}.$$ (8.110)

The different cases $k = 0, \pm 1$, correspond to spatial sections which are Euclidean ($k = 0$), have positive spatial curvature ($k = +1$) or negative curvature ($k = -1$). The topological index k is called the *curvature index*. However, the spatial geometry is *closed* if $k = +1$ but *can be either open or closed* if $k = 0$ or -1. The popular belief that the Universe is open if $k = 0$ or -1 is incorrect [see e.g. M. Lachièze-Rey, J.P. Luminet, (1994)].

8.5   Other metrics of astrophysical interest

The gravitational collapse of a star to a black hole of mass M, charge Q and angular momentum J results in the metric

$$ds^2 = \frac{\Delta}{\rho^2}\left[dt - \frac{J}{M}\sin^2\theta d\varphi\right]^2 - \frac{\sin^2\theta}{\rho^2}\left[\left(r^2 + \frac{J^2}{M^2}\right)d\varphi - \frac{J}{M}dt\right]^2$$
$$-\frac{\rho^2}{\Delta}dr^2 - \rho^2 d\theta^2,$$ (8.111)

where

$$\Delta \equiv r^2 - 2Mr + \frac{J^2}{M^2} + Q^2$$ (8.112a)

$$\rho^2 = r^2 + \frac{J^2}{M^2}\cos^2\theta.$$ (8.112b)

Equation (8.111) gives the *Kerr–Newman* metric. Its properties are clearly summarised by N. Straumann (1984) [see also C.W. Misner *et al.* (1973)]. It contains some special cases found earlier [$Q = J = 0$ (Schwarzschild); $Q = 0$ (Kerr); $J = 0$ (Reissner–Nordström)]. This metric plays an important role when discussing rotating black holes.

Another class of metrics, of cosmological interest this time, consists of models which are homogeneous but *anisotropic*, such as those of the form

$$ds^2 = dt^2 - X^2(t)dx^2 - Y^2(t)dy^2 - Z^2(t)dz^2, \tag{8.113}$$

with $X(t)$, $Y(t)$ and $Z(t)$ functions of time representing the expansion of the universe in the x, y and z directions. This form characterises *models of Bianchi type I*. An interesting special case, the *Kasner (in 1921) solution*, is given by

$$X(t) = X_0 t^{p_1}, \quad Y(t) = Y_0 t^{p_2}, \quad Z(t) = Z_0 t^{p_3}, \tag{8.114}$$

with

$$p_1 + p_2 + p_3 = p_1^2 + p_2^2 + p_3^2 = 1. \tag{8.115}$$

This solution represents an *anisotropically expanding universe* – a volume element dV grows in time as $\sqrt{|g|}d^3x = td^3x$. One or more dimensions contract while the others expand. Homogeneous and isotropic models are studied in detail in the book by M.P. Ryan and L.C. Shepley (1975); a less general study of these universes, but one devoting more space to the Kasner solution and its relation to the "mixmaster" model [see C.W. Misner (1969)] can be found in C.W. Misner, K.S. Thorne, J.A. Wheeler (1973).

8.6 The linearised Einstein equations

In many cases the space–time manifold V^4 and its metric $g^{(0)}_{\mu\nu}$ are known and we wish to consider "small" perturbations $h_{\mu\nu}$ of the metric, which may or may not be induced by "small" perturbations of the energy–momentum tensor:

$$\begin{cases} g_{\mu\nu} = g^{(0)}_{\mu\nu} + h_{\mu\nu} \\ |h_{\mu\nu}| \ll |g^{(0)}_{\mu\nu}|, |h_{\mu\nu}|^2 \ll |h_{\mu\nu}|. \end{cases} \tag{8.116}$$

The "small" perturbations $h_{\mu\nu}$ obey the *linearised* Einstein equations if we neglect terms[18] of $O\left(|h_{\mu\nu}|^2\right)$. The first problem is to know if this decomposition is valid *in every coordinate system*. Put another way, if we change coordinates, are the new $h_{\mu\nu}$ still "small"? In general they are not, and if we wish to maintain the inequalities $|h_{\mu\nu}| \ll |g^{(0)}_{\mu\nu}|$ and $|h_{\mu\nu}|^2 \ll |h_{\mu\nu}|$ we have to *restrict* the allowed coordinate changes. In this section we shall limit ourselves to the case where the metric $g^{(0)}_{\mu\nu}$ is that of Minkowski space, i.e. $\eta_{\mu\nu}$:

$$g_{\mu\nu} = \eta_{\mu\nu} + h_{\mu\nu}. \tag{8.117}$$

The second problem concerns the raising and lowering of indices. It is clear that to

---

[18] And their derivatives; for example, $\left|\partial_\mu h_{\alpha\beta}\right| \gg \left|\partial_\mu h_{\alpha\beta}\right|^2$; etc.

make a coherent approximation *we must use the metric* $\eta_{\mu\nu}$ *to raise and lower indices.* Hence quantities with indices are tensors with respect to Lorentz transformations and not with respect to general coordinate changes. Note, however, an important difference concerning the inverse $g^{\mu\nu}$ of $g_{\mu\nu}$. We have

$$g^{\mu\nu} \neq \eta^{\mu\alpha}\eta^{\nu\beta}g_{\alpha\beta} \tag{8.118}$$

and moreover

$$g^{\mu\nu} = \eta^{\mu\nu} - h^{\mu\nu} + O\left(h_{\alpha\beta}^2\right), \tag{8.119}$$

with

$$h^{\mu\nu} \equiv \eta^{\mu\alpha}\eta^{\nu\beta}h_{\alpha\beta}. \tag{8.120}$$

Under these conditions and with these conventions, the Christoffel symbols $\Gamma_{\alpha\beta}^{\mu}$, the Ricci tensor $R_{\mu\nu}$ and the Einstein tensor $G_{\mu\nu}$ become

$$\Gamma_{\alpha\beta}^{\mu} = \frac{1}{2}\eta^{\mu\lambda} \cdot \left[\partial_\alpha h_{\beta\lambda} + \partial_\beta h_{\alpha\lambda} - \partial_\lambda h_{\alpha\beta}\right] \tag{8.121}$$

$$R_{\mu\nu} = \frac{1}{2}\left[\partial_{\mu\nu}h - \partial_{\mu\alpha}h_\nu^{\;\alpha} - \partial_{\nu\alpha}h_\mu^{\;\alpha} + \Box h_{\mu\nu}\right] \tag{8.122}$$

$$G_{\mu\nu} = \frac{1}{2}\left[\partial_{\mu\nu}h - \partial_{\mu\alpha}h_\nu^{\;\alpha} - \partial_{\nu\alpha}h_\mu^{\;\alpha} + \Box h_{\mu\nu}\right] - \eta_{\mu\nu}\left[\Box h - \partial_{\alpha\beta}h^{\alpha\beta}\right]. \tag{8.123}$$

The latter equation can be simplified by setting

$$\bar{h}_{\mu\nu} \equiv h_{\mu\nu} - \frac{1}{2}\eta_{\mu\nu}h, \tag{8.124}$$

where h is the trace of $h_{\mu\nu}$,

$$h \equiv h^\mu_{\;\mu} = \eta^{\alpha\beta}h_{\alpha\beta}; \tag{8.125}$$

and we get

$$G_{\mu\nu} = \frac{1}{2}\left\{\Box\bar{h}_{\mu\nu} + \eta_{\mu\nu}\partial_{\alpha\beta}\bar{h}^{\alpha\beta} - \partial_{\alpha(\mu}\bar{h}_{\nu)}^\alpha\right\}. \tag{8.126}$$

Hence the linearised Einstein equations become

$$\Box\bar{h}_{\mu\nu} + \eta_{\mu\nu}\partial_{\alpha\beta}\bar{h}^{\alpha\beta} - \partial_{\alpha(\mu}\bar{h}_{\nu)}^\alpha = 16\pi G T_{\mu\nu}. \tag{8.127}$$

Note that for this equation to be self-consistent, the energy–momentum tensor $T_{\mu\nu}$ must also be of order $h_{\mu\nu}$. It follows that the conservation law $\nabla_\mu T^{\mu\nu} = 0$ reduces to the Lorentzian relation $\partial_\mu T^{\mu\nu} = 0$. As $G_{\mu\nu}$ has zero divergence, the left hand side of (8.126) is also divergenceless (linearised Bianchi identities), which implies $\partial_\mu T^{\mu\nu} = 0$. These four conditions restrict the possible coordinate changes, as we now discuss.

Consider a coordinate change of the general type

$$x^\mu \to x^\mu + \varepsilon^\mu(x) = x'^\mu \tag{8.128}$$

where $\varepsilon^\mu(x)$ is an arbitrary function subject to the restriction that $h'_{\mu\nu}$ is a "small" quantity, i.e. the relations (8.116) still hold. The transformation (8.128) for a 4-vector A^μ is

$$A^{\mu'} = \frac{\partial x^{\mu'}}{\partial x^\mu} A^\mu$$
$$= \left[\eta^{\mu'}{}_\mu + \partial_\mu \varepsilon^{\mu'}(x) \right] A^\mu; \tag{8.129}$$

and similarly the perturbation $h_{\mu\nu}$ of the metric $\eta_{\mu\nu}$ in the new coordinate system is

$$h'_{\mu\nu} = h_{\mu\nu} - \partial_{(\mu}\varepsilon_{\nu)}, \tag{8.130}$$

so that if $h'_{\mu\nu}$ and $h_{\mu\nu}$ are to be small, this must also hold for the derivatives $\partial_\mu \varepsilon_\nu(x)$: *not all* coordinate tranformations conserve the "smallness" of $h_{\mu\nu}$. The transformations (8.128) are called *gauge transformations* as they are analogous to similar transformations in electromagnetism[19].

Under a gauge transformation $\bar{h}_{\mu\nu}$ transforms as

$$\bar{h}'_{\mu\nu} = \bar{h}_{\mu\nu} - \partial_{(\mu}\varepsilon_{\nu)} + \eta_{\mu\nu}\partial_\alpha \varepsilon^\alpha. \tag{8.131}$$

Here we should recall that *all* the indices are Lorentzian, and as a result the Lorentz invariance of the calculation is manifest throughout.

Equation (8.127) is analoguous to the Maxwell equation (5.24) for the electromagnetic potential; moreover it is invariant under gauge (coordinate) transformations (8.128). The condition[20]

$$\partial_\alpha \bar{h}^{\alpha\beta} = 0, \tag{8.132}$$

called the *"Lorentz gauge"*, restricts the allowed coordinate changes. Adopting this gauge immediately gives

$$\partial_\lambda \bar{h}'^{\mu\lambda} = \partial_\lambda \bar{h}^{\mu\lambda} - \Box \varepsilon^\mu = 0, \tag{8.133}$$

leading to $\Box \varepsilon^\mu = 0$. This condition is characteristic of *harmonic coordinates*, i.e. coordinates preserving the Lorentz condition (8.132).

With this gauge condition, the linearised Einstein equations become

$$\Box \bar{h}_{\mu\nu} = 16\pi G T_{\mu\nu}, \tag{8.134}$$

with the general solution

$$\bar{h}_{\mu\nu}(t, x) = 4G \int d^3x' \frac{T_{\mu\nu}\left(x^0 - |\mathbf{x} - \mathbf{x}'|, \mathbf{x}'\right)}{|\mathbf{x} - \mathbf{x}'|}, \tag{8.135}$$

which is very similar to the *retarded solution* for the electromagnetic 4-potential[21] in tems of the 4-current. Analogously, the relation (8.135) shows explicitly the propagation of the action of gravity from a point $\mathbf{x}'$ of the source to a point $\mathbf{x}$, which takes place at the velocity of light $c = 1$.

---

[19] See e.g. S. Weinberg (1972) p. 257; C.W. Misner, K.S. Thorne, J.A. Wheeler (1973), p. 440.
[20] This condition is sometimes called the "De Donder gauge" or "harmonic gauge".
[21] See e.g. A.O. Barut (1964) or C. Itzykson, J.P. Zuber (1980) etc.

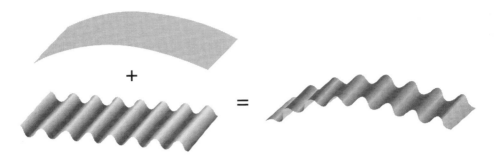

Fig. 8.5. A gravitational wave can be regarded as a small perturbation propagating on the space–time manifold.

The linearised Einstein equations are important in many ways. First, they allow simple calculations and direct interpretations of weak gravitational perturbations to the space–time of special relativity, such as in the solar system. Also, Eq. (8.134) with $T_{\mu\nu} \equiv 0$ shows the existence of *gravitational radiation*, which can propagate even without a source:

$$\Box \bar{h}_{\mu\nu} = 0. \tag{8.136}$$

In other words, even in a vacuum the gravitational field could no longer vanish; Einstein's equation $G_{\mu\nu} = 0$ has solutions differing from $g_{\mu\nu} = \eta_{\mu\nu}$.

8.7   Gravitational radiation

The linearised Einstein theory presented above shows us that small gravitational perturbations propagate at the velocity of light in Minkowski space. One can also show that small perturbations propagate in any curved space [see e.g. A. Papapetrou (1974) or C. W. Misner, K.S. Thorne, J.A. Wheeler (1973)] as illustrated in **Fig. 8.5**.

Here we shall limit ourselves to small perturbations from Minkowski space, which provides enough insight[22].

The importance of gravitational radiation in astrophysics comes from the fact that every physical phenomenon causes its emission in varying degrees. There is no gravitationally neutral matter: the gravitational collapse of a star or the motion of a planet about a star both produce gravitational waves. The binary pulsar PSR 1913+16 has already provided proof of the existence of these waves: two neutron stars (one a pulsar) are spiralling towards each other at a rate corresponding exactly to the loss of energy in gravitational waves as calculated in general relativity. Although these waves have not yet been observed in the laboratory – their amplitude is extremely small –

---

[22] A first, very simple, approach can be found in P.C.W. Davies (1980); a more advanced treatment is given by K. Thorne (1983), while a general review with a very complete bibliography can be found in K. Thorne (1989); for discussion of other problems connected with gravitational radiation see N. Deruelle & T. Piran (1983) and B. Carter & J.B. Hartle (1987).

experiments have been running in most parts of the world for more than a quarter of a century [J. Weber (1960) (1961) (1969)], and it seems likely that sensitivities will improve to the point of detection in the comparatively near future. This will open the way to a genuine gravitational wave astronomy, which will allow us to observe the first instants of the Universe and give access to many other phenomena which are currently hidden from us.

(1) Consider the relation (8.136) and the gauge condition (8.132)

$$\begin{cases} \Box \bar{h}_{\mu\nu} = 0 \\ \partial_\mu \bar{h}^{\mu\nu} = 0, \end{cases} \tag{8.137}$$

characterising the propagation of free gravitational waves in Minkowski space. Plane waves, of the form

$$\bar{h}_{\mu\nu} = A_{\mu\nu} e^{-ik.x} + A_{\mu\nu}^* e^{ik.x}, \tag{8.138}$$

are solutions of the system (8.137), given an amplitudes $A_{\mu\nu}$ (which is a symmetric Lorentz tensor) such that

$$k^2 A_{\mu\nu} = 0, \tag{8.139a}$$

$$k_\mu A^{\mu\nu} = 0. \tag{8.139b}$$

As $A_{\mu\nu}$ is symmetric it has ten *a priori* independent components. However, the four gauge conditions give four additional constraints, reducing the number of independent components to six. Gauge invariance

$$x^\mu \rightarrow x'^\mu = x^\mu + \varepsilon^\mu(x), \tag{8.140}$$

(four conditions) finally reduces these to two. We show this briefly. We note first that if we wish to preserve the plane-wave character of $\bar{h}_{\mu\nu}$, the functions $\varepsilon^\mu(x)$ must reduce to

$$\varepsilon^\mu(x) = i\xi^\mu e^{-ik\cdot x} - i\xi^{\mu*} e^{ik\cdot x}, \tag{8.141}$$

using the relation $\Box \varepsilon^\mu(x) = 0$ [cf. Eq. (8.133)], where the ξ^ν are infinitesimal functions of k. The gauge transformation (8.140) then gives [cf. Eq. (8.131)]

$$A_{\mu\nu} \rightarrow A'_{\mu\nu} = A_{\mu\nu} + k_{(\mu} \xi_{\nu)} - \eta_{\mu\nu} k_\alpha \xi^\alpha, \tag{8.142}$$

but since $A'_{\mu\nu}$ still obeys the gauge condition, we have also

$$0 = k_\mu A'^{\mu\nu} = k^2 \xi^\nu = 0, \tag{8.143}$$

which is little more than an expression of $\Box \varepsilon^\mu(x) = 0$. We now use these properties to show explicitly how the number of independent components of $A_{\mu\nu}$ can be reduced to two for a plane wave propagating in the $0z$ direction, i.e. with wave vector (where $k \equiv |\mathbf{k}|$)

$$k^\mu = (k, 0, 0, k) \quad \text{or} \quad k_\mu = (k, 0, 0, -k). \tag{8.144}$$

This clearly does not affect the generality of the proof as it is always possible to find a coordinate system such that (8.144) holds. Under these conditions, and using (8.139b), we have

$$A_{\mu 0} = A_{\mu 3} \qquad (\mu = 0, 1, 2, 3). \tag{8.145}$$

The transformation (8.142) then gives

$$\left\{ \begin{array}{l} A'^{00} = A^{00} + k(\xi^0 + \xi^3) \\ A'^{01} = A^{01} + k\xi^1 \\ A'^{02} = A^{02} + k\xi^2 \end{array} \right. \qquad \left\{ \begin{array}{l} A'^{11} = A^{11} + k(\xi^0 - \xi^3) \\ A'^{12} = A^{12} \\ A'^{22} = A^{22} + k(\xi^0 - \xi^3). \end{array} \right. \tag{8.146}$$

This linear system has six equations for four unknowns $(\xi^0, \xi^1, \xi^2, \xi^3)$ which we wish to determine so as to simplify the form of $A_{\mu\nu}$ (note that $A_{\mu\nu}$ and $A'_{\mu\nu}$ represent *the same* physical situation). This system shows that the component A_{12} is unaffected by a gauge transformation, so we can regard it as one of the independent components of $A_{\mu\nu}$; this shows moreover that the combination $A^{11} + A^{22}$ also has this property:

$$A'^{11} + A'^{22} = A^{11} + A^{22}; \tag{8.147}$$

so that ξ^1 and ξ^2 are uniquely determined:

$$k\xi^1 = A'^{01} - A^{01}, \qquad k\xi^2 = A'^{02} - A^{02}. \tag{8.148}$$

Finally, we can always *choose* the 4-vector ξ^μ so that

$$A'^{00} = A'^{01} = A'^{02} = 0 = A'^{11} + A'^{22}. \tag{8.149}$$

In this gauge $A_{\mu\nu}$ thus becomes

$$\|A_{\mu\nu}\| = \begin{vmatrix} 0 & 0 & 0 & 0 \\ 0 & A^{11} & A^{12} & 0 \\ 0 & A^{12} & -A^{22} & 0 \\ 0 & 0 & 0 & 0 \end{vmatrix} \tag{8.150}$$

while $\bar{h}_{\mu\nu}$ takes the form

$$h_{\mu\nu} = \operatorname{Re}\left\{ [\alpha e_{\mu\nu}(+) + \beta e_{\mu\nu}(\times)] e^{ik.x} \right\}, \tag{8.151}$$

where α and β are two arbitrary complex functions of k and $e_{\mu\nu}(+)$ and $e_{\mu\nu}(\times)$ are the two polarization matrices

$$\|e_{\mu\nu}(+)\| = \begin{vmatrix} 0 & 0 & 0 & 0 \\ 0 & 1 & 0 & 0 \\ 0 & 0 & -1 & 0 \\ 0 & 0 & 0 & 0 \end{vmatrix} \tag{8.152}$$

$$\|e_{\mu\nu}(\times)\| = \begin{vmatrix} 0 & 0 & 0 & 0 \\ 0 & 0 & 1 & 0 \\ 0 & 1 & 0 & 0 \\ 0 & 0 & 0 & 0 \end{vmatrix}. \tag{8.153}$$

This gauge is called the TT gauge, TT standing for "Transverse-Traceless".

(2) The two polarization modes (8.152) and (8.153) of a gravitational wave must now be interpreted physically by examining the effects of such a wave on test particles. Such effects can only be detected by measuring them *with respect to another test particle*. We thus write down the equation of geodesic deviation in the TT gauge for the distance between two such particles.

A simpler approach [J. Foster, J.D. Nightingale (1979)] is to consider a test particle at the origin and another at $\{\eta^\mu = (0, \eta)\}$, the two particles being at rest before the wave passes. They are therefore separated by

$$\ell = |g_{\mu\nu}\eta^\mu\eta^\nu|^{1/2}$$
$$= \left[(\delta_{ij} - h_{ij})\eta^i\eta^j\right]^{1/2}, \tag{8.154}$$

which, with the coordinate change

$$\eta^i \rightarrow \kappa^i = \eta^i + \frac{1}{2}h^i_j\eta^j, \tag{8.155}$$

can be rewritten in the form

$$\ell = [\delta_{ij}\kappa^i\kappa^j]^{1/2} + 0(\|h_{ij}\|^2), \tag{8.156}$$

showing that κ can be regarded as a position vector in the usual space.

If the vector κ is parallel to the z axis we have $\kappa^3 = \eta^3$, since $h^3_j = 0$ [see Eq. (8.150)]: if the separation of the two test particles is parallel to the propagation direction of the wave their distance is unaffected by its passage; *a gravitational wave is purely transverse.*

If the two particles lie in a plane orthogonal to the wave direction, i.e. κ^i given by

$$\|\kappa^i\| = \|\eta^i\| - \frac{1}{2}\alpha\cos[k(t-z)].\|\begin{vmatrix} 1 & 0 \\ 0 & -1 \end{vmatrix}\eta^i\| \tag{8.157}$$

for the $+$ polarisation mode (α is taken as real and positive for simplicity) and by

$$\|\kappa^i\| = \|\eta^i\| - \frac{1}{2}\alpha\cos[k(t-z)].\|\begin{vmatrix} 0 & 1 \\ 1 & 0 \end{vmatrix}\eta^i\| \tag{8.158}$$

for the $\times$ mode, we see that the motion of the two particles is given by **Fig. 8.6.**

The 4-vector η^μ joining the two test particles obeys the equation of geodesic deviation, which when linearised implies

$$\frac{d^2\eta^\mu}{d\tau^2} = R^\mu{}_{00j}\eta^i_o, \tag{8.159}$$

where we have chosen the reference frame in which one particle is at rest at the origin (i.e. $u^\mu = (1, \mathbf{0})$) so that $\Gamma^\mu_{00} = 0$. Equation (8.159) then reduces (up to terms of second order) to

$$\frac{d^2\eta^i}{d\tau^2} \sim R^i{}_{00j}\eta^j_o, \tag{8.160}$$

which can be rewritten as

$$\frac{d^2\eta^i}{dt^2} = -\frac{1}{2}\frac{\partial^2 h^i_j}{\partial t^2}\eta^j_o, \tag{8.161}$$

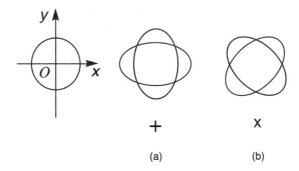

Fig. 8.6. The two oscillation modes of a gravitational wave are illustrated by the variations of the distances from the origin (where a test particle is situated) of a ring of test particles (mode $+$: (a); mode $\times$: (b)).

where we have used the linearised Riemann tensor expressed in the TT gauge. Considering a wave propagating in the positive z-direction, i.e. with

$$\begin{cases} h_{11} = -h_{22} = \Phi_+(t-z) \\ h_{12} = h_{21} = \Phi_\times(t-z), \end{cases} \tag{8.162}$$

we note immediately that there is no oscillation parallel to the z-direction: $h_j^3 = 0$, and we thus have $\ddot{\eta}^3 = 0$. Gravitational waves are *transverse waves*. Moreover equation (8.161) has the approximate solutions

$$\begin{pmatrix} \eta^1 \\ \eta^2 \end{pmatrix} \simeq \begin{pmatrix} \eta_0^1 \\ \eta_0^2 \end{pmatrix} - \frac{1}{2}\Phi_+(t-z)\begin{pmatrix} 1 & 0 \\ 0 & -1 \end{pmatrix}\begin{pmatrix} \eta_0^1 \\ \eta_0^2 \end{pmatrix} \tag{8.163}$$

$$\begin{pmatrix} \eta^1 \\ \eta^2 \end{pmatrix} \simeq \begin{pmatrix} \eta_0^1 \\ \eta_0^2 \end{pmatrix} - \frac{1}{2}\Phi_\times(t-z)\begin{pmatrix} 0 & 1 \\ 1 & 0 \end{pmatrix}\begin{pmatrix} \eta_0^1 \\ \eta_0^2 \end{pmatrix} \tag{8.164}$$

found by approximate integration of equation (8.161), assuming that the test particles are at rest with separation η_0^i before the passage of the wave. These two relations, specialised to the case of plane periodic waves, show the effect of the two polarisation modes on test particles (see **Fig. 8.6**).

(3) The preceding discussion suggests the possibility of detecting gravitational waves: in principle all we need to do is to measure the relative displacement of two test particles. This can be done in many ways; we mention first the gravitational wave detector developed by J. Weber (1961), (1969), (1980), which is a *resonant detector* (see **Fig. 8.7**). In practice, it is clear that one does not use a system such as that in **Fig. 8.7**, but other oscillating systems. Weber's antenna consisted of a heavy metal cylinder whose oscillation modes could be excited by the passage of gravitational waves.

Let us consider the simple antenna shown in **Fig. 8.7**. We note first that if a wave propagates parallel to the axis the antenna does not react, as the waves are transverse. The antenna is therefore *directional*, with the largest reaction obtained when the wave

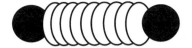

Fig. 8.7. Resonant detector. Two test particles linked by a spring form a gravitational wave detector.

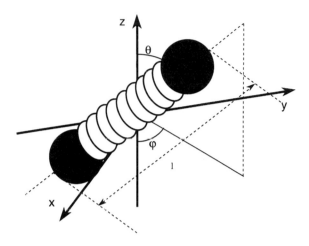

Fig. 8.8. Geometry of the resonant detector of Fig. 8.7. (The wave propagates along the z axis.)

propagates perpendicular to the separation of the two masses. We shall restrict ourselves to this case.

With the notation used above, the geodesic deviation equation for the separation of the two masses is

$$\frac{d^2}{dt^2}\delta\eta^i + \gamma\frac{d}{dt}\delta\eta^i + \omega_0^2\delta\eta^i \sim -\frac{1}{2}\frac{\partial^2}{\partial t^2}h^i_j \cdot \eta_0^j + O(h^2) \qquad (8.165)$$

where ω_0 is the resonant frequency of the antenna and γ is the damping coefficient of the system. Here $\delta\eta^i$ is the variation of the distance between the two masses after the passage of the wave, while η_0^i is this distance before the wave passes. Since $\eta^i = \eta_0^i + \delta\eta^i$ and $\delta\eta^i$ is first order in h, we have dropped second-order terms in the right hand side of (8.165). This equation is written in a local coordinate system for one of the masses (in the TT gauge); the interested reader should consult C. Misner, K. Thorne and J.A. Wheeler (1973) for a fuller discussion.

As an example, consider a plane wave with polarisation $+$ and frequency ω propagating along the z axis (with the detector oriented as in **Fig. 8.8**)

$$\begin{cases} h_{11} = -h_{22} = A_+e^{-i\omega(t-z)} \\ h_{12} = \quad h_{21} = 0. \end{cases} \qquad (8.166)$$

Inserting (8.166) into the harmonic oscillator equation (8.165) for the detector, and

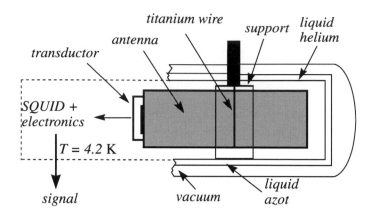

Fig. 8.9. Schematic diagram of the Weber detector. A heavy aluminium cylinder is suspended by a cable and its oscillations in response to gravitational waves are measured using a transducer. Later versions (Rome) are placed in a low temperature enclosure [after E. Amaldi *et al.* (1986)].

considering only the separation $\delta\eta$ of the two masses, we get

$$\frac{d^2}{dt^2}\delta\eta + \gamma\frac{d}{dt}\delta\eta + \omega_0^2\delta\eta = -\frac{1}{2}\omega^2\ell A_+ e^{-iwt}\sin^2\theta\cos 2\varphi, \qquad (8.167)$$

and the amplitude of the stationary solution is

$$\delta\eta(\omega) = \frac{\frac{1}{2}\omega^2\ell A_+\sin^2\theta.\cos 2\varphi}{\omega^2 - \omega_0^2 + i\gamma\omega}. \qquad (8.168)$$

We see that, as with any oscillator, the bandwith of the gravitational antenna controls what signal is detected; either one sharply peaked about the resonant frequency ω_0 (with a high quality factor $Q \equiv \omega_0/2\gamma$) or a much wider spectrum, with reduced amplitude. In practice the choice of quality factor is determined by the kind of signal (random, peaked, etc.) that one hopes to detect [see e.g. C. Misner, K. Thorne, J.A. Wheeler (1973) or K. Thorne (1980) (1983) (1989)].

The first detector constructed on these lines was that of J. Weber (1960), a "bar" – a cylinder – of aluminium weighing about 1.5 tons, whose modes of excitation responded to the passage of a wave (**Fig. 8.9**). The deformation was measured using a transducer[23]. In its fundamental mode the cylinder could be regarded as a harmonic oscillator.

Such measurements are extremely difficult to make because of the very low energy of a gravitational wave: a very intense wave, such as given by a supernova explosion in the Galactic centre, would give a typical wave energy of order 10^{-25} J; the wave amplitude is of order 10^{-18} and the displacement $\delta\eta$ several orders of magnitude smaller than the size of a proton! The extremely small amplitude length variations of the Weber bar demand very careful measurement of the transducer signals, and complex analyses

---

[23] i.e. a device tranforming mechanical deformations into electric signals. These may be piezoelectric ceramics, capacitors, etc.

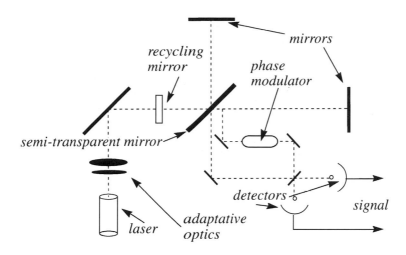

Fig. 8.10. The Michelson interferometer used as a gravitational wave detector. To get a large signal requires very long arms. This is achieved by multiple reflections. [After J.Y. Vinet *et al.* (1990).]

for the extraction of the signal from the accompanying noise (thermal noise in the cylinder[24], noise from the electronics, etc.). To reduce these effects[25] the antenna must be cooled [see e.g. E. Amaldi *et al.* (1986)] as much as possible (in practice to the temperature of liquid helium, i.e. ~ 4 K) and/or one can use a large monocrystal.

Whatever improvements are made, one soon reaches the "quantum limit": the energies involved in the passage of a wave are so weak that the energy change of the bar is of order a few phonons. If E_{grav} is the energy given to the bar by the wave, we must have $E_{\text{grav}} > \omega_0$. This energy is of order

$$E_{\text{grav}} \sim \frac{1}{2} M v^2, \tag{8.169}$$

where M is the antenna mass and v is the velocity induced by the wave, i.e.

$$v \sim \omega h \ell. \tag{8.170}$$

so we get finally

$$h_{\text{min}} > (\ell^2 \omega_0 M)^{-1/2}. \tag{8.171}$$

An ideal antenna has $h_{\text{min}} > 10^{-21}$. It may be possible to surpass the quantum limit, as the uncertainty principle allows us to measure a quantum variable with arbitrary precision provided that the conjugate variable is essentially unconstrained.

Another type of detector currently being studied consists of a large interferometer

---

[24] At the ambient temperature the motion of the bar resulting from thermal noise is of order $x_{\text{noise}} \sim (2 k_B T / M \omega_0^2)^{1/2} \sim 10^{-14}$ cm, and the corresponding energy is of order 10^{-23} J ...
[25] See e.g. the articles by V.B. Braginsky, J. Weber, W.M. Fairbank, W.M. Fairbank *et al.* in B. Bertotti (1974); see also F. Everitt, W.W. Hansen (1975).

(Michelson or Fabry–Perot). A Michelson interferometer (see **Fig. 8.10**) involves three mirrors of which two are linked: two of the mirrors are fixed to large masses, freely suspended at the ends of the two arms of the interferometer (see *Chapter 1*). A semi-transparent mirror, standard in this type of apparatus, is designed to reflect (and transmit) a coherent light beam emitted by a laser. The path length difference of the two beams thus gives rise to detectable interference fringes when they are recombined. When a gravitational wave crosses the detector, one of the arms will contract and the other expand, changing the optical path length of the beams and thus displacing the fringes. However, given the very small mirror displacements, one needs very long arms. For example, to detect a wave with a frequency of 1 kHz requires arms of order 100 km. This is achieved by using multiple reflections from secondary mirrors which lead to distances of order kilometers.

(4) We now ask how gravitational waves are emitted, and with what amplitude. Consider matter characterised by an energy–momentum tensor $T_{\mu\nu}$, sufficiently compact that at infinity space–time can be taken as Minkowskian, and diffuse enough that the linerised Einstein equations hold. Then in the Lorentz gauge (8.134) becomes

$$\Box \bar{h}_{\mu\nu} = 16\pi G T_{\mu\nu}, \tag{8.134}$$

and its *retarded* solution (containing only outgoing waves)

$$\bar{h}_{\mu\nu}(t, \mathbf{x}) = 4G \int d^3 x' \frac{T_{\mu\nu}(t - |\mathbf{x} - \mathbf{x}'|, \mathbf{x}')}{|\mathbf{x} - \mathbf{x}'|} \tag{8.135}$$

is valid. Strictly speaking we should add to $T_{\mu\nu}$ the contribution of the gravitational waves themselves, but we may neglect this term.

Denoting by $r \equiv |\mathbf{x}|$ the distance from the origin, and assuming that matter is concentrated there, the solution (8.135) can be approximated by

$$\bar{h}_{\mu\nu}(t, r) \sim \frac{4G}{r} \int d^3 x' T_{\mu\nu}(t - r, \mathbf{x}'), \tag{8.172}$$

far from the source, i.e. in the *wave zone*. We note first that $\bar{h}_{0\mu}$ is constant in time and does not contribute to the gravitational waves: from conservation of the energy–momentum tensor we have (*Chapter 3*)

$$\int d\Sigma_\mu T^{\mu\nu} = \text{const.}$$

$$= \int d_3 x' T^{0\mu}.$$

We may thus limit ourselves to the *spatial* part of $\bar{h}_{\mu\nu}$, i.e. $\bar{h}_{ij}$. Now using the conservation relation $\partial_\mu T^{\mu\nu} = 0$ in the explicit form

$$\partial_0 T^{00} + \partial_k T^{0k} = 0 \tag{8.173a}$$

$$\partial_0 T^{i0} + \partial_k T^{ik} = 0, \tag{8.173b}$$

some algebra gives

$$\bar{h}_{ij}(t,r) = \frac{2G}{r}\frac{d^2}{dt^2}\int d^3x \rho(t-r,\mathbf{x})x_i x_j, \tag{8.174}$$

where ρ is the mass density of the source ($\rho \equiv T_{00}$) and we have assumed only slow motions of the source. This assumption amounts to (i) neglecting the energy of internal motions, and (ii) considering only the contribution of the mass density. Thus ρ is the proper mass density. The relation (8.174) was first given in 1918 by Einstein.

Relation (8.174) can be derived as follows [J. Foster, J.D. Nightingale (1989)]. From the identity

$$\int d^3x \partial_k [T^{ik}x^j] = \int d^3x \partial_k T^{ik}x^j + \int d^3x T^{ij}, \tag{8.175}$$

whose left hand side vanishes (Gauss's theorem and the spatial confinement of the source), and using the conservation relation (8.173b), we get

$$\int d^3x T^{ij} = \frac{1}{2}\frac{d}{dt}\int d^3x(T^{io}x^j + T^{jo}x^i). \tag{8.176}$$

Gauss's theorem also gives

$$\int d^3x \partial_k[T^{ok}x^i x^j] = 0, \tag{8.177}$$

so that the right hand side of (8.176) is

$$\frac{d}{dt}\int d^3x(T^{oi}x^j + T^{oj}x^i) = \frac{d^2}{dt^2}\int d^3x T^{00}x^i x^j, \tag{8.178}$$

where we have again used the conservation relation (8.173a). This finally gives (8.174).

The expression (8.174) for $\bar{h}_{ij}$ brings in the second time derivative, which appears as the second order moment, i.e. the *quadrupole moment*, of the mass distribution of the source. Let us study this a little more closely by comparing it with the usual case of electromagnetic waves, which bring in the second derivative of the *dipole* moment of the charge distribution of the source. We note first that we can pass from the Coulomb law to the law of universal attraction by the substitution $e^2 \to -m^2$, so the radiated electromagnetic power [see J.D. Jackson (1962)]

$$\mathcal{E} = \frac{2}{3}e^2\gamma^2 = \frac{2}{3}\left(\frac{d^2}{dt^2}\mathbf{d}\right)^2, \tag{8.179}$$

where γ is the acceleration and $\mathbf{d}$ the dipole moment,

$$\mathbf{d} = \sum_\ell e\mathbf{x}_\ell, \tag{8.180}$$

suggests that the power radiated by a gravitational source will be very similar. However, this is not in fact correct, as the mass dipole moment

$$\mathbf{d} = \sum_\ell m_\ell \mathbf{x}_\ell, \tag{8.181}$$

satisfies $\ddot{\mathbf{d}} = 0$! In fact

$$\frac{d}{dt}\mathbf{d} = \sum_\ell m_\ell \frac{d}{dt}\mathbf{x}_\ell = \text{total momentum} = \text{const.},$$

so that the derivative vanishes. We can easily show that the conservation of angular momentum implies that the gravitational analogue of magnetic dipole radiation also vanishes [see C. Misner, K. Thorne, J.A. Wheeler (1973)]. Gravitational radiation is thus at least *quadrupolar*: a spherically symmetric source, even one which varies in time, does not emit any gravitational radiation.

(5) From the reasoning above it follows that almost all phenomena produce gravitational waves! However, we can only expect significant production from astrophysical objects: supernova explosions, binary systems, collisions of stellar objects, etc. Estimates of gravitational wave production in a given physical situation rely purely on theoretical models of astrophysical objects. For example, it is assumed that a neutron star forms by implosion[26] of the core of a supernova which elsewhere explodes; this requires specific models which bring in many physical effects whose description is not always clear (equation of state of dense matter, etc.), and which so far have not led to the prediction of neutron star formation. All this is discussed in detail[27] by K. Thorne (1980).

Figure 8.11, taken from K. Thorne (1989), gives an idea of the expected amplitudes, using currently accepted astrophysical models, for various gravitational wave frequencies. Among the phenomena considered are (i) explosions (supernovae, neutron star glitches, gravitational collapse[27] of a star to a black hole, coalescence of a binary system, *etc.*), (ii) periodic systems (binary systems, rotation of distorted neutron stars or white dwarfs etc.), (iii) a random background of primordial gravitational waves (i.e. created in the Big Bang, or those formed in the gravitational collapse of possible population III stars[28]. More detailed diagrams for each of these phenomena can be found in K. Thorne (1987) and references therein.

We can find an order of magnitude estimate of the amplitude of a gravitational wave of frequency ω by using (8.174). We have

$$h \sim \frac{2G}{r}\omega^2 MR^2\varepsilon, \tag{8.182}$$

where the second time derivative gives the factor ω^2, for a plane wave of frequency ω, where R is the size of the source and M its mass. The factor ε describes the *asymmetry* of the source. Now $\omega R \sim v$ is a characteristic velocity for the internal motions of the source, and

$$v^2 \sim 2\frac{GM}{R}, \tag{8.183}$$

---

[26] The core collapse must be asymmetric if it is to produce gravitational waves.
[27] A simple review is given by M. Tinto (1988).
[28] Assumed to be formed before galaxies.

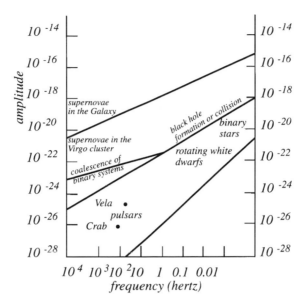

Fig. 8.11. Estimates of the amplitudes of gravitational waves as a function of frequency for various astrophysical processes [from K. Thorne (1989)].

so we get the crude estimate [M. Tinto (1988)]

$$h \sim \varepsilon \left(\frac{R_{\text{Sch.}}}{R} \right) \cdot \left(\frac{R_{\text{Sch.}}}{r} \right) \sim \varepsilon \left(\frac{R_{\text{Sch.}}}{r} \right) \cdot \left(\frac{v}{c} \right)^2. \tag{8.184}$$

This relation clearly shows what conditions must be obeyed by gravitational wave sources in order to be effective emitters: their sizes must be as close as possible to the Schwarzschild radius ($R \gtrsim R_{\text{Sch.}}$); they must be very relativistic and as asymmetric as possible.

(6) Currently the best proof of the existence of gravitational waves does not come from direct detection but from study of the binary pulsar PSR 1913+16 [J.H. Taylor, J.M. Weisberg (1982) (1989)]: these are so precise that the energy loss to gravitational radiation can be evaluated and compared with the predictions of Einstein's (1918) formula. We shall only give a simple discussion here, which is not accurate enough for the study of the binary pulsar.

The first equation is the energy of the emitted waves. Using the field equations (8.137) obeyed by the components $\bar{h}_{\mu\nu}$ (in the TT gauge) the corresponding energy–momentum tensor is (see *Appendix C*) proportional to[29]

$$T_{\alpha\beta}^{\text{wave}} \propto \frac{1}{2} \partial_\alpha \bar{h}_{\mu\nu} \cdot \partial_\beta \bar{h}^{\mu\nu}, \tag{8.185}$$

---

[29] We have neglected the term in $\eta_{\alpha\beta} L$; this vanishes for combinations of plane waves obeying the equations of motion.

where the constant of proportionality varies as G^{-1} for dimensional reasons. This relation shows that the energy in a gravitational wave is *a priori* of second order and is therefore negligible in linearised theory. The second order Einstein equations for free waves are

$$G_{\mu\nu} = G^{(1)}_{\mu\nu} + G^{(2)}_{\mu\nu} = 0, \qquad (8.186)$$

and show that $G^{(2)}_{\mu\nu}$ can be regarded as $-16\pi G$ times the energy–momentum tensor of the linearised gravitational wave. We finally get [see e.g. S. Weinberg (1972)]

$$T^{\alpha\beta}_{\text{wave}} = \frac{1}{32\pi G} \partial^\alpha \bar{h}_{\mu\nu} \cdot \partial^\beta \bar{h}^{\mu\nu}, \qquad (8.187)$$

which can be rewritten in the form

$$T^{\alpha\beta}_{\text{wave}} = \frac{k^\alpha k^\beta}{16\pi G}[|\alpha|^2 + |\beta|^2] \qquad (8.188)$$

for a plane wave with wave vector k^α of type (8.151).

At this point it is important to note that the earlier expressions (8.185) and (8.187) for $T^{\mu\nu}_{\text{wave}}$ have no direct physical meaning; in fact gravitational energy[30] cannot *a priori* be localised: no gravitational force is exerted at a single point! Thus we should replace the relations above by *averages* over several wavelengths; so (8.187) should be

$$T^{\alpha\beta}_{\text{wave}} = \frac{1}{32\pi G} < \partial^\alpha \bar{h}_{\mu\nu} \cdot \partial^\beta \bar{h}^{\mu\nu} > . \qquad (8.189)$$

The energy density (or flux, i.e. the energy per unit time and area, up to a factor c) for a plane wave propagating along the x axis can finally be written as

$$T^{01}_{\text{wave}} = \frac{1}{16\pi G}\left[\left(\frac{d}{dt}h_{23}\right)^2 + \frac{1}{4}\left(\frac{d}{dt}h_{22} - \frac{d}{dt}h_{33}\right)^2\right]. \qquad (8.190)$$

Now using the approximate expression for h_{ij} given by (8.172) or (8.174) and integrating the energy flux over a surface surrounding the source, we get[31] the *radiated power* in gravitational waves:

$$\frac{dE}{dt} = -\frac{G}{45}\frac{d^3 Q_{ij}}{dt^3} \cdot \frac{d^3 Q^{ij}}{dt^3}, \qquad (8.191)$$

where Q_{ij} is the quadrupole moment defined by

$$Q_{ij} \equiv \int d^3 \rho(x)[3x_i x_j - x^2 \delta_{ij}]. \qquad (8.192)$$

This is Einstein's (1918) formula, the details of which can be found in e.g. L. Landau

---

[30] We have seen above that a gravitational wave excites oscillation modes of a detector: it thus communicates energy and momentum. A gravitational wave therefore possesses an energy content.

[31] The proof only holds for a system with negligible tensions, which is weakly self-gravitating, and whose velocities are small compared with that of light. Moreover, even when extended to sources not obeying these conditions, there are delicate problems in applying the formalism to the binary pulsar PSR 1913+16, since there $GM/Rc^2 \sim 0.2$. [We are indebted for these remarks to L. Blanchet and T. Damour, whose article in the collection edited by N. Deruelle and T. Piran (1983) should be consulted.]

and E. Lifschitz (1962). The relation (8.190) shows that for an oscillating system with frequency ω, the power goes as ω^6.

An important example of a gravitational wave emitter is a binary system, such as the binary pulsar PSR 1913+16. Here we shall limit ourselves to two pointlike stars of masses m_1 and m_2 in a circular orbit about their common centre of mass. We thus have a system with a time-dependent quadrupole moment, which therefore emits gravitational waves. As a result the two stars slowly spiral towards each other and ultimately coalesce. We examine this more closely. If ℓ_1, ℓ_2 denote the distances of the two stars from their centre of mass (at the origin), the time-dependent parts of the the various components of the quadrupole moment can be found without difficulty from the equations giving the trajectories of the two stars,

$$\begin{cases} x_1 & = -x_2 = \ell \cos \omega t \\ y_1 & = -y_2 = \ell \sin \omega t, \end{cases} \tag{8.193}$$

where ℓ is their distance apart and where ω is the angular velocity, i.e.

$$\omega = \left(\frac{GM}{\ell^3}\right)^{1/2} \tag{8.194}$$

$(M = m_1 + m_2)$ resulting from the balance of gravity and centrifugal force for circular motion. We thus find

$$\begin{cases} Q_{11} = -Q_{22} = \dfrac{3}{2}\mu\ell^2 \cos 2\omega t + \text{const.} \\[2mm] Q_{12} = \quad Q_{21} = \dfrac{3}{2}\mu\ell^2 \sin 2\omega t + \text{const.} \end{cases} \tag{8.195}$$

where μ is the reduced mass,

$$\mu = \frac{m_1 m_2}{m_1 + m_2}. \tag{8.196}$$

Now inserting (8.195), after differentiating three times, into (8.191), we get

$$\left|\frac{dE}{dt}\right| = \frac{32}{5}\frac{G^4}{c^5}\frac{M^3\mu^2}{\ell^5}, \tag{8.197}$$

where we have given the factors of c explicitly. Noting that (8.194) gives the orbital period $T = 2\pi/\omega$, we get

$$\frac{1}{T}\frac{dT}{dt} = \frac{3}{2}\frac{1}{\ell}\frac{d\ell}{dt}$$

$$= -\frac{3}{2}\frac{1}{E}\frac{dE}{dt} = -\frac{96}{5}\frac{G^3}{c^5}\frac{M^2\mu}{\ell^4}, \tag{8.198}$$

where we have used the fact that the system energy is

$$E = -\frac{1}{2}\frac{G\mu M}{\ell}. \tag{8.199}$$

Integrating the differential equation (8.198) for $\ell(t)$, we see that as t tends to t_0, $\ell(t)$

tends to zero, i.e. the two stars coalesce, when

$$t_0 = \frac{5}{256} \frac{c^5}{G^3} \frac{\ell_0^4}{M^2 \mu},$$ (8.200)

with ℓ_0 the initial distance of the two stars. When relation (8.198) is applied to the binary pulsar PSR 1913+16, the agreement is quite good: correcting for the nonzero eccentricity of the orbit, the agreement is exact! *This demonstrates the existence of gravitational waves.*

Exercises

8.1   Consider Einstein's equations in a space–time with 2+1 dimensions. How many independent components does the Riemann tensor $^{(3)}R_{ijkl}$ have? How many does its contraction $^{(3)}R_{ij}$ have? Deduce that a 2+1 dimensional space–time with $^{(3)}R_{ij} = 0$ is necessarily flat.

8.2   Show that the Einstein equations can be put into the form

$$R_{\mu\nu} = 8\pi G \left[T_{\mu\nu} - \frac{1}{2} g_{\mu\nu} T^{\lambda}{}_{\lambda} \right].$$

Write down the Einstein equations with cosmological coordinates in Schwarzschild coordinates and solve them, to find

$$ds^2 = \left(1 - \frac{2GM}{r} - \frac{1}{3} \lambda r^2 \right) dt^2$$ (E8.1)

$$= \left(1 - \frac{2GM}{r} - \frac{1}{3} \lambda r^2 \right)^{-1} dr^2 - r^2 d\Omega^2.$$ (E8.2)

Find the perihelion advance of a planet using this metric. Knowing that the perihelion advance for Mercury is $43.11'' \pm 0.45''$, what limits can you place on λ?

8.3   (i) Calculate the Riemann tensor for the most general spherically symmetric metric, and

(ii) Deduce for the Schwarzschild metric written in standard coordinates that this tensor is not singular at the point $r = 2GM$.

8.4   Using Birkhoff's theorem, show that a test particle inside a sphere feels no gravitational force.

8.5   The geometry of a spherically symmetric star is described by the metric

$$ds^2 = e^{2\phi} dt^2 - e^{2\Lambda} dr^2 - r^2 d\Omega^2,$$

and the matter is assumed to be isentropic (i.e. $d\mathcal{S} = 0$). The energy–momentum tensor for the fluid of which the star is made has the perfect fluid form

$$T^{\mu\nu} = (\rho + P) u^{\mu} u^{\nu} - P g^{\mu\nu}.$$

(i) Show that the conservation of $T^{\mu\nu}$, i.e. $\nabla_\mu T^{\mu\nu} = 0$, implies

$$d[\rho/n] = -P d[1/n],$$

where n is the baryon density of the star (note that $\nabla_\mu J^\mu = 0$), i.e.

$$d\rho = \frac{\rho + P}{n} dn.$$

(ii) Write the Euler equation explicitly for the fluid (to get

$$\left(\rho + P \right) u^\nu \nabla_\mu u_\nu + \nabla_\mu P + u^\nu \nabla_\nu P . u_\mu = 0 \right).$$

(iii) For a static star show that

$$\frac{dP(r)}{dr} = -[\rho(r) + P(r)] \frac{\partial}{\partial r} \phi(r).$$

8.6 Show that in a Friedman universe the conservation of the energy–momentum tensor, $\nabla_\mu T^{\mu\nu} = 0$, implies

$$d(\rho R^3) + P dR^3 = 0.$$

8.7 Show that in a synchronous reference frame $\sigma_{\mu\nu} = 0$ requires

$$\partial_0 g_{ij} = \frac{1}{3} g_{ij} \partial_\mu u^\mu.$$

8.8 Consider Minkowski space $\mathcal{M}^4$ with its usual metric

$$ds^2 = dt^2 - d\mathbf{x}^2$$

(i) Write this metric in the hyperbolic coordinate system $(\tau, r, \theta, \varphi)$ with $t = \tau \mathrm{ch}\chi$, $r = \tau \mathrm{sh}\chi$, $r/t = \mathrm{th}\chi$ and $t^2 - r^2 = \tau^2$.

(ii) Identify the metric obtained with the metric of a Friedman universe with the scale factor $R(\tau) = \tau (\tau > 0)$. What are the worldlines of the galaxies? Find the photon trajectories.

(iii) What is the curvature of this space–time (the Milne universe?) What energy–momentum tensor does it correspond to?

8.9 Let $A_{\mu\nu}$ be the amplitude of a gravitational wave $\bar{h}_{\mu\nu}$,

$$\bar{h}_{\mu\nu} = \mathcal{R}e \, A_{\mu\nu} e^{ik.x},$$

in the TT gauge (the wave propagates along the z axis) and let $R(\chi)$ be a rotation through an angle χ about the z axis. Write down $R(\chi)$ explicitly and find the form of $A_{\mu\nu}$ after this transformation.

8.10 Show that in the TT gauge $\Gamma^\mu_{00} = 0$ and $\Gamma^\mu_{0\nu} = \frac{1}{2} \dot{h}^\mu_\nu$.

8.11 Write the geodesic equation for a test particle acted upon by a gravitational wave in the TT gauge. What conditions may we draw?

8.12 Prove relations (8.159) to (8.161).

8.13   Calculate the effect of a gravitational wave of frequency ω and transverse polarisation ($\times$) propagating in the $z > 0$ direction on an antenna with the geometry of **Fig. 8.8**.

8.14   Prove the relation (8.174) following the argument of (8.175)–(8.177) [use the conservation relations (8.173) and Gauss's theorem].

8.15   Show that the tensor $T^{\mu\nu}_{\text{wave}}$ [Eq. (8.187)] is gauge invariant.

8.16   Calculate the Einstein tensor $G_{\mu\nu}$ to second order in h and ∂h. Deduce the form of $T^{\text{wave}}_{\mu\nu}$ [ref: S. Weinberg (1972)].

Appendix A Tensors

In *Chapter 3* we "defined" tensors empirically, without giving them a precise mathematical meaning. We thus laid down the precise positioning of the indices and the notation. These are completely logical if we examine tensors from a rather more mathematical point of view.

More detailed treatments can be found in various books[1].

In this appendix we shall briefly review the notion of the *dual* of a vector space, then give a mathematical definition of tensors, particularly *affine* tensors. Finally we shall indicate what criteria decide if a quantity behaves as a tensor.

1   Dual of a vector space

Let $\mathbf{E}$ be any vector space on C ($\mathbf{R}^n, \mathbf{C}^n, \mathscr{M}$, etc.). A *linear form* on $\mathbf{E}$ is a linear application of $\mathbf{E}$ onto C.

Example 1: $f^i(x) = x^i$ is a linear form on C^n.

Example 2: Let $\mathbf{E}$ be the vector space of numerical functions which are continuous on [0,1]. The following applications are linear forms:

$$x(t) \longrightarrow x(t_0) \text{ with } t_0 \in [0, 1]$$

$$x(t) \longrightarrow \int_0^1 x(t)dt$$

Example 3: Let $\mathbf{E} = C^n$ and $A \in \mathscr{L}(\mathbf{E}, \mathbf{E})$ be a linear application of $\mathbf{E}$ onto itself. Then Tr. A is a linear form.

Example 4: Let $\mathbf{E}$ be the vector space of convergent numerical sequences. The application associating a sequence $\{s_n\}$ with its limit s is a linear form.

Definition: *The vector space $\mathscr{L}(\mathbf{E}, C)$ of linear forms on $\mathbf{E}$ is called the dual space of $\mathbf{E}$ and is denoted $\mathbf{E}^*$. If dim. $\mathbf{E} = n$ then dim. $\mathbf{E}^* = n$.*

Now let $\{\mathbf{e}_i\}_{i=1,2,\dots,n}$ be a basis of $\mathbf{E}$, assumed to have dimension n. Now define the linear forms $\mathbf{e}^{*1}, \mathbf{e}^{*2}, \dots, \mathbf{e}^{*n}$ by

$$\mathbf{e}^{*i}(\mathbf{e}_j) = \delta^i{}_j$$

---

[1]   See e.g. A. Lichnérowicz, (1955, 1960).

[note that for each **x** of **E**, $\mathbf{e}^{*i}(\mathbf{x}) = x^i$, where x^i is the contravariant component of **x** in the basis $\{\mathbf{e}^i\}$]; we can then easily show that $\{\mathbf{e}^{*i}\}$ is a basis for $\mathbf{E}^*$, the dual basis of $\{\mathbf{e}_i\}$.

Let $\{\mathbf{e}_i\}$ and $\{\mathbf{e}'_i\}$ be two bases of **E** and let $\mathscr{R}$ be the transformation matrix

$$\mathbf{e}'_i = R_i{}^j \mathbf{e}_j;$$

if **x** is a vector of **E**, we have $\mathbf{x} = x^i \mathbf{e}_i = x'^j \mathbf{e}'_j$; which immediately implies

$$x^i = R_j{}^i x'^j.$$

In other words, the contravariant components of **x** in the new basis are found using the matrix $\mathscr{R}^{-1T}$, where T denotes the transpose.

If dim.**E** $=n$, it is easy to show that any linear form on **E**, assumed to be equipped with a scalar product $<, >$, can be written in the form

$$\mathbf{y}^*(\mathbf{x}) = <\mathbf{y}, \mathbf{x}>,$$

where **x, y** are elements of **E** and $\mathbf{y}^*$ is an element of $\mathbf{E}^*$, uniquely associated with **y**.

2   Tensor products of vector spaces

Let **E** and **F** be two vector spaces of finite dimensions: dim.**E**=n and dim.**F**=m. We define a vector space of mn dimensions, written $\mathbf{E} \otimes \mathbf{F}$, by supplying $\mathbf{E} \times \mathbf{F}$ with an *ad hoc* structure. With the pair $(\mathbf{x}, \mathbf{y})$ with **x** in **E** and **y** in **F**, we associate an element of $\mathbf{E} \otimes \mathbf{F}$, written $\mathbf{x} \otimes \mathbf{y}$, with the correspondence having the following properties:

(i)

$$\mathbf{x} \otimes (\mathbf{y}_1 + \mathbf{y}_2) = \mathbf{x} \otimes \mathbf{y}_1 + \mathbf{x} \otimes \mathbf{y}_2$$
$$(\mathbf{x}_1 + \mathbf{x}_2) \otimes \mathbf{y} = \mathbf{x}_1 \otimes \mathbf{y} + \mathbf{x}_2 \otimes \mathbf{y}$$

(for all vectors $\mathbf{x}, \mathbf{x}_1, \mathbf{x}_2$, of **E** and $\mathbf{y}, \mathbf{y}_1, \mathbf{y}_2$ of **F**)

(ii) $\alpha(\mathbf{x} \otimes \mathbf{y}) = \alpha\mathbf{x} \otimes \mathbf{y} = \mathbf{x} \otimes \alpha\mathbf{y}$

(iii) if $\{\mathbf{e}_i\}$ and $\{\mathbf{f}_j\}$ are bases of **E** and **F** respectively, then the nm elements

$$\mathbf{e}_i \otimes \mathbf{f}_j \equiv \varepsilon_{ij}$$

form a basis of $\mathbf{E} \otimes \mathbf{F}$.

We can thus write

$$\mathbf{x} \otimes \mathbf{y} = x^i y^j \mathbf{e}_i \otimes \mathbf{f}_j = x^i y^j \varepsilon_{ij};$$

so that the $x^i y^j$ are the *contravariant components* of the tensor $\mathbf{x} \otimes \mathbf{y}$ in the base ε_{ij}. Of course, we can form the tensor product of a vector space with the dual of another vector space. Thus, considering the vector spaces **E**, $\mathbf{E}^*$, **F** and $\mathbf{F}^*$, we can form the following tensor products

$$\mathbf{E} \otimes \mathbf{F}, \mathbf{E} \otimes \mathbf{F}^*, \mathbf{E}^* \otimes \mathbf{F}^*, \mathbf{E}^* \otimes \mathbf{F}, \mathbf{E} \otimes \mathbf{E}^*, \dots$$

whose elements will have the general forms

$$\mathbf{T} = T^{ij}\mathbf{e}_i \otimes \mathbf{f}_j$$
$$\mathbf{T}' = T'^i{}_j\mathbf{e}_i \otimes \mathbf{f}^{*j}$$
$$\mathbf{T}'' = T''_{ij}\mathbf{e}^{*i} \otimes \mathbf{f}^{*j}$$
$$\mathbf{T}''' = T'''_i{}^j\mathbf{e}^{*i} \otimes \mathbf{f}_j$$
$$\mathbf{T}'''' = T''''^i{}_j\,\mathbf{e}_i \otimes \mathbf{e}^{*j}$$
$$\cdots \qquad \cdots$$

We thus see rather better what the position of the indices means for a tensor of rank 2.

Having defined the tensor product of two vector spaces, it is now possible to form tensor products of arbitrary numbers of vector spaces. For example, if **E**, **F** and **G** are three vector spaces of finite dimension, we can form $\mathbf{E} \otimes \mathbf{F}$ and then $(\mathbf{E} \otimes \mathbf{F}) \otimes \mathbf{G}$, which by construction is $\mathbf{E} \otimes \mathbf{F} \otimes \mathbf{G}$. The elements of this vector space are tensors of rank 3, of the form

$$\mathbf{T} = T^{ijk}\mathbf{e}_i \otimes \mathbf{f}_j \otimes \mathbf{g}_k.$$

An *affine tensor* attached to a vector space **E** is an element of the vector space

$$\underbrace{\mathbf{E} \otimes \mathbf{E} \otimes \ldots \otimes \mathbf{E}}_{p \text{ times}} \otimes \underbrace{\mathbf{E}^* \otimes \mathbf{E}^* \otimes \ldots \otimes \mathbf{E}^*}_{q \text{ times}} \equiv T;$$

this is a tensor of rank $p + q$ which is p times contravariant and q times covariant. We shall confine ourselves to affine tensors in the following: in special relativity we consider only affine tensors in Minkowski space $\mathscr{M}$. Further, we have dim.$\mathbf{T} = n^{p+q}$.

3 Criteria for being a tensor

We now study the effect of *changes of basis* for the components of an affine tensor; let $T^{\mu_1\mu_2\cdots\mu_p}$ be its components: the tensor considered is thus an element of $\mathbf{E}^{\otimes p}$. Let $\{\mathbf{e}_i\}$ and $\{\mathbf{e}'_i\}$ be two different bases of **E**, and $\mathscr{R}$ the transformation matrix, with elements $R_i{}^j$,

$$\mathbf{e}'_i = R_i{}^j\mathbf{e}_j.$$

The basis $\mathbf{E}^{\otimes p}$, i.e. $\{\mathbf{e}_{i_1} \otimes \mathbf{e}_{i_2} \otimes \ldots \otimes \mathbf{e}_{i_p}\}$ becomes

$$\mathbf{e}'_{i_1} \otimes \mathbf{e}'_{i_2} \otimes \ldots \otimes \mathbf{e}'_{i_p} = R_{i'_1}{}^{j_1} R_{i'_2}{}^{j_2} \ldots R_{i'_p}{}^{j_p}\mathbf{e}_{j_1} \otimes \mathbf{e}_{j_2} \otimes \ldots \otimes \mathbf{e}_{j_p}$$

which can also be written

$$\varepsilon'_{i_1 i_2 \ldots i_p} = R_{i'_1}{}^{j_1} R_{i'_2}{}^{j_2} \ldots R_{i'_p}{}^{j_p}\varepsilon_{j_1 j_2 \ldots j_p}$$

in an obvious notation. Expressing the tensor **T** in the two bases $\{\varepsilon_{i_1 i_2 \ldots i_p}\}$ and $\{\varepsilon'_{i_1 i_2 \ldots i_p}\}$ leads to

$$\mathbf{T} = T^{\mu_1 \mu_2 \ldots \mu_p} \varepsilon_{i_1 i_2 \ldots i_p}$$
$$= T'^{\nu_1 \nu_2 \ldots \nu_p} \varepsilon'_{i_1 i_2 \ldots i_p}$$

and finally to

$$T'^{\nu_1 \nu_2 \ldots \nu_p} = (R^{-1})_{\mu_1}{}^{\nu_1} (R^{-1})_{\mu_2}{}^{\nu_2} \ldots (R^{-1})_{\mu_p}{}^{\nu_p} T^{\mu_1 \mu_2 \ldots \mu_p}.$$

Introducing the notation

$$(R^{-1})_\mu{}^\nu \equiv R^\mu{}_\nu;$$

the transformation law for the contravariant components of the tensor **T** can be rewritten in the simpler form

$$T'^{\mu_1 \mu_2 \ldots \mu_p} = R^{\mu_1}{}_{\nu_1} R^{\mu_2}{}_{\nu_2} \ldots R^{\mu_p}{}_{\nu_p} T^{\nu_1 \nu_2 \ldots \nu_p}.$$

It is easy to show that the covariant components transform as

$$T'_{\mu_1 \mu_2 \ldots \mu_p} = R_{\mu_1}{}^{\nu_1} R_{\mu_2}{}^{\nu_2} \ldots R_{\mu_p}{}^{\nu_p} T_{\nu_1 \nu_2 \ldots \nu_p};$$

and the mixed components, i.e. contra- and co-variant, of a tensor are written

$$T'^{\mu_1 \mu_2 \ldots}_{\nu_1 \nu_2 \ldots} = R^{\mu_1}{}_{\alpha_1} R^{\mu_2}{}_{\alpha_2} \ldots R_{\nu_1}{}^{\beta_1} R_{\nu_2}{}^{\beta_2} \ldots T^{\alpha_1 \alpha_2 \ldots}_{\beta_1 \beta_2 \ldots}.$$

A very useful notation is given by writing

$$T'^{\mu_1 \mu_2 \ldots}_{\nu_1 \nu_2 \ldots} \equiv T^{\mu'_1 \mu'_2 \ldots}_{\nu'_1 \nu'_2 \ldots}$$
$$(R^{-1})^\mu{}_\alpha \equiv R^{\mu'}{}_\alpha$$
$$R_\mu{}^\alpha \equiv R^\mu{}_{\alpha'};$$

with these conventions we have

$$R^\mu{}_{\alpha'} R^{\alpha'}{}_\nu = \delta^\mu{}_\nu$$

and

$$R^{\mu'}{}_\alpha R^\alpha{}_{\nu'} = \delta^{\mu'}{}_{\nu'}.$$

The transformation law for the mixed components of a tensor, of rank 2 for example, is thus

$$T^{\mu'}{}_{\nu'} = R^{\mu'}{}_\alpha R^\beta{}_{\nu'} T^\alpha{}_\beta.$$

The primed indices correspond to the components of **T** in the new basis.

We thus easily get [see A. Lichnérowicz (1960)] the following criteria for being a tensor.

Theorem: *For a set of m^q quantities $T^{i_1 i_2 \ldots i_q}$ defined in a basis $\{\mathbf{e}_{i_1} \otimes \mathbf{e}_{i_2} \otimes \ldots \otimes \mathbf{e}_{i_q}\}$ of the space $\mathbf{E}^{\otimes q}$ (dim. $\mathbf{E} = m$) to be regarded as the contravariant components of a tensor, it is necessary and sufficient that under any change of basis they transform according to the rules given above.*

Corollary: *For a set of m^q quantities $T^{i_1 i_2 \ldots i_q}$ defined in a basis $\{\mathbf{e}_{i_1} \otimes \mathbf{e}_{i_2} \otimes \ldots \otimes \mathbf{e}_{i_q}\}$ of the space $\mathbf{E}^{\otimes q}$ (dim. $\mathbf{E} = m$) to be regarded as the contravariant components of a tensor, it is necessary and sufficient that for any vectors $\mathbf{x}_{(1)}, \mathbf{x}_{(2)}, \ldots, \mathbf{x}_{(q)}$ of $\mathbf{E}$, the quantity $T^{i_1 i_2 \ldots i_q} x_{(1)i_1} x_{(2)i_2} \ldots x_{(q)i_q}$ remains invariant under any change of basis.*

Appendix B Exterior Differential Forms

We limit ourselves pragmatically to elementary notions in this appendix; these are for use in the integration of various physical quantities (currents, energy–momentum tensor, etc.). Nevertheless, differential forms are used in all sorts of physical problems from thermodynamics to mechanics, including electromagnetism and fluid mechanics. More advanced treatments can be found in the books by H. Flanders (1963), A. Lichnérowicz (1960) and B. Schutz (1985).

We first give a simple idea of this notion. We consider $\mathbf{R}^3$ with coordinates $\{x^i\}_{i=1,2,3}$. Consider a vector field of components A^i; a tensor field of components T^{ij}; and finally another tensor field of components Q^{ijk}. Consider now the following integrals[1]

$$I = \int_{\mathscr{C}} A_i dx^i, \tag{B.1}$$

$$J = \int_{\mathscr{S}} T_{ij} dx^i dx^j, \tag{B.2}$$

$$K = \int_{\mathscr{V}} Q_{ijk} dx^i dx^j dx^k. \tag{B.3}$$

The first integral, I, is a curvilinear integral along the curve $\mathscr{C}$, and its integrand,

$$\omega = A_i dx^i, \tag{B.4}$$

is a *differential form of degree 1*. The integral J is a surface integral over $\mathscr{S}$. However, a general surface integral in $\mathbf{R}^3$ has the form

$$\int_{\mathscr{S}} P\, dx\, dy + Q\, dy\, dz + R\, dx\, dz; \tag{B.5}$$

this means that we must have in some sense

$$dx\, dx = dy\, dy = dz\, dz = 0, \tag{B.6}$$

and on the other hand that terms of the type $dy\, dx$, etc... must be expressible as functions of $dx\, dy$, etc.

---

[1] At this point we are not preoccupied with the positions of the various indices.

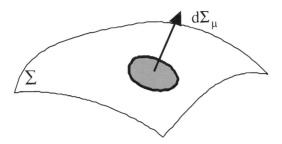

Fig. B.1. Oriented surface element in $\mathbf{R}^2$.

To make this somewhat more precise, consider two coordinates ξ^1 and ξ^2 on the surface $\mathscr{S}$, thus described parametrically as

$$x = x(\xi^1, \xi^2), \quad y = y(\xi^1, \xi^2), \quad z = z(\xi^1, \xi^2), \tag{B.7}$$

so that the integral (B.5) can be rewritten in the form

$$\int A_{12} d\xi^1 d\xi^2, \tag{B.8}$$

where A_{12} is a function of ξ^1 and ξ^2 found from (B.7) and P, Q, R, and the integral (B.8) is over a domain of $\mathbf{R}^2$. The integrand $d\xi^1 d\xi^2$ of the latter integral represents an *oriented* surface element of $\mathbf{R}^2$ (see **Fig. B.1**). Put another way, if $\mathbf{e}_1$ and $\mathbf{e}_2$ represent basis vectors (tangent to the coordinate curves ξ^1 and ξ^2), we have

$$d\mathbf{S} = \mathbf{e}_1 d\xi^1 \wedge \mathbf{e}_2 d\xi^2 = \mathbf{e}_1 \wedge \mathbf{e}_2 d\xi^1 d\xi^2 \tag{B.9}$$

where $\wedge$ denotes the usual vector product in $\mathbf{R}^3$. It follows that if we exchange the roles of the vectors $\mathbf{e}_1 d\xi^1$ and $\mathbf{e}_2 d\xi^2$, $d\vec{S}$ changes sign but not modulus. Moreover, this property is by construction invariant under coordinate changes on the surface $\mathscr{S}$. We see finally that we must set

$$d\xi^1 d\xi^2 = -d\xi^2 d\xi^1, \tag{B.10}$$

which is evidently compatible with the property (B.6). Then the integral (B.8) can be rewritten as

$$\int A_{ij} d\xi^i \wedge d\xi^j, \tag{B.11}$$

where we have used the notation

$$d\xi^i d\xi^j \equiv d\xi^i \wedge d\xi^j, \tag{B.12}$$

and A_{ij} now appears as an *antisymmetric tensor*. Similarly, the integral J can be rewritten in the form

$$J = \int_{\mathscr{S}} T_{ij} dx^i \wedge dx^j. \tag{B.13}$$

We return now to the general form (B.5) which we usually write as

$$\int_{\mathscr{S}} \vec{V} \cdot d\vec{S}, \tag{B.14}$$

where $\vec{V}$ is a vector of $\mathbf{R}^3$ with components (P, Q, R). As dS_i can always be written as

$$dS_i = \frac{1}{2!} \varepsilon_{ijk} d\xi^j \wedge d\xi^k, \tag{B.15}$$

it is clear that (B.11) can also be put into the form (B.14) by setting

$$A_{jk} = \frac{1}{2} \varepsilon_{ijk} V^i; \tag{B.16}$$

this is only possible because A_{jk} is an antisymmetric tensor of $\mathbf{R}^3$.

We note finally that if we make a coordinate change of the type

$$\xi^1 \longleftrightarrow \xi^2, \tag{B.17}$$

which changes the orientation of the basis $\{\mathbf{e}_1, \mathbf{e}_2\}$, then dS_i does not change sign (the pseudo-tensor ε_{ijk} changes sign as well as $d\xi^i \wedge d\xi^j$) and the second-order differential form making up the integrand is invariant.

What happens to the integrand of (B.3)? In view of the remarks above we can write it as

$$Q_{ijk} dx^i \wedge dx^j \wedge dx^k. \tag{B.18}$$

Now the differential form $dx^i \wedge dx^j \wedge dx^k$ is completely antisymmetric in the indices (i, j, k). If we wish to obtain a scalar for K, we must have Q_{ijk} proportional to the pseudo-tensor ε_{ijk}:

$$Q_{ijk} = \varphi \varepsilon_{ijk}. \tag{B.19}$$

Thus, the third-order differential form (B.18) is proportional to the volume element

$$\varepsilon_{ijk} dx^i \wedge dx^j \wedge dx^k = 2 d^3 x. \tag{B.20}$$

We note in passing that we can start from a form of degree zero (i.e. a scalar function f), and by differentiation obtain a form of degree one, e.g.

$$\alpha = df = \partial_i f \cdot dx^i. \tag{B.21}$$

We shall use the ideas above in a pragmatic way, particularly the *exterior calculus* defined by the properties of the "vector product" $\wedge$ in $\mathbf{R}^3$, defined below for $\mathbf{R}^n$.

1   Exterior calculus

Consider a vector space E over $\mathbf{R}$ with n dimensions and the tensor product $E^{\otimes p}$. We construct the space $E^{\wedge p}$ of p-vectors: we set $E^{\wedge 0} \equiv \mathbf{R}$ and $E^{\wedge 1} \equiv E$. The *exterior product* $\wedge$ of two vectors $\mathbf{x}$ and $\mathbf{y}$ of E is defined to have the following properties:

$$
\begin{aligned}
&\text{(i)} && \mathbf{x} \wedge \mathbf{y} = -\mathbf{y} \wedge \mathbf{x} \\
&\text{(ii)} && \mathbf{x} \wedge \mathbf{x} = 0 \\
&\text{(iii)} && (\lambda \mathbf{x}_1 + \mu \mathbf{x}_2) \wedge \mathbf{y} = \lambda(\mathbf{x}_1 \wedge \mathbf{y}) + \mu(\mathbf{x}_2 \wedge \mathbf{y}) \\
&\text{(iv)} && \mathbf{x} \wedge (\lambda' \mathbf{y}_1 + \mu' \mathbf{y}_2) = \lambda'(\mathbf{x} \wedge \mathbf{y}_1) + \mu'(\mathbf{x} \wedge \mathbf{y}_2).
\end{aligned}
\tag{B.22}
$$

In terms of components, with

$$
\mathbf{x} = \mathbf{e}_i x^i, \mathbf{y} = \mathbf{e}_i y^i \quad (i = 1, 2, \dots, n),
\tag{B.23}
$$

we thus have

$$
\mathbf{x} \wedge \mathbf{y} = x^i y^j \mathbf{e}_i \wedge \mathbf{e}_j
\tag{B.24}
$$

$$
= \sum_{i<j} (x^i y^j - x^j y^i) \mathbf{e}_i \wedge \mathbf{e}_j.
\tag{B.25}
$$

The space $E^{\wedge 2}$ is thus defined as the set of all linear combinations of the form

$$
\sum \alpha_{ij} \mathbf{e}_i \wedge \mathbf{e}_j.
$$

Relation (B.25) then shows that $E^{\wedge 2}$ is *identical to the vector space of antisymmetric second-order tensors.*

$E^{\wedge p}$ can be defined similarly and is identical to the antisymmetric tensors of order p; its dimension is thus C_n^p and a basis is

$$
\mathbf{e}_{i_1 i_2 \cdots i_p} \equiv \mathbf{e}_{i_1} \wedge \mathbf{e}_{i_2} \wedge \cdots \wedge \mathbf{e}_{i_p},
$$

with $i_1 < i_2 < \cdots < i_p$.

Now consider the spaces $E^{\wedge p}$ and $E^{\wedge q}$. If $\alpha \in E^{\wedge p}$ and $\beta \in E^{\wedge q}$, we can define $\alpha \wedge \beta \in E^{\wedge(p+q)}$ (which will be 0 if and only if $p + q > n$) with the following properties, where the operation $\wedge$ is such that

$$
\begin{aligned}
&\text{(i)} && \alpha \wedge \beta \quad \text{is distributive with respect to addition} \\
&\text{(ii)} && \alpha \wedge (\beta \wedge \gamma) = (\alpha \wedge \beta) \wedge \gamma \quad \text{(associativity)} \\
&\text{(iii)} && \beta \wedge \alpha = (-)^{pq} \alpha \wedge \beta.
\end{aligned}
\tag{B.26}
$$

The exterior product $\wedge$ thus maps from the space $E^{\wedge p} \times E^{\wedge q}$ to the space $E^{\wedge(p+q)}$. A linear operation A of E on itself transforms an element $\alpha \in E^{\wedge p}$

$$
\alpha = \alpha^{i_1 i_2 \cdots i_p} \mathbf{e}_{i_1} \wedge \mathbf{e}_{i_2} \wedge \cdots \wedge \mathbf{e}_{i_p},
\tag{B.27}
$$

into an element

$$
\alpha' = \alpha^{i_1 i_2 \cdots i_p} A_{i_1}{}^{j_1} A_{i_2}{}^{j_2} \cdots A_{i_p}{}^{j_p} \mathbf{e}_{j_1} \wedge \mathbf{e}_{j_2} \wedge \cdots \wedge \mathbf{e}_{j_p},
\tag{B.28}
$$

where the indices $(j_1, j_2, \cdots j_p)$ of the $A_i{}^j$ must be antisymmetric:

$$
\begin{aligned}
A_{i_1}{}^{j_1} \cdot A_{i_2}{}^{j_2} \cdots A_{i_p}{}^{j_p} \mathbf{e}_{j_1} \wedge \mathbf{e}_{j_2} \wedge \cdots \wedge \mathbf{e}_{j_p} \\
= \det A \cdot \mathbf{e}_{i_1} \wedge \mathbf{e}_{i_2} \wedge \cdot \wedge \mathbf{e}_{i_p}.
\end{aligned}
\tag{B.29}
$$

The passage from (B.27) to (B.28) is effected by a change of basis: $\mathbf{e}_i \to A_i{}^j \mathbf{e}_j$.

2   Differential forms

(1) In $\mathbf{R}^n$ consider the infinitesimal element dx^i ($i = 1, 2, \ldots, n$) and a covector $A_i(x)$. By definition, the quantity

$$\omega = A_i dx^i, \tag{B.30}$$

is a *differential form of degree one*, the dx^i providing a basis for the vector space formed by the ω: $dx^i \equiv \mathbf{e}^i$. We can therefore study the exterior algebra of differential forms and thus produce the *second degree* form from two quantities analogous to the ones above:

$$
\begin{aligned}
\alpha &= \omega_1 \wedge \omega_2 \\
&= A_i^{(1)} dx^i \wedge A_j^{(2)} dx^j \\
&= A_i^{(1)} \cdot A_j^{(2)} dx^i \wedge dx^j \qquad (i, j = 1, 2, \ldots, n) \\
&= \frac{1}{2} [A_i^{(1)} A_j^{(2)} - A_i^{(2)} A_j^{(1)}] dx^i \wedge dx^j.
\end{aligned}
\tag{B.31}
$$

Quite generally, a differential form of degree p ($p \le n$) can be written in the form

$$\alpha = A_{i_1 i_2 \cdots i_p} dx^{i_1} \wedge dx^{i_2} \wedge \cdots \wedge dx^{i_p}, \tag{B.32}$$

where $A_{i_1 i_2 \cdots i_p}$ is an antisymmetric tensor.

(2) If we change the coordinate system in $\mathbf{R}^n$, we have

$$dx^i = \frac{\partial x^i}{\partial x^{i'}} dx^{'i} \equiv A_{i'} dx^{i'}, \tag{B.33}$$

so that the form ω (B.30) now becomes

$$\omega = A_i dx^i = A_i \frac{\partial x^i}{\partial x^{i'}} dx^{i'} = A_{i'} dx^{i'} \tag{B.34}$$

and the general form α (B.31)

$$
\begin{aligned}
\alpha &= A_{i_1 i_2 \cdots i_p} dx^{i_1} \wedge dx^{i_2} \wedge \cdots \wedge dx^{i_p} \\
&= A_{i'_1 i'_2 \cdots i'_p} dx^{i'_1} \wedge dx^{i'_2} \wedge \cdots \wedge dx^{i'_p}.
\end{aligned}
\tag{B.35}
$$

These relations show by identification that

$$A_{i'} = \frac{\partial x^i}{\partial x^{i'}} A_i$$

and that

$$A_{i'_1 i'_2 \cdots i'_p} = \frac{\partial x^{i_1 i_2 \cdots i_p}}{\partial x^{i'_1 i'_2 \cdots i'_p}} A_{i_1 i_2 \cdots i_p}, \tag{B.36}$$

where we have set

$$\frac{\partial x^{i_1 i_2 \cdots i_p}}{\partial x^{i'_1 i'_2 \cdots i'_p}} \equiv \det \left[\frac{\partial x^{i_k}}{\partial x^{i'_\ell}} \right], \tag{B.37}$$

which is the *Jacobian* of the transformation $\{x\} \to \{x'\}$. In passing from the basis

$dx^{i_1} \wedge \cdots \wedge dx^{i_p}$ to the basis $dx^{i'_1} \wedge \cdots \wedge dx^{i'_p}$, we must *antisymmetrise* the coefficient, which gives rise to the determinant above.

As an example, consider $\mathbf{R}^2$ in Cartesian coordinates (x,y) and the second-order form

$$
\begin{aligned}
\mathscr{V} &= \frac{1}{2!}\varepsilon_{ij}dx^i \wedge dx^j \\
&= dx \wedge dy,
\end{aligned}
\tag{B.38}
$$

i.e. the usual volume element in $\mathbf{R}^2$. In polar coordinates (r, θ),

$$
\begin{cases}
x^1 = r \cos \theta \equiv x \\
x^2 = r \sin \theta \equiv y;
\end{cases}
\tag{B.39}
$$

we thus have

$$
\begin{cases}
dx^1 = dr \cos \theta - r \sin \theta d\theta \\
dx^2 = dr \sin \theta + r \cos \theta d\theta;
\end{cases}
\tag{B.40}
$$

or

$$
\begin{aligned}
dx \wedge dy &= (dr \cos \theta - r \sin \theta d\theta) \wedge (dr \sin \theta + r \cos \theta d\theta) \\
&= r \cos^2 \theta dr \wedge d\theta - r \sin^2 d\theta \wedge dr \\
&= rdr \wedge d\theta = rdr.d\theta,
\end{aligned}
\tag{B.41}
$$

where we have used the antisymmetry properties of the exterior product.

(3) Consider in $\mathbf{R}^3$ a parametrically defined surface S:

$$
\begin{cases}
x = x(\xi^1, \xi^2) \\
y = y(\xi^1, \xi^2) \\
z = z(\xi^1, \xi^2)
\end{cases}
\tag{B.42}
$$

and any differential form, e.g.

$$
\omega = Adx \wedge dy + Bdx \wedge dz + Cdy \wedge dz,
\tag{B.43}
$$

where A, B and C are functions of (x, y, z). If we constrain the point (x, y, z) to lie in the surface S, the form ω reduces to

$$
\begin{aligned}
\omega &= A\left[x(\xi^1, \xi^2), y(\xi^1, \xi^2), z(\xi^1, \xi^2)\right] \\
&\quad \times \left\{\frac{\partial x}{\partial \xi^1}d\xi^1 + \frac{\partial x}{\partial \xi^2}d\xi^2\right\} \wedge \left\{\frac{\partial y}{\partial \xi^1}d\xi^1 + \frac{\partial y}{\partial \xi^2}d\xi^2\right\} + \cdots
\end{aligned}
\tag{B.44}
$$

$$
\begin{aligned}
&= \left\{A\left[x(\xi^1, \xi^2), y(\xi^1, \xi^2), z(\xi^1, \xi^2)\right] \cdot \left\{\frac{\partial x}{\partial \xi^1}\frac{\partial y}{\partial \xi^2} - \frac{\partial y}{\partial \xi^1}\frac{\partial x}{\partial \xi^2}\right\} + \cdots\right\} \\
&\quad \times d\xi^1 \wedge d\xi^2 \equiv \phi(\xi^1, \xi^2)d\xi^1 \wedge d\xi^2.
\end{aligned}
\tag{B.45}
$$

The latter form is said to be *induced* by ω on S. Consider another example where the form ω is of order three:

$$
\omega = \psi(x, y, z)dx \wedge dy \wedge dz.
\tag{B.46}
$$

If the point (x, y, z) is constrained to lie in S, ω becomes

$$\omega = \psi\left[x(\xi^1, \xi^2), y(\xi^1, \xi^2), z(\xi^1, \xi^2)\right] \times \left\{\frac{\partial x}{\partial \xi^1} d\xi^1 + \frac{\partial x}{\partial \xi^2} d\xi^2\right\}$$

$$\wedge \left\{\frac{\partial y}{\partial \xi^1} d\xi^1 + \frac{\partial y}{\partial \xi^2} d\xi^2\right\} \wedge \left\{\frac{\partial z}{\partial \xi^1} d\xi^1 + \frac{\partial z}{\partial \xi^2} d\xi^2\right\} = 0, \tag{B.47}$$

a result we could have predicted. Assume now that ω is a first-order form:

$$\omega = A_i(x, y, z) dx^i. \tag{B.48}$$

We have as above

$$\omega = A_i\left[x(\xi^1, \xi^2), y(\xi^1, \xi^2), z(\xi^1, \xi^2)\right]\left\{\frac{\partial x^i}{\partial \xi^1} d\xi^1 + \frac{\partial x^i}{\partial \xi^2} d\xi^2\right\}. \tag{B.49}$$

This result goes through easily for the case of a surface of p dimensions embedded in $\mathbf{R}^n$, for forms of order *less than or equal* to p.

3   Volume element: dual forms

(1) Consider in $\mathbf{R}^n$ with any coordinates $\{x^1, x^2, \ldots, x^n\}$ the n-order form

$$\omega = \phi(x^1, x^2, \ldots, x^n) dx^1 \wedge dx^2 \wedge \cdots \wedge dx^n. \tag{B.50}$$

In another coordinate system $\{y^1, y^2, \ldots, y^n\}$, it becomes

$$\omega = \psi(y^1, y^2, \ldots, y^n) dy^1 \wedge dy^2 \wedge \cdots \wedge dy^n, \tag{B.51}$$

where the functions $\phi(x)$ and $\psi(y)$ are related by

$$\phi(x^1, x^2, \ldots, x^n) = \psi(y^1, y^2, \ldots, y^n) \det\left[\frac{\partial y^i}{\partial x^j}\right]. \tag{B.52}$$

On the other hand, the metric tensor g_{ij} transforms as

$$g'_{ij} = \frac{\partial x^k}{\partial y^i} \cdot \frac{\partial x^\ell}{\partial y^j} g_{k\ell}, \tag{B.53}$$

so we also have

$$|g'| = \left[\det \frac{\partial x^k}{\partial y^\ell}\right]^2 \cdot |g|. \tag{B.54}$$

It follows that the differential form

$$\eta \equiv \sqrt{|g|} dx^1 \wedge dx^2 \wedge \cdots \wedge dx^n, \tag{B.55}$$

retains the same form under all coordinate changes (which preserve the orientation of the coordinate lines). The above form clearly reduces to the usual volume element $dV = dx^1 \wedge dx^2 \wedge \cdots \wedge dx^n$ in Cartesian coordinates.

The volume element η can be rewritten in a manifestly covariant way using the totally antisymmetric (pseudo-)tensor $\sqrt{|g|}\varepsilon_{i_1 i_2 \dots i_n}$ defined by

$$\varepsilon_{i_1 i_2 \dots i_n} = \begin{cases} +1 & \text{if } (i_1, i_2, \dots, i_n) \text{ is an even permutation of } (1, 2, \dots, n) \\ -1 & \text{if an odd permutation} \\ 0 & \text{otherwise.} \end{cases} \tag{B.56}$$

We thus get the desired manifestly covariant expression for η:

$$\eta = \frac{1}{n!} \sqrt{|g|}\varepsilon_{i_1 i_2 \dots i_n} dx^{i_1} \wedge dx^{i_2} \wedge \dots \wedge dx^{i_n}. \tag{B.57}$$

(2) Now consider a p-dimensional surface S, embedded in $\mathbf{R}^n$. How do we construct a surface element for a general S? First, we need a differential form of order p, and thus the form $dx^{i_1} \wedge dx^{i_2} \wedge \dots \wedge dx^{i_p}$. Also the order of the dx^i cannot matter. The only totally antisymmetric tensor we have is $\varepsilon_{i_1 i_2 \dots i_n} \sqrt{|g|}$. It follows that the only form of order p with the desired properties is proportional to

$$d\Sigma_{i_{p+1} i_{p+2} \dots i_n} = \frac{1}{p!} \sqrt{|g|}\varepsilon_{i_1 i_2 \dots i_n} dx^{i_1} \wedge dx^{i_2} \wedge \dots \wedge dx^{i_p}. \tag{B.58}$$

For an ordinary two-dimensional surface in $\mathbf{R}^3$ this differential form is, in Cartesian coordinates,

$$d\Sigma_i = \frac{1}{2!} \sqrt{1}\, \varepsilon_{ijk}\, dx^j \wedge dx^k, \tag{B.59}$$

or

$$\begin{cases} d\Sigma_1 = dy \wedge dz \\ d\Sigma_2 = dx \wedge dz \\ d\Sigma_3 = dx \wedge dy. \end{cases} \tag{B.60}$$

Thus, in $\mathbf{R}^3$, $d\Sigma_i$ indeed reduces to the usual surface element. To be more specific we take the case of a sphere of radius 1 and use coordinates x and y. We have

$$z = \pm \left[1 - x^2 - y^2 \right]^{1/2},$$

and if, for example, we consider the northern hemisphere ($+$ sign) we have

$$dz = \frac{-1}{[1 - x^2 - y^2]^{1/2}} \{x dx + y dy\},$$

which, when inserted into the expressions above for the $d\Sigma_i$, gives

$$d\Sigma_1 = \frac{x}{\sqrt{1 - x^2 - y^2}} dx \wedge dy$$

$$d\Sigma_2 = \frac{y}{\sqrt{1 - x^2 - y^2}} dx \wedge dy$$

$$d\Sigma_3 = \frac{z}{\sqrt{1 - x^2 - y^2}} dx \wedge dy.$$

We have rewritten z in the expression for $d\Sigma_3$ so that the symmetry of the $d\Sigma_i$ is more apparent. We thus finally write for the sphere

$$d\Sigma_i = n^i \sqrt{g_s} dx \wedge dy,$$

where n^i is the unit normal and g_s is the determinant of the metric on S, i.e.

$$g_s = \frac{1}{1 - x^2 - y^2} > 0.$$

(3) Consider a form of order p in $\mathbf{R}^n$,

$$\omega = \frac{1}{p!} A_{i_1 i_2 \ldots i_p} dx^{i_1} \wedge dx^{i_2} \wedge \cdots \wedge dx^{i_p}, \tag{B.61}$$

where $A_{i_1 i_2 \ldots i_p}$ is completely antisymmetric and the indices take all values from 1 to n. With this form we associate a form of order $(n - p)$, called the *dual form*, defined by

$$^*\omega \equiv \frac{1}{p!} \sqrt{|g|} \varepsilon_{i_1 i_2 \ldots i_n} A^{i_1 i_2 \ldots i_p} dx^{i_{p+1}} \wedge dx^{i_{p+2}} \wedge dx^{i_n}, \tag{B.62}$$

where the indices are raised, as always using the metric tensor g^{ij}:

$$A^{i_1 i_2 \ldots i_p} = g^{i_1 j_1} g^{i_2 j_2} \ldots g^{i_p j_p} A_{j_1 j_2 \ldots j_p}. \tag{B.63}$$

Defining the *scalar product of two differential forms* of the same order p by

$$< \alpha, \beta > \equiv \frac{1}{p!} \alpha_{i_1 i_2 \ldots i_p} \beta^{i_1 i_2 \ldots i_p}, \tag{B.64}$$

it is easy to verify the following properties:

$$\alpha \wedge {}^*\beta = \beta \wedge {}^*\alpha = < \alpha, \beta > \eta \tag{B.65}$$

$$< {}^*\alpha, {}^*\beta > = \text{sign}\,(g) < \alpha, \beta > \tag{B.66}$$

$$< \alpha \wedge \beta, \eta > = < {}^*\alpha, \beta > \qquad [d^{\circ}(\alpha) = n - p, \ d^{\circ}(\alpha) = p]. \tag{B.67}$$

For α of order p, $^*\alpha$ is of order $n - p$, so it follows that $^{**}\alpha$ is also a form of degree p. Using the properties of the Levi–Civita pseudo-tensor $\varepsilon_{i_1 i_2 \ldots i_n}$, we can show that

$$^{**}\alpha = (-1)^{p(n-p)} \text{sign}\,(g)\alpha. \tag{B.68}$$

For example, if $n = 4$ and $p = 2$, the form

$$F = \frac{1}{2!} F_{\mu\nu} dx^{\mu} \wedge dx^{\nu} \tag{B.69}$$

has the dual form

$$^*F = \frac{1}{2!} \varepsilon_{\mu\nu\alpha\beta} F^{\alpha\beta} dx^{\mu} \wedge dx^{\nu}, \tag{B.70}$$

and we have

$$^{**}F = -F. \tag{B.71}$$

4   Differentiation of forms

(1) We have already considered a simple example of differentiation, namely of a form of order 0, i.e. a scalar function. Differentiation gave a form of order 1. We shall now define an operation d which associates a form of order $p+1$ with one of order p. This operation has the following properties:

$$\begin{cases} \text{(i)} & d(\alpha + \beta) = d\alpha + d\beta \qquad [d^\circ(\alpha) = d^\circ(\beta)] \\ \text{(ii)} & d(\alpha \wedge \beta) = d\alpha \wedge \beta + (-1)^p \alpha \wedge d\beta \qquad [d^\circ(\alpha) = p] \\ \text{(iii)} & d(d\alpha) = 0. \end{cases} \qquad \text{(B.72)}$$

The first property is *linearity*, while the second is the analogue of Leibniz's rule, taking account of the property (B.26), which must be preserved. The last property can be regarded as a condition of compatibility with the fact that in $\mathbf{R}^n$ forms of order $n+1$ vanish identically. Consider a form of the type

$$\omega = \Phi(x^1, x^2, \dots, x^n) dx^1 \wedge dx^2 \wedge \cdots \wedge dx^n. \qquad \text{(B.73)}$$

We have

$$d\omega = d\Phi \wedge dx^1 \wedge dx^2 \wedge \cdots \wedge dx^n + \Phi \sum_{i=1}^{i=n} dx^1 \wedge \cdots d^2 x^i \wedge \cdots \wedge dx^n. \qquad \text{(B.74)}$$

As $d\omega = 0$ and $d\Phi \wedge dx^1 \wedge \cdots = 0$, it follows that

$$\sum_{i=1}^{i=n} dx^1 \wedge \cdots \wedge d^2 x^i \wedge \cdots \wedge dx^n \equiv 0 ; \qquad \text{(B.75)}$$

which must hold for every n, so that we must impose $d(dx^i) = 0$ for all i. As a result it is easy to show that $d(d\alpha) = 0$. This property is called *"Poincaré's lemma"*.

The three conditions above are very natural and one can show that the operation d exists and is unique [see H. Flanders (1963)].

We easily verify that

$$d(\Phi d\omega) = d\Phi \wedge d\omega, \qquad \text{(B.76)}$$

and that for a form $\omega = A_{i_1 \dots i_p} dx^{i_1} \wedge dx^{i_2} \wedge \cdots \wedge dx^{i_p}$, we have

$$\begin{aligned} d\omega &= \frac{1}{p!} \partial_k A_{i_1 i_2 \dots i_p} dx^k \wedge dx^{i_1} \wedge dx^{i_2} \wedge \cdots \wedge dx^{i_p} \\ &= (p+1) \partial_{[k} A_{i_1 i_2 \dots i_p]} dx^k \wedge dx^{iq_1} \wedge dx^{i_2} \wedge \cdots \wedge dx^{i_p}. \end{aligned} \qquad \text{(B.77)}$$

As an example, consider in $\mathscr{M}^n$ the first-order form

$$\mathscr{A} = A_\mu dx^\mu,$$

and take $d\mathscr{A}$. We then have

$$\begin{aligned} d\mathscr{A} &= \partial_\nu A_\mu \, dx^\nu \wedge dx^\mu \\ &= 2\partial_{[\nu} A_{\mu]} dx^\nu \wedge dx^\mu. \end{aligned}$$

(2) Consider a differential form ω such that

$$\omega = d\alpha. \tag{B.78}$$

From the definition of the operation d we have $d\omega = 0$. ω is a *closed* form. A differential form of the type represented by such an ω is called *exact*. We can ask, given a closed form ω, i.e. one such that $d\omega = 0$, whether it is exact, i.e. is $\omega = d\alpha$? In fact this is true, but only *locally* [see H. Flanders (1963)]. Moreover, the form α is clearly not unique, as any form α' such that

$$\alpha' = \alpha + d\gamma \qquad (\gamma: \text{arbitrary form}), \tag{B.79}$$

is also a solution.

(3) We call *coderivation* δ the operation defined by

$$\delta \equiv {}^*d^*. \tag{B.80}$$

It satisfies $\delta^2 = 0$, as can be seen directly.

5   Maxwell's equations in differential forms

Consider the following forms[2]:

$$\mathscr{F} = \frac{1}{2!}F_{\mu\nu}dx^\mu \wedge dx^\nu \tag{B.81}$$

$$\mathscr{J} = J_\mu dx^\mu, \tag{B.82}$$

and let us apply the operation d to the form $\mathscr{F}$; we get

$$d\mathscr{F} = \frac{1}{2!}\partial_{[\lambda}F_{\mu\nu]}dx^\lambda \wedge dx^\mu \wedge dx^\nu = 0; \tag{B.83}$$

since $\partial_{[\lambda}F_{\mu\nu]} = 0$, by virtue of Maxwell's equations. We also have

$$^*\mathscr{F} = \frac{1}{2!}\varepsilon_{\mu\nu\alpha\beta}F^{\alpha\beta}dx^\mu \wedge dx^\nu, \tag{B.84}$$

so that after some algebra we get

$$d^*\mathscr{F} = 4\pi{}^*\mathscr{J}, \tag{B.85}$$

using the second set of Maxwell equations; again, taking the dual of the last relation

$$^*d^*\mathscr{F} \equiv \delta\mathscr{F} = 4\pi\mathscr{J}. \tag{B.86}$$

Thus, Maxwell's equations are equivalent to the following relations between differential forms:

$$\begin{cases} d\mathscr{F} = 0 \\ \delta\mathscr{F} = 4\pi\mathscr{J}. \end{cases} \tag{B.87}$$

---

[2] More details can be found in C. Misner, K. Thorne, J. Wheeler (1973).

Thus the form $\mathscr{F}$ is *closed*. It is therefore locally *exact*: there exists a form $\mathscr{A}$ such that

$$\mathscr{F} = d\mathscr{A}. \tag{B.88}$$

As $\mathscr{F}$ is a form of order 2, $\mathscr{A}$ is a form of degree 1:

$$\mathscr{A} = A_\mu dx^\mu, \tag{B.89}$$

and we thus have

$$\mathscr{F} = \frac{1}{2!} F_{\mu\nu} dx^\mu \wedge dx^\nu = dA_\nu \wedge dx^\nu \tag{B.90}$$

$$= \partial_\mu A_\nu \, dx^\mu \wedge dx^\nu = \frac{1}{2} \left[\partial_\mu A_\nu - \partial_\nu A_\mu \right] dx^\mu \wedge dx^\nu, \tag{B.91}$$

i.e. $F_{\mu\nu} = \partial_\nu A_\nu - \partial_\nu A_\mu$. We see also that the non-uniqueness of the form $\mathscr{A}$ expresses the gauge properties of the 4-potentials. Thus $\mathscr{A}$ and $\mathscr{A}'$ give rise to the same field $\mathscr{F}$ if and only if $\mathscr{A} - \mathscr{A}' = d\phi$, where ϕ is a scalar function. The form $\mathscr{A}$ thus obeys the relation

$$\delta d\mathscr{A} = 4\pi \mathscr{J} . \tag{B.92}$$

6   Integration of differential forms – Stokes' theorem

We consider $\mathbf{R}^n$ and a differential form ω of order $p \leq n$. It is clear that such a form can be integrated only over a surface S^q of dimension $q \leq p$ of $\mathbf{R}^n$; the symbol

$$\int_{S^q} \omega \equiv I, \tag{B.93}$$

has the following sense: (i) the form ω is first constrained to S^q, making it a form of order q; (ii) S^q is decomposed into disjoint parts each described by well-defined coordinates; (iii) on each of these parts – let Σ be one of them –

$$I_\Sigma \equiv \int_{\Sigma \subset S^q} f(\xi^1, \xi^2, \ldots, \xi^q) d\xi^1 d\xi^2 \ldots d\xi^q, \tag{B.94}$$

where $\{\xi^1, \xi^2, \ldots, \xi^q\}$ is a coordinate system which describes Σ. Clearly I, or I_Σ, does not depend on the coordinates chosen.

An important result is *Stokes' theorem*. Let ω be a differential form and $\mathscr{V}$ a volume of $\mathbf{R}^n$; and $\partial\mathscr{V}$ the boundary of $\mathscr{V}$: $\dim[\partial\mathscr{V}] = \dim[\mathscr{V}] - 1$. Stokes theorem is the relation

$$\int_{\mathscr{V}} d\omega = \int_{\partial\mathscr{V}} \omega. \tag{B.95}$$

For example, in $\mathbf{R}^3$, if $\mathscr{V}$ is a finite volume bounded by a surface $S \equiv \partial\mathscr{V}$, we have

$$\int_{\mathscr{V}} d \left[\omega_{ij} dx^i \wedge dx^j \right] = \int_S \omega_{ij} dx^i \wedge dx^j, \tag{B.96}$$

or

$$\int_{\mathscr{V}} \partial_{[k}\omega_{ij]}dx^k \wedge dx^i \wedge dx^j = \int_S \omega_{ij}dx^i \wedge dx^j. \tag{B.97}$$

As ω_{ij} is totally antisymmetric we can associate with it a vector A_k [where i, j and k are an even permutation of $(1, 2, 3)$], so that the preceding relation takes the familiar form

$$\int_{\mathscr{V}} div.\vec{A}\ d^3x = \int_S \vec{A}.d\vec{S}. \tag{B.98}$$

In Minkowski space, the form $\omega = J^\mu d\Sigma_\mu$ gives

$$\int_{\mathscr{V}} d[J^\mu d\Sigma_\mu] = \int_{\partial\mathscr{V}} J^\mu d\Sigma_\mu, \tag{B.99}$$

a formula analogous to the preceding one in $\mathbf{R}^3$, which can be rewritten as

$$\int_{\mathscr{V}} \nabla_\mu J^\mu \eta = \int_{\partial\mathscr{V}} J^\mu d\Sigma_\mu. \tag{B.100}$$

Appendix C Variational Form of the Field Equations

The usual equations of motion of a particle or a system of particles can be deduced from a *variational principle* [principle of least action leading either to Lagrange's or Hamilton's equations, depending on the variables used, see H. Goldstein (1980) or L. Landau and E. Lifschitz (1960)], in both the Newtonian and relativistic cases [see A.O. Barut (1965) or J.L. Anderson (1967)] with some subtleties concerning the constraint $u_\mu u^\mu = 1$ in the latter case [G. Kalman (1961); A. Peres, N. Rosen (1960)].

In the same way, the equations satisfied by the *fields* (continuous systems with an infinite number of degrees of freedom), whatever tensor nature they may have, can often be deduced from variational principles. This is true of the equations of electromagnetism, for example, but not the equation for heat transfer.

There are many analytic procedures for describing the motion of a particle or the evolution of a field. There is no *a priori* reason to confine oneself to differential equations or second order partial differential equations.

The main advantage of a variational formalism is that it allows one to find the *conserved quantities* in the motion (i.e. the first integrals) and directly to exploit the symmetries of the physical problem considered; these two aspects are connected, as we shall see. We shall introduce such a formalism here only because it allows us to define the energy and momentum of a field very simply.

Consider a field φ, with components $\varphi^{..}$ (the dots denote arbitrary tensor indices), and let $\mathscr{L}[\varphi(x), \partial\varphi(x)]$ be the *Lagrangian density*, related to the total action S of the field by

$$S = \int d^4x\, \mathscr{L}[\varphi(x), \partial\varphi(x)], \tag{C.1}$$

where the integral is extended over a space–time volume enclosed between two given spacelike surfaces. Now write that the action integral is minimised for the fields which exist in nature; let $\delta S = 0$, or

$$\delta S = \int d^4x\, \delta\mathscr{L}[\varphi(x), \partial\varphi(x)] = 0. \tag{C.2}$$

Introducing the local variation

$$\delta\varphi(x) = \varphi'(x) - \varphi(x), \tag{C.3}$$

between two infinitesimally neighbouring fields φ and φ', we get successively

$$\delta S = \int d^4x \{ \mathscr{L}(\varphi + \delta\varphi, \partial\varphi + \delta\partial\varphi) - \mathscr{L}(\varphi, \partial\varphi) \}$$

$$= \int d^4x \left[\frac{\partial\mathscr{L}}{\partial\varphi} \delta\varphi + \frac{\partial\mathscr{L}}{\partial(\partial_\mu\varphi)} \partial_\mu(\delta\varphi) \right]. \tag{C.4}$$

Note that there are summations over the implicit indices in φ, $\partial/\partial(\partial_\mu\varphi)$, etc. We have again used the relation $\partial\delta\varphi = \delta\partial\varphi$, which can easily be proved. Integrating the last term of equation (C.4) by parts, we get

$$\int d^4x \left\{ \left[\frac{\partial\mathscr{L}}{\partial\varphi} - \partial_\mu \frac{\partial\mathscr{L}}{\partial(\partial_\mu\varphi)} \right] \delta\varphi + \int_\Sigma d\Sigma_\mu \left[\frac{\partial\mathscr{L}}{\partial\partial_\mu\varphi} \delta\varphi \right] \right\} = 0, \tag{C.5}$$

where we have used the fact that the variations $\delta\varphi$ are *assumed* to vanish at the limits of integration: the integrated term comes from the last term of (C.5) using Stokes' theorem,

$$\int d^4x \partial_\mu \left[\frac{\partial\mathscr{L}}{\partial\partial_\mu\varphi} \delta\varphi \right] = \int_\Sigma d\Sigma_\mu \frac{\partial\mathscr{L}}{\partial\partial_\mu\varphi} \delta\varphi,$$

where the term on the right hand side vanishes since $\delta\varphi = 0$ on the surface of integration.

As the latter equation has to hold whatever the domain of integration and whatever the variation $\delta\varphi$, we deduce that φ must obey the *Euler–Lagrange* equations:

$$\frac{\partial\mathscr{L}}{\partial\varphi} - \partial_\mu \frac{\partial\mathscr{L}}{\partial(\partial_\mu\varphi)} = 0. \tag{C.6}$$

As an example, we consider the scalar field, whose Lagrangian is

$$\mathscr{L}_{\text{scal}} = \frac{1}{2} \left[\partial_\mu\varphi \cdot \partial^\mu\varphi - m^2\varphi^2 \right] + 4\pi J\varphi; \tag{C.7}$$

the Euler–Lagrange equations are then

$$[\Box + m^2]\varphi = 4\pi J, \tag{C.8}$$

which is the *Klein–Gordon equation* coupled to the source[1] J. We can similarly verify that Maxwell's equations, written in the form (5.24), can be recovered from the Lagrangian

$$\mathscr{L}_{\text{em}}[A] = -\frac{1}{4}(\partial^\nu A^\mu - \partial^\mu A^\nu) \cdot (\partial_\nu A_\mu - \partial_\mu A_\nu) \tag{C.9}$$

$$= -\frac{1}{4} F_{\mu\nu} \cdot F^{\mu\nu}; \tag{C.10}$$

this Lagrangian is invariant under gauge transformations.

We now turn to the symmetries of space–time, and in particular the invariance of physics under spatio-temporal translations (homogeneity of space and uniformity in time). In fact, the energy and momentum of a physical system – classical or quantum,

---

[1] The factor 4π is purely conventional.

relativistic or not, consisting of particles or fields – are *defined* as *constants of the motion* associated with these symmetries; energy with time uniformity and momentum with spatial homogeneity. We therefore impose invariance of the action S under infinitesimal space–time translations:

$$x'^{\mu} = x^{\mu} + \varepsilon^{\mu}. \tag{C.11}$$

It is clear that if the limits of integration for the action are not also varied the system will not be invariant under the transformation (C.11) or under other space–time variations (e.g. Lorentz transformations, dilatations). This can be done [see A.O. Barut (1965)] but it is simpler to follow C. Itzykson and J.P. Zuber (1980), who introduce a variation ε^{μ} into (C.11) which depends on the space–time point, i.e. $\varepsilon^{\mu} = \varepsilon^{\mu}(x)$. Under these conditions, the variations of $\phi(x)$ and $\partial_{\mu}\phi(x)$ become

$$\delta\phi(x) = \varepsilon^{\mu}(x)\partial_{\mu}\phi(x) \tag{C.12}$$

$$\begin{aligned}\delta\partial_{\mu}\phi(x) &= \partial_{\mu}\phi(x+\varepsilon) - \partial_{\mu}\phi(x) \\ &= \partial_{\mu}\left[\varepsilon^{\nu}(x)\partial_{\nu}\phi(x)\right] \\ &= \varepsilon^{\nu}(x)\partial_{\mu\nu}\phi(x) + \partial_{\nu}\phi(x)\cdot\partial_{\mu}\varepsilon^{\nu}(x).\end{aligned} \tag{C.13}$$

Substituting these two relations into the variation of the action integral and integrating by parts the term in $\partial_{\mu}\varepsilon^{\nu}(x)$, we get

$$\delta S = \int d^4x\, \varepsilon^{\nu}(x)\left\{\partial_{\nu}L - \partial_{\mu}\left[\frac{\partial L}{\partial(\partial_{\mu}\phi)}\partial_{\nu}\phi\right]\right\} = 0, \tag{C.14}$$

which must hold whatever the function $\varepsilon^{\nu}(x)$ is. It follows that the quantity

$$T^{\mu\nu} \equiv \frac{\partial L}{\partial(\partial_{\mu}\phi)}\partial^{\nu}\phi - \eta^{\mu\nu}L \tag{C.15}$$

is conserved, i.e.

$$\partial_{\mu}T^{\mu\nu} = 0. \tag{C.16}$$

This is the *energy–momentum tensor* of the field φ. This tensor represents in some sense a density of energy and momentum for the field. The associated *global* quantity P^{μ} is the total energy–momentum contained in the field φ. P^{μ} is conserved in time, and thus has the same value on any spacelike surface. We can show this using Stokes' theorem: we have

$$P^{\mu}(\Sigma) = \int_{\Sigma} d\Sigma_{\nu}\, T^{\mu\nu}; \tag{C.17}$$

and we would like to show that $P^{\mu}(\Sigma)$ is independent of Σ, or

$$\frac{\delta}{\delta\Sigma}P^{\mu}(\Sigma) = 0. \tag{C.18}$$

In other words, if we change the surface Σ by an amount $\delta\Sigma$, we should like to show that

$$P^{\mu}(\Sigma + \delta\Sigma) = P^{\mu}(\Sigma).$$

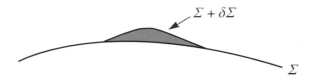

Fig. C.1. Conserved quantities. These are independent of the surface over which one integrates the corresponding density. To see this it suffices to apply Stokes' theorem to the volume enclosed between the (arbitrary) surfaces Σ and $\Sigma + \delta\Sigma$.

We write this relation explicitly and integrate $T^{\mu\nu}$ over the spacelike surface defined by Σ and $\Sigma + \delta\Sigma$ (see **Fig. C.1**). Stokes' theorem then gives

$$\int d\Sigma_\nu \, T^{\mu\nu} = \int d^4x \partial_\nu \, T^{\mu\nu} \tag{C.19}$$

(the domain of these two integrals is shown shaded in the figure). As the second integral vanishes [cf. Eq. (C.15) above], the first must also do so, and the integrals over the surfaces Σ and $\Sigma + \delta\Sigma$ are equal.

In general the *canonical* energy–momentum tensor (that found from the Lagrangian) *is not symmetric*. However, one can always make it symmetric by adding a term which is a 4-divergence. [see A.O. Barut (1965)]. Just as the Lagrangian of a mechanical system is defined only up to a total derivative (i.e. up to a derivative along the motion), the Lagrangian of a field is defined only up to divergenceless terms. To see this, we only have to add such a term to any Lagrangian; in varying the action we used an integration by parts, whereas here we will find integrated terms which vanish since the variations of the fields φ are taken to be zero over the boundaries of the integration region.

We consider two examples of energy–momentum tensors, those of the scalar field and the electromagnetic field. We find in the two cases the results

$$T_{\mu\nu\,\text{scal}} = \partial_\mu\varphi \cdot \partial_\nu\varphi - \frac{1}{2}\eta_{\mu\nu}\left[(\partial\varphi)^2 - m^2\varphi + 8\pi J\varphi\right] \tag{C.20}$$

$$T_{\mu\nu\,\text{em}} = F^\lambda_\mu \cdot F_{\lambda\nu} + \frac{1}{4}\eta_{\mu\nu}F_{\lambda\sigma} \cdot F^{\lambda\sigma}.$$

Explicit calculation of $T_{00\text{em}}$ gives

$$T_{00\text{em}} = \frac{1}{2}[\mathbf{E}^2 + \mathbf{B}^2], \tag{C.21}$$

which is just the usual expression for the electromagnetic energy density, while T^{0i} is given by

$$T^{0i} = (\mathbf{E} \wedge \mathbf{B})^i, \tag{C.22}$$

which is the Poynting vector.

Appendix D The Concept of a Manifold

The notion of a manifold, more precisely that of a differentiable manifold, is fundamental in general relativity. This notion generalises that of a surface embedded in a Euclidean or pseudo-Euclidean space. Moreover, it applies to many other kinds of mathematical object, such as continuous (Lie) groups, for example the rotation or Lorentz groups, the phase space of a mechanical system of N particles (i.e. the space of the $\{q^i, p_i\}_{i=1,2,...,N}$, where the q^i are generalised coordinates and the p_i are the canonically conjugate momenta), etc.

Before defining a manifold and enumerating its properties, we try to acquire some intuition in simple cases, such as that of the usual sphere S^2 embedded in $\mathbf{R}^3$.

Thus, we consider a sphere, of unit radius, defined parametrically by the relations

$$\begin{cases} x = \cos\varphi \cdot \sin\theta \\ y = \sin\varphi \cdot \sin\theta \\ z = \cos\theta. \end{cases} \qquad (D.1)$$

A point of the sphere S^2 is thus fixed by giving the two coordinates (θ, φ), and the relations (D.1) constitute a mapping of S^2 onto $\mathbf{R}^3$, i.e. an embedding of the sphere in Euclidean space. We could, however, regard this rather differently: with a point of S^2 we associate the coordinates (θ, φ), i.e. a point of a space $\mathbf{R}^2$ (see **Fig. D.1**). We examine this more closely. What are the coordinates of the north pole of the sphere? It is easy to check that this point has coordinates $(0, \varphi)$ where φ is *arbitrary*! In other words, the north pole is represented not by a point of the plane (θ, φ), but by a segment of a straight line. To this first anomaly of the coordinate system we may add another. Consider an arbitrary point of the meridian $\varphi = 0$, i.e. a point with coordinates $(0, \theta)$: this can also be represented by the point $(2\pi, \theta)$ in the space $\mathbf{R}^2$ of pairs (θ, φ). If we want to obtain a representation of S^2 in $\mathbf{R}^2$, we must eliminate these pathologies, and thus exclude the boundaries which give problems. It follows that only a part of the sphere – S^2 minus a semi-great-circle from one pole to the other – is correctly represented by an *open* set (such as a rectangle without its boundaries) of $\mathbf{R}^2$: to each point of S^2 there corresponds one and only one point of this open set, and reciprocally.

Such a representation (i.e. giving an open set of S^2 and a mapping onto $\mathbf{R}^2$) constitutes a *local chart* of the sphere. The chart is incomplete, as we have just seen:

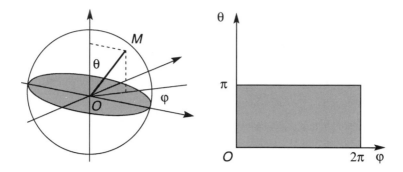

Fig. D.1. Representation of the sphere S^2 **in** $\mathbf{R}^2$. Although the spherical coordinates (θ, φ) obey $0 \leq \theta \leq \pi$ and $0 \leq \varphi \leq 2\pi$, we must limit the variations of θ and φ so as to exclude the boundaries. Otherwise a point of the meridian $\varphi = 0$ would be represented by two points of $\mathbf{R}^2$ and the north (south) pole would be represented by a segment of a straight line. Only a part of the sphere is represented by this coordinate system, in an open set of $\mathbf{R}^2$: the rectangle shown in the figure, minus its boundary.

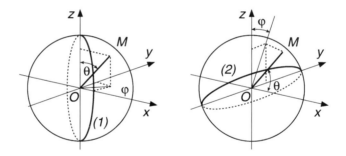

Fig. D.2. With the use of standard polar coordinates, we represent the sphere S^2 minus the meridian denoted by (1). Using another polar system (shown on the figure with the x axis chosen as the new "z axis") we describe S^2 minus the meridian (2). The second coordinate system thus allows us the describe the points of (1).

an open set of the sphere (shown above) is correctly represented by an open rectangle in $\mathbf{R}^2$.

If we want to represent *all* of the sphere, we need another chart; we could for example use different polar coordinates, whose representation would break down for a different semi-great-circle (see **Fig. D.2**) which does not intersect the first one. The two charts together would allow us to represent the *whole* sphere. This is called an *atlas*. Of course these two coordinate systems must give consistent results in regions of S^2 where both are defined.

At this point we could argue that we have not chosen a good coordinate system. We therefore study examples of other systems. We consider first the local chart given by

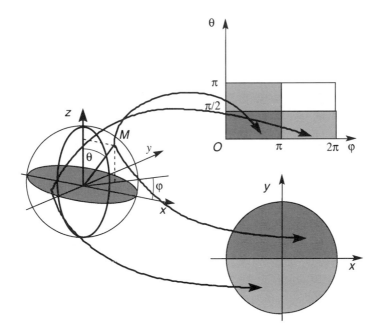

Fig. D.3. Comparison of two local charts on the sphere S^2. Polar coordinates allow us to represent the sphere in the open set $\mathcal{V}_1$ of $\mathbf{R}^2$ (mapping Ψ_1) and the coordinates (x,y) in the open set $\mathcal{V}_2$ of $\mathbf{R}^2$. The common regions of S^2 are shown double-hatched in $\mathcal{V}_1$ and $\mathcal{V}_2$.

the open hemisphere

$$|x| < 1, \quad |y| < 1, \quad z = +\sqrt{1 - x^2 - y^2}, \tag{D.2}$$

and the mapping of S^2 on to $\mathbf{R}^2$ defined by the coordinates (x,y). In $\mathbf{R}^2$ the image of the hemisphere is the open disc

$$x^2 + y^2 < 1 \tag{D.3}$$

(see **Fig. D.3**). To cover the sphere completely we have to use other local charts. We could for example use besides (D.2) the open sets

$$(i)\ x > 0, \ (ii)\ y > 0, \tag{D.4}$$

which are also open hemispheres, and to which we should add $S^2 - (\text{D.2}) - (\text{D.4})$ (open). This example shows that an *atlas* of the sphere must involve several charts. It can in fact be proved that at least *two* charts are needed to cover the entire sphere. This situation is completely general; given any surface, there is *a priori* no coordinate system which covers it completely.

It is interesting to compare two coordinate systems: in a region of S^2 where they overlap they must give the same results, i.e. we must be able to pass from one system to the other.

Figure D.3 illustrates a case where the first local chart $\{\mathcal{V}_1, \Psi_1\}$ is given by the open hemisphere $y > 0$ *and* polar coordinates (θ, φ) constrained by $0 < (\theta, \varphi) < \pi$, while

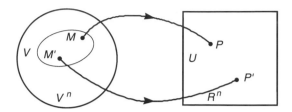

Fig. D.4. A manifold. To a point M of the open subset V of the manifold $\mathscr{V}^n$ there corresponds a point P of $\mathbf{R}^n$; the open set V corresponds to the open set U of $\mathbf{R}^n$ and conversely. If M and M' are neighbouring points of U in $\mathbf{R}^n$, their counterparts P and P' in $\mathbf{R}^n$ are also neighbours, and conversely.

the second local chart, i.e. $\{\mathscr{V}_2, \Psi_2\}$, is given by the open hemisphere $z > 0$ *and* the projections (x, y). Of course, in the intersection of the two hemispheres (i.e. the quarter sphere $y, z > 0$), the two coordinate systems are related by

$$\begin{cases} x = \sin\theta \cdot \cos\varphi \\ y = \sin\theta \cdot \sin\varphi \end{cases} \tag{D.5}$$

where θ and φ are constrained by

$$\begin{cases} 0 < \theta < \pi/2 \\ 0 < \varphi < \pi. \end{cases} \tag{D.6}$$

The maps of S^2 onto $\mathbf{R}^2$ defining the local charts must have an important property: given two neighbouring points of S^2 we wish to find that the representative points of $\mathbf{R}^2$ are close also. Conversely, neighbouring points of the chart should represent neighbouring points of S^2. In other words, the mapping Ψ of S^2 onto $\mathbf{R}^2$ must be continuous as well as its inverse. Open sets of $\mathbf{R}^2$ must correspond to open sets of S^2 and *vice versa*; Ψ is a *homeomorphism*.

1   Differentiable manifolds

The example above is very instructive in several ways, and can be generalised to surfaces of any dimension embedded in $\mathbf{R}^n$. However, the notion of a manifold goes beyond this: a manifold is defined *intrinsically*, independently of any possible embedding in $\mathbf{R}^n$. It is an independent mathematical object. Thus, a manifold differs from a vector space in several regards. There is no origin or scalar multiplication, addition, parallelism at distance, or constant vector field. The linear structure of a vector space is lost in favour of changes of curvilinear coordinates and the associated covariance.

   An n-dimensional manifold $\mathscr{V}^n$ is a set which is locally homeomorphic to an open set of $\mathbf{R}^n$. An element of any open set of an *n*-dimensional manifold is represented by a point of $\mathbf{R}^n$ (i.e. by n parameters) which is an element of an open subset. Conversely, to an element of an open set of $\mathbf{R}^n$ corresponds one and only one point of an open set of the

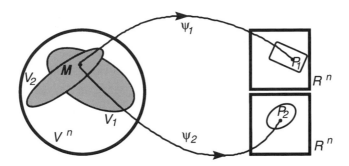

Fig. D.5. The two open sets V_1 and V_2 of $\mathscr{V}^n$ have nonzero intersection. A point of the intersection is thus represented in two different ways in $\mathbf{R}^n$: Ψ_1 gives coordinates $\{\xi^i\}$ and Ψ_2 gives coordinates $\{\eta^i\}$ for the same point M.

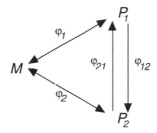

Fig. D.6. Coordinate changes φ_2 corresponding to the diagram above, i.e. to $\varphi = \Psi_2 o \Psi_1^{-1}$.

manifold; moreover this mapping is bicontinuous (**Fig. D.4**). Intuitively this means that $\mathscr{V}^n$ is locally "of the same type" as $\mathbf{R}^n$; $\mathscr{V}^n$ has the same topology – the same notion of continuity – as $\mathbf{R}^n$. The numbers $\xi^1(M), \xi^2(M), \ldots, \xi^n(M)$ are the *coordinates* of M. As before, given an open set V of $\mathscr{V}^n$ and a mapping Ψ, injective and bicontinuous *(a homeomorphism)*, the pair $\{V, \Psi\}$ is called a chart. Similarly, a denumerable family of open sets V_i of $\mathscr{V}^n$ covering the manifold *and* homeomorphisms Ψ_i (i.e. a denumerable family of *local charts* $\{V_i, \Psi_i\}$) is called an *atlas*.

Given two local charts $\{V_1, \Psi_1\}$ and $\{V_2, \Psi_2\}$, where V_1 and V_2 have common points, it is clear that the latter have two possible representations – two different coordinate systems – in $\mathbf{R}^n$ (**Fig. D.5**). If $\{\xi^i\}$ and $\{\eta^i\}$ are the coordinates of a point M common to V_1 and V_2, the passage from one system to the other involves a transformation (**Fig. D.6**) φ,

$$\varphi = \Psi_2 o \Psi_1^{-1}. \tag{D.7}$$

Let P_1 be a point of $\mathscr{V}_1$ representing a point M of $\mathscr{V}^n$: $P_1 = \Psi_1(M)$. In V_2, the representative point of M, i.e. P_2, is $\Psi_2(M)$: $P_2 = \Psi_2(M)$. If φ is the *coordinate change* sought, i.e. such that $P_2 = \varphi(P_1)$, it is clear that we have $\varphi_2(M) = \varphi[\Psi_1(M)]$, giving (D.7).

A differentiable manifold is a manifold such that all possible mappings[1] $\varphi_{ij} = \Psi_j o \Psi_i^{-1}$ *are differentiable.* In practice this means that coordinate changes are differentiable functions.

These definitions allow one to define differential forms, integration, etc.

---

[1] Of course, $\varphi_{ii} = I$, the identity map.

References

E.A. Abbott, *Flat Land, A Romance of Many Dimensions* [Barnes and Nobles, New York, 1963].

A. Abragam, *The Mössbauer Effect* [Gordon and Breach, New York,1964].

W.S. Adams, The Relativity Displacement of the Spectral Lines in the Companion of Sirius [*Proc. Nat. Acad. Sci. (USA)* **11**, 382, 1925; and *Observatory*, **49**, 88, 1925].

E.G. Adelberger, B.R. Heckel, G. Smith, Y. Su, H.E. Swanson, Eötvös Experiments, Lunar Ranging and the Strong Equivalence Principle [*Nature,* **347**, 261, 1990a].

E.G. Adelberger, C.W. Stubbs, B.R. Heckel, Y. Su, H.E. Swanson, G. Smith, J.H. Gundlach, Testing the Equivalence Principle in the Field of the Earth: Particle Physics at masses below $1\mu eV$? [*Phys. Rev.* **D42**, 3267, 1990b].

T. Alvager, F.J.M. Farley, J. Kjellman, I. Wallin, Test of the Second Postulate of Special Relativity in the GeV Region [*Phys. Letts.* **12**, 260, 1964].

E. Amaldi, C. Cosmelli, G.V. Pallotino, G. Pizzella, R. Rapagnani, F. Ricci, Preliminary Results on the Operation of a 2270 Kg Cryogenic Gravitational – Wave Antenna with a Resonant Capacitive Transducer and a d.c. SQUID Amplifier [*Nuovo Cim.* **9C**, 829, 1986].

J.D. Anderson, P.B. Esposito, W. Martin, C.L. Thornton, D.O. Muhleman, Experimental Test of General Relativity Using Time-Delay Data from Mariner 6 and Mariner 7 [*Ap. J.* **200**, 221, 1975].

J.L. Anderson, *Principles of Relativity Physics* [Academic Press, New York, 1967].

J.N. Bahcall, W. Sargent, M. Schmidt, An Analysis of the Absorption Spectrum of 3C191 [*Ap. J.* **149**, 11, 1967].

J.N. Bahcall, M. Schmidt, Does the Fine–Structure Constant Vary with Cosmic Time? [*Phys. Rev. Letts.* **19**, 1294, 1967].

J. Bailey, K. Borer, F. Combley, H. Drum, F. Farley, J. Field, W. Flegel, P. Hattersley, F. Krienen, F. Lange, E. Picasso, W. von Rüden, The Anomalous Magnetic Moment of Positive and Negative Muons [*Phys. Letts.* **68B**, 191, 1977].

J.M. Barnothy, *History of Gravitational Lenses and the Phenomena they Produce,* in *Gravitational Lenses,* J.M. Moran, J.N. Hewitt, K.Y. Lo, eds. [Springer, Berlin, 1989].

A.O. Barut, *Electrodynamics and Classical Theory of Fields* [McMillan, New York, 1964, reprinted 1965].

J.W. Beams, Finding a Better Value for G [*Phys. Today*, May 1971].

J. Bekenstein, Fine–structure constant: is it really a constant? [*Phys. Rev.* **D25,** 1527, 1982].

J.D. Bekenstein, *The Missing Light Puzzle: A Hint about Gravitation?* in *Proceedings of the 2nd Canadian Conference on General Relativity and Relativity Astrophysics*; A. Coley, C. Dyer and T. Tupper, eds. [World Scientific, Singapore, 1988].

L. Bel, *Rigid Motion Invariance of Newtonian and Einstein's Theories of General Relativity,* in *Recent Developments in Gravitation,* E. Verdaguer, J. Garriga, J. Céspedes, eds. [World Scientific, Singapore, 1990].

F. Belinfante, On the Current and the Density of the Electric Charge, the Energy, the Linear Momentum, and the Angular Momentum of Arbitrary Fields [*Physica* **7,** 449, 1940].

B. Bertotti, *Gravitational Experiments* [Ecole d'Été de Varenna, 1972; Acad. Press; New York, 1974].

B. Bertotti, D. Brill, R. Krotkov, *Experiments on Gravitation,* in *Gravitation: An Introduction to Current Research,* L. Witten, ed. [J. Wiley, New York, 1962].

J.E. Blamont, F. Roddier, Precise Observations of the Profile of the Fraunhoffer Strontium Resonance Line. Evidence for the Gravitational Redshift on the Sun [*Phys. Rev. Letts.* **7,** 437, 1961].

D. Bohm, *The Special Theory of Relativity* [Benjamin, New York, 1965].

H. Bondi, *Cosmology* [Cambridge University Press, Cambridge, 1960].

H. Bondi, *Some Special Solutions of the Einstein Equations, part I: Special Relativity,* in *Lectures on General Relativity,* A. Trautman, H. Bondi and F.A.E. Pirani, eds. [Prentice Hall, Englewood Cliffs, 1965].

M. Born, E. Wolf, *Principles of Optics* [Pergamon Press, Oxford, 1975].

R.R. Bourassa, R. Kantowski, The Theory of Gravitational Lenses [*Ap. J.* **195,** 13, 1975].

R.R. Bourassa, R. Kantowski, T.D. Norton, The Spheroidal Gravitational Lens [*Ap. J.* **185,** 747, 1975].

M.G. Bowler, *Gravitation and Relativity* [Pergamon Press, Oxford, 1976].

V.B. Braginsky, V.I. Panov, Verification of the Equivalence of Inertial and Gravitational Mass [*Soviet Phys. J.E.T.P.* **34,** 463, 1971].

C. Brans, R.H. Dicke, Mach's Principle and a Relativistic Theory of Gravitation [*Phys. Rev.* **124,** 925, 1961].

J. Brault, Gravitational Redshift of Solar Lines [*Bull. Am. Phys. Soc.* **8,** 28, 1963].

K. Brecher, Is the Speed of Light Independant of the Velocity of the Source? [*Phys. Rev. Letts.* **39,** 1051, 1977].

A. Brillet, J.L. Hall, Improved Laser Test of the Isotropy of Space [*Phys. Rev. Letts.* **42,** 549, 1979].

A. Brillet, J.L. Hall, *An Improved Test of the Isotropy of Space Using Laser Techniques,* in *Proceedings of the Second Marcel Grossmann Meeting on General Relativity,* R. Ruffini, ed. [North Holland, Amsterdam, 1982].

B.C. Brown, G.E. Masek, T. Maung, E.S. Miller, H. Ruderman, W. Vernon, Experimental Comparison of the Velocities of eV (Visible) and GeV Electromagnetic Radiation [*Phys. Rev. Letts.* **30,** 763, 1973].

W. Buchel (trans. I.M. Freeman), Why is Space Three-dimensional? [*Am. J. Phys.* **37,** 1222, 1969].

M. Bunge, Mach's Critique of Newtonian Mechanics [*Am. J. Phys.* **34,** 585, 1966].

W.L. Burke, Multiple Gravitational Imaging by Distributed Masses [*Ap. J.* **244,** L1, 1981].

E. Byckling, K. Kajantie, *Particle Kinematics* [J. Wiley, New York, 1973].

V. Canuto, Atomic and Gravitational Clocks [*Int. J. Theor. Phys.* **28,** 1005, 1990].

V. Canuto, I. Goldmann, Atomic and Gravitational Clocks [*Nature* **296,** 709, 1982a].

V. Canuto, I. Goldmann, *The Strong Equivalence Principle and its Violation,* in *Early Evolution of the Universe and its Present Structure,* G.O. Abell, G. Chincarini, eds.; IAU Symposium 104 [Reidel, Dordrecht, 1982b].

V. Canuto, I. Goldmann, Astrophysical Consequences of a Violation of the Strong Equivalence Principle [*Nature,* **304,** 311, 1983].

S.M. Carroll, W.H. Press, E.L. Turner, The Cosmological Constant [*Ann. Rev. Astron. Astrophys.,* **30,** 499, 1992]

B. Carter, *Mathematical Foundations of the Theory of Relativistic Stellar and Black Hole Configuration,* in 1986 Cargèse Summer School *"Gravitation in Astrophysics",* B. Carter, J.B. Hartle, eds. [Plenum Press, 1987].

B. Carter, Outer Curvature and Conformal Geometry and Embedding [*J. Geom. Phys.* **8,** 53, 1992].

J.P. Cedarholm, G.F. Bland, B.L. Havens, C.H. Townes, New Experimental Test of Special Relativity [*Phys. Rev. Letts.* **1,** 342, 1958].

D.C. Champeney, G.R. Isaak, A.M. Khan, [*Phys. Letts.* **7,** 241, 1963].

T.E. Chupp, R.J. Hoare, R.A. Loveman, E.R. Obeiza, J.M. Richardson, M.E. Wagshul, Results of a New Test of Local Lorentz Invariance: A Search for Mass Anisotropy in ^{21}Ne [*Phys. Rev. Letts.* **63,** 1541, 1989].

E.E. Clark, The Uniform Transparent Gravitational Lens [*Month. Not. R. Astr. Soc.* **158,** 233, 1972].

G.M. Clemence, The Relativity Effect in Planetary Motions [*Rev. Mod. Phys.* **19,** 361, 1947].

B. Cohen, *Les Origines de la Physique Moderne* [Petite Bibliothèque Payot n° 21; Paris, 1960].

J.H. Cooke, R. Kantowski, Time Delays for Multiply Imaged Quasars [*Ap. J.* **195,** L11, 1975].

J.L. Coolidge, *A History of Geometrical Methods* [Dover; New York, 1963].

T. Damour, Strong Field Effects in General Relativity [*Helv. Acta Phys.* **59,** 292, 1986].

T. Damour, G.W. Gibbons, J.H. Taylor, Limits on the Variability of G Using Binary-Pulsar Data [*Phys. Rev. Letts.* **61,** 1151, 1988].

T. Damour, G. Schäfer, New Tests of the Strong Equivalence Principle Using Binary-Pulsar Data [*Phys. Rev. Letts.* **66,** 2549, 1991].

B. Davies, Elementary Theory of Perihelion Precession [*Am. J. Phys.* **51,** 909, 1983].

P.C.W. Davies, Time Variation of the Coupling Constants [*J. Phys.* **A5,** 1296, 1972].

P.C.W. Davies, *The Search for Gravity Waves* [Cambridge University Press, Cambridge, 1980].

N. Deruelle, T. Piran, eds. *Gravitational Radiation* [NATO Advanced Institute, Les Houches 1982; North Holland, Amsterdam, 1983].

S. Deser, Self-Interaction and Gauge Invariance [*Gen. Relat. Grav.* **1,** 9, 1970].

S. Deser, *The Gravitational Field* [Lectures notes; Orsay, 1971].

W. De Sitter, *Phys. Z.* **14,** 429, 1267, 1913 [cited by K. Brecher (1977)].

R.H. Dicke, The Eötvös Experiment [*Scientific Am.,* p. 84, Dec. 1961].

R.H. Dicke, *The Theoretical Significance of Experimental Relativity* [Gordon and Breach; New York (1965)].

R.H. Dicke, The Oblateness of the Sun and Relativity [*Science* **184,** 419, 1974].

R.H. Dicke, H.M. Goldenberg, Solar Oblateness and General Relativity [*Phys. Rev. Letts.* **18,** 313, 1967].

R.H. Dicke, H.M. Goldenberg, The Oblateness of the Sun [*Ap. J. Sup.* **27,** 131, 1974].

A.D. Dolgov, Ya. B. Zeldovich, Cosmology and Elementary Particles [*Rev. Mod. Phys.* **53,** 1, 1981].

R.W.P. Drever, A Search for Anisotropy of Inertial Mass Using a Full Precession Technique [*Phil. Mag.* **6,** 683, 1961].

F.J. Dyson, Time Variation of the Charge of the Proton [*Phys. Rev. Letts.* **19,** 1291, 1967].

D.H. Eckart, C. Jekeli, A.R. Lazarewicz, A.J. Romaides, R. Sands, Tower Gravity Experiment: Evidence for Non Newtonian Gravity [*Phys. Rev. Letts.* **60,** 2567, 1988].

A.S. Eddington, *The Mathematical Theory of Relativity* [Cambridge UP, London, 1922].

N. Efimov, *Géométrie Supérieure* [Mir, Moscow, 1981].

J. Ehlers, *General Relativity and Kinetic Theory,* in *General Relativity and Cosmology* R.K. Sachs, ed., Varenna Course, 1969 [Academic Press, New York, 1971].

J. Ehlers, *The Nature and Structure of Spacetime* in *The Physicist's Conception of Nature,* J. Mehra, ed. [Reidel Publ. Co., Dordrecht, Holland, 1973].

A. Einstein, *Zur Allgemeinen Relativitätstheorie* [Preuss. Akad. Wissen., Berlin, 778, 1915a].

A. Einstein, *Zur Allgemeinen Relativitätstheorie* [Preuss. Akad. Wissen., Berlin, 799, 1915b].

A. Einstein, Lens-Action of a Star by the Deviation of Light in the Gravitational Field [*Science* **84,** 506, 1936].

L.P. Eisenhart, *Riemannian Geometry* [Dover, New York, 1966].

J. Eisenstaedt, Histoire et Singularités de la Solution de Schwarzschild (1915–1923) [*Arch. Hist. Exact Sci.* **27**, 157, 1982].

J. Eisenstaedt, La Relativité Générale à l'Etiage (1925–1955) [*Arch. Hist. Exact Sci.* **35**, 115, 1986].

J. Eisenstaedt, Trajectoires et Impasses de la Solution de Schwarzschild [*Arch. Hist. Exact Sci.*, **37**, 275, 1988].

J. Eisenstaedt, *Cosmology: A Space for Thought on General Relativity*, in *Foundations of Big Bang Cosmology*, F.W. Meyerstein, ed. [World Scientific, Singapore, 1989].

G.F.R. Ellis, *Relativistic Cosmology*, in *General Relativity and Cosmology*, R.K. Sachs, ed.; Varenna course, 1969; [Academic Press, New York, 1971].

G.F.R. Ellis, *Cosmology and Verifiability* [Boston Colloquium for the Philosophy of Science, Nov. 1973].

R. Eötvös, Über die Anziehung der Erde auf Verschiedene Substanzen [*Math. Naturwiss. Ber. aus. Ungarn* **8**, 65, 1889].

R. Eötvös, D. Pekàr, E. Fekete, Beiträge zum Gesetzte der Proportionalität von Trägheit und Gravität [*Annalen d. Phys.* **68**, 11, 1922].

F. Everitt, W.W. Hansen, *Gravitation, Relativity and Precise Experimentation* in *Proceedings of the First Marcel Grossmann Meeting on General Relativity*, R. Ruffini, ed.; Trieste, 1975 [North Holland, Amsterdam, 1975].

W.M. Fairbanks, *The Use of Low Temperature Technology in Gravitation Experiments*, in *Varenna Summer School (1972), Experimental Gravitation*, B. Bertotti, ed. [Academic Press, New York, 1974].

S. Feuer, *Einstein et le Conflit des Générations* [Ed. Complexe, Bruxelles, 1978]. *Einstein and the Generation of Science* [Basic Books, 1974].

R.P. Feynman, *Lectures on Gravitation* [lecture notes by F.B. Morinigo, W.G. Wagner; Cal. Tech. 1962–1963].

M. Fierz, W. Pauli, Relativistic Wave Equations of Arbitrary Spin in an Electromagnetic Field [*Proc. Roy. Soc.* **A173**, 211, 1939].

L. Flamm, Beitrage zur Einsteinschen Gravitationstheorie [*Physik. Zeit.* **17**, 448, 1916].

H. Flanders, *Differential Forms* [Academic Press; New York, 1963].

V. Fock, *The Theory of Space–Time and Gravitation* [Pergamon; Oxford, 1966].

E.B. Fomalont, R.A. Sramek, The Deflection of Radio Waves by the Sun [*Comm. Astrophys.* **7**, 19, 1977].

J. Foster, J.D. Nightingale, *A Short Course in General Relativity* [Longman, London, 1979].

J.G. Fox, Experimental Evidence for the Second Postulate of Special Relativity [*Am. J. Phys.* **30**, 297, 1962].

J.G. Fox, Evidence against Emission Theories [*Am. J. Phys.* **33**, 1, 1965].

Th. Frankel, *Gravitational Curvature* [Freeman, San Francisco, 1979].

A. Friedman, Über die Krümmung des Raumes [*Zeits. Phys.* **10**, 377, 1922].

A. Friedman, Über die Möglichkeit einer Welt mit konstanter negativer Krümmung des Raumes [*Zeits. Phys.* **21,** 326, 1924].

G. Gatewood, C. Gatewood, A Study of Sirius [*Ap. J.* **225,** 191, 1978].

G. Gatewood, J. Russel, Astrometric Determination of the Gravitational Redshift of Van Maanen 2 (EG5) [*Ap. J.* **79,** 815, 1974].

G.T. Gillies, The Newtonian Gravitational Constant: an Index of Measurements in *Rapport du Bureau International des Poids et Mesures BITM–83/1*; Sèvres, 1983, [*Metrologia* **24** Sup., I, 1987].

H. Goldstein, *Classical Mechanics* [Addison-Wesley, Reading, 1980].

M.L. Good, K_0 and the Equivalence Principle [*Phys. Rev.* **121,** 311, 1961].

J.L. Greenstein, A. Boksenberg, R. Carswell, K. Shortridge, The Rotation and Gravitational Redshift of White Dwarfs [*Ap. J.* **212,** 186, 1977].

J.L. Greenstein, J.B. Oke, H.L. Shipman, Effective Temperature, Radius, and Gravitational Redshift of Sirius B [*Ap. J.* **169,** 563, 1971].

J.L. Greenstein, V. Trimble, The Einstein Redshift of White Dwarfs [*Ap. J.* **149,** 283, 1967].

S.R. de Groot, W.A. Van Leeuwen, Ch. G. Van Weert, *Relativistic Kinetic Theory* [North Holland, Amsterdam, 1980].

D.J. Grove, J.G. Fox, e/m for 385 MeV Protons [*Phys. Rev.* **90,** 378, 1953].

S.N. Gupta, Gravitation and Electromagnetism [*Phys. Rev.* **96,** 1683, 1954].

S.N. Gupta, Einstein's and Other Theories of Gravitation [*Rev. Mod. Phys.* **29,** 337, 1957].

S.N. Gupta, *Quantum Theory of Gravitation,* in *Recent Developments in General Relativity* [Pergamon Press, New York, 1962].

Z.G.T. Gviragossián, G.B. Rothbart, M.R. Yearian, R.A. Gearheart, J.J. Murray, Relative Velocity Measurements of Electrons and Gamma Rays at 15 GeV. [*Phys. Rev. Letts.* **34,** 355, 1975].

R. Hagedorn, *Relativistic Kinematics* [W.A. Benjamin, New York, 1964].

E.R. Harrison, *Cosmology, the Science of the Universe* [Cambridge University Press, Cambridge, 1981].

M.P. Haugan, C.M. Will, Weak Interactions and Eötvös Experiments [*Phys. Rev. Letts.* **37,** 1, 1976].

M.P. Haugan, C.M. Will, Modern Tests of Special Relativity [*Phys. Today,* May 1987, p. 69].

P. Havas, Four-Dimensional Formulation of Newtonian Mechanics and their Relations in the Special and General Theory of Relativity [*Rev. Mod. Phys.* **36,** 938, 1964].

R.W. Hellings, P.J. Adams, J.D. Anderson, M.S. Keesey, E.L. Lau, E.M. Standish, V.M. Canuto, I. Goldman, Experimental Test of the Variability of G Using Viking Lander Ranging Data [*Phys. Rev. Letts.* **51,** 1609, 1983].

J. Hershey, Astrometric Study of the Sproul Plate Series on Van Maanen's Star, Including Gravitational Redshift [*A. J.* **83,** 197, 1978].

D. **Hilbert**, *Die Grundlagen der Physik* [Nachr. Gesell. Wissensch., Göttingen, 395, 1915].

D. **Hilbert**, S. **Cohn–Vossen**, *Geometry and the Imagination* [Chelsea, London, 1952].

B. **Hoffmann**, H. **Dukas** *Albert Einstein, Creator and Rebel* [Viking Press, New York, 1972].

S.C. **Holding**, F.D. **Stacey**, G.J. **Tuck**, Gravity in Mines – An Investigation of Newton's Law [*Phys. Rev.* **D33**, 3487, 1986].

J.P. **Hsu**, Analysis of Weak Interactions and Eötvös Experiments [*Phys. Rev.* **D17**, 3164, 1978].

V.W. **Hughes**, H.G. **Robinson**, V. **Beltran–Lopez**, Upper Limits for the Anisotropy of Inertial Mass from Nuclear Resonance Experiments [*Phys. Rev. Letts.* **4**, 342, 1960].

R.A. **Hulse**, J.H. **Taylor**, Discovery of a Pulsar in a Binary System [*Ap. J.* **195**, L51, 1975].

L. Kh. **Ingel'**, Gravitational Lenses and Velocities faster than Light [*Sov. Astron. Letts.* **1**, 54, 1975].

J.M. **Irvine**, The Constancy of the Laws of Physics in the Light of Prehistoric Nuclear Reactors [*Contemp. Phys.* **24**, 427, 1983].

G.R. **Isaak**, The Mössbauer Effect, Application to Relativity [*Phys. Bull.* **21**, 250, 1970].

C. **Itzykson**, J.P. **Zuber**, *Quantum Field Theory* [McGraw Hill, New York, 1980].

J.D. **Jackson**, *Classical Electrodynamics* [Wiley, New York, 1962].

M. **Jammer**, *Concepts of Space* [Harvard University Press; Cambridge (Mass.), 1954].

M. **Jammer**, *Concepts of Mass* [Harvard University Press; Cambridge (Mass.), 1961].

M. **Jammer**, *Some Fundamental Problems in the Special Theory of Relativity* in *"Problems in the Foundations of Physics"*, 1977 Varenna Summer School; Toraldo di Francia, ed. [North Holland, Amsterdam, 1979].

G. **Kalman**, Lagrangian Formalism in Relativistic Dynamics [*Phys. Rev.* **123**, 384, 1961].

R.J. **Kennedy**, E.M. **Thorndike**, Experimental Establishment of the Relativity of Time, [*Phys. Rev.* **42**, 400, 1932].

C.W. **Kilmister**, *General Theory of Relativity* [Pergamon Press; Oxford, 1973].

H. von **Klüber**, The Determination of Einstein's Light-Deflection in the Gravitational Field of the Sun [*Vistas in Astronomy* **3**, 47, 1960].

L. **Koester**, Verification of the Equivalence of Gravitational and Inertial Mass for the Neutron [*Phys. Rev.* **D14**, 907, 1976].

R.H. **Kraichnan**, Special-Relativistic Derivation of Generally Covariant Gravitation Theory [*Phys. Rev.* **98**, 1118, 1955].

L.B. **Kreuzer**, Experimental Measurement of Equivalence of Active and Passive Gravitational Mass [*Phys. Rev.* **169**, 1007, 1968].

K. **Kuroda**, H. **Hirakawa**, Experimental Test of the Law of Gravitation [*Phys. Rev.* **D32**, 342, 1985].

M. Lachièze-Rey, J.P. Luminet, A Primer in Cosmic Topology [*Phys. Reports* **254,** 135, 1995.]

S.K. Lamoreaux, J.P. Jacobs, B.R. Heckel, J.F. Raab, E.N. Fortson, New Limits on Spatial Anisotropy from Optically Pumped ^{201}Hg and ^{199}Hg [*Phys. Rev. Letts.* **57,** 3125, 1986].

L. Landau, E. Lifschitz, *Mechanics* [Pergamon Press, Oxford, 1960].

L. Landau, E. Lifschitz, *The Classical Theory of Fields* [Pergamon Press, Oxford, 1962].

L. Landau, E. Lifschitz, *Quantum Mechanics, Non-relativistic Theory* [Pergamon Press, Oxford, 1965].

M. Le Bellac, J.M. Lévy–Leblond, Galilean Electromagnetism [*Nuovo Cim.* **14B,** 217, 1973].

J.M. Lévy–Leblond, Mécanique Quantique des Forces de Gravitation et Stabilité de la Matière [*J. Phys. (Paris),* Colloque C3; **30,** 43, 1969 (Suppl.)].

A. Lichnérowicz, *Eléments de Calcul Tensoriel* [A. Colin; Paris, 1955].

A. Lichnérowicz, *Algèbre et Analyse Linéaire* [Masson, Paris, 1960].

S. Liebes, Gravitational Lenses [*Phys. Rev.* **B133,** 835, 1964]

A.P. Lightman, W.H. Press, R.H. Price, S.A. Teukolsky, *Problem Book in Relativity and Gravitation* [Princeton University Press, Princeton, 1975].

F. Link, *Eclipse Phenomena in Astronomy* [Springer, Berlin, 1969].

D.R. Long, Why do we Believe Newtonian Gravitation at Laboratory Dimensions? [*Phys. Rev.* **D9,** 850, 1974].

D.R. Long, Experimental Determination of the Gravitational Inverse Square Law [*Nature* **260,** 417, 1976].

M.J. Longo, Tests of Relativity from SN1987A [*Phys. Rev.* **D36,** 3276, 1987].

H. Mandelberg, L. Witten, Experimental Verification of the Relativistic Doppler Effect, [*J. Opt. Soc. Am.* **52,** 529, 1962].

R.W. Mandl, Letter to A. Einstein (1935, unpublished).

R. Mansouri, R.U. Sexl, A Test Theory of Special Relativity: I Simultaneity and Clock Synchronization [*Gen. Rel. Grav.* **8,** 497, 1977a].

R. Mansouri, R.U. Sexl, A Test Theory of Special Relativity: II First Order Tests [*Gen. Rel. Grav.* **8,** 515, 1977b].

R. Mansouri, R.U. Sexl, A Test Theory of Special Relativity: III Second-Order Tests [*Gen. Rel. Grav.* **8,** 809, 1977c].

M. Maurette, Fossil Nuclear Reactors [*Ann. Rev. Nucl. Sci.* **26,** 319, 1976].

S. Mavridès, *L'Univers Relativiste* [Masson, Paris, 1973].

J. Mehra, *Einstein, Hilbert and the Theory of Gravitation* [Reidel, 1974].

H. Minkowski, *Space and Time,* in *The Principle of Relativity* [Dover, New York, 1908].

N. Mio, K. Tsubono, H. Hirakawa, Experimental Test of the Law of Gravitation at Small Distances [*Phys. Rev.* **D36,** 2321, 1987].

C.W. Misner, Mixmaster Universe [*Phys. Rev. Letts.* **22,** 1071, 1969].

C.W. Misner, K.S. Thorne, J.A. Wheeler, *Gravitation* [Freeman, San Francisco, 1973].

J.M. Moran, J.N. Hewitt, K.Y. Lo, eds.; *Gravitational Lenses* [Springer, Berlin, 1989].

D.O. Muhleman, R.D. Ekers, E.B. Fomalont, Radio Interferometric Test of the General Relativistic Light Bending near the Sun [*Phys. Rev. Letts.* **24**, 1377, 1970].

E. Namer, *L'Affaire Galilée* [Archives Julliard, Paris, 1975].

C. Neumann, *Über des newtonische Prinzip der Fernwirkung* [Leipzig. 1896].

D. Newman, G.W. Ford, A. Rich, E. Sweetman, Precision Experimental Verification of Special Relativity [*Phys. Rev. Letts.* **40**, 1355, 1978].

E. Newman, J.N. Goldberg, Measurements of Distance in General Relativity [*Phys. Rev.* **114**, 1391, 1959].

G. Nordström, Zur Theorie der Gravitation vom Standpunkt des Relativitäts Prinzips [*Annalen d. Phys.* **42**, 533, 1913].

K. Nordtvedt, Equivalence Principle for Massive Bodies: I Phenomenology [*Phys. Rev.* **169**, 1014, 1968a]; II Theory [*Phys. Rev.* **169**, 1017, 1968b].

K. Nordtvedt, Testing Relativity with Laser Ranging to the Moon [*Phys. Rev.* **170**, 1186, 1968c].

J.E. Norvath, H. Vucetich, Oklo Phenomenon and the Principle of Equivalence [*Phys. Rev.* **D37**, 931, 1988].

B.E.J. Pagel, On the Limits to Past Variability of the Proton–Electron Mass Ratio Set by Quasar Absorption Line Redshifts [*Mon. Not. R. Astr. Soc.* **179**, 81P, 1977].

S. Pakvasa, W.A. Simmons, T.J. Weiler, Test of Equivalence Principle for Neutrinos and Antineutrinos [*Phys. Rev.* **D39**, 1761, 1989].

A. Palatini, Deduzione Invariantiva delle Equazioni Gravitazionali dal Principio di Hamilton [*Rendic. Circ. Mat., Palermo,* **43**, 203, 1919].

V.I. Panov, V.N. Frontov, The Cavendish Experiment at Large Distance [*Soviet Phys. J.E.T.P.* **50**, 852, 1980].

A. Papapetrou, *Lectures in General Relativity* [Reidel, Dordrecht, 1974].

J.A. Peacock, *Review of Gravitational-Lens Theory,* in *Quasars and Gravitational Lenses* [Proceedings of the 24th Liège International Astrophysical Colloquium, 1983].

P.J.E. Peebles, *Principles of Physical Cosmology* [Princeton University Press, Princeton, 1993].

P.J. Peebles, R.H. Dicke, Cosmology and the Radioactive Decay Ages of Terrestrial Rocks and Meteorites [*Phys. Rev.* **128**, 2006, 1962].

A. Peres, Constancy of the Fundamental Electric Charge [*Phys. Rev. Letts.* **19**, 1293, 1967].

A. Peres, N. Rosen, Covariant Formalism for Particle Dynamics [*Nuovo Cim.* **18**, 144, 1960].

H. Poincaré, *La Science et l'Hypothèse* [1904, reprinted by Flammarion, Paris, 1968].

D.M. Popper, Redshift in the Spectrum of 40 Eridani B [*Ap. J.* **120**, 316, 1954].

R.V. Pound, G.A. Rebka, Apparent Weight of Photons [*Phys. Rev. Letts.* **4**, 337, 1960].

R.V. Pound, J.L. Snider, Effect of Gravity on Gamma Radiation [*Phys. Rev.* **B140**, 788, 1964].

J.D. Prestage, J.J. Bollinger, W.M. Itano, D.J. Wineland, Limits for Spatial Anisotropy by Use of Nuclear-Spin-Polarized $^9Be^+$ Ions [*Phys. Rev. Letts.* **54,** 2387, 1985].

R.D. Reasenberg, I.I. Shapiro, P.E. MacNeil, R.B. Goldstein, J.C. Breidenthal, J.C. Brenkle, D.L. Cain, T.M. Kaufman, T.A. Komarek, A.I. Zygielbaum, Viking Relativity Experiment: Verification of Signal Retardation by Solar Gravity [*Ap. J.* **234,** L219, 1979].

E. Recami, R. Mignani, Classical Theory of Tachyons [*Riv. Nuovo Cim.* **4,** 209, 1974].

S. Refsdal, The Gravitational Lens Effect [*Month. Not. R. Astr. Soc.* **128,** 295, 1964a].

S. Refsdal, On the Possibility of Determining Hubble's Parameter and the Masses of Galaxies from the Gravitational Lens Effect [*Month. Not. R. Astr. Soc.* **128,** 307, 1964b].

S. Refsdal, On the Possibility of Testing Cosmological Theories from the Gravitational Lens Effect [*Month. Not. R. Astr. Soc.* **132,** 101, 1966].

H. Reichenbach, *The Philosophy of Space and Time* [Dover, New York, reprinted 1958].

W. Rindler, Counter Example to the Lenz–Schiff Argument [*Am. J. Phys.* **36,** 540, 1968].

W. Rindler, *Essential Relativity* [Springer-Verlag, Berlin, 1977].

H.P. Robertson, Kinematics and World-Structure [*Ap. J.* **82,** 284, 1935].

H.P. Robertson, Kinematics and World-Structure [*Ap. J.* **83,** 187, 1936].

H.P. Robertson, Relativistic Cosmology [*Rev. Mod. Phys.* **5,** 62, 1933].

H.P. Robertson, Postulate versus Observation in the Special Theory of Relativity [*Rev. Mod. Phys.* **21,** 378, 1949].

H.P. Robertson, in *Space Age Astronomy*, A.J. Deutsch, W.B. Klemperer, eds. [Academic Press, New York, 1962].

H.P. Robertson, T.W. Noonan, *Relativity and Cosmology* [W.B. Saunders, Philadelphia, 1968].

D.S. Robertson, W.E. Carter, W.H. Dillinger, New Measurements of Solar Gravitational Deflection of Radio Signals Using VLBI [*Nature* **349,** 768, 1991].

P. Roll, R. Krotkov, R. Dicke, The Equivalence of Inertial and Passive Gravitational Mass [*Ann. Phys. (NY)* **26,** 442, 1964].

R.D. Rose, H.M. Parker, R.A. Lowry, A.R. Kuhlthau, J.W. Beams, Determination of the Gravitational Constant G [*Phys. Rev. Letts.* **23,** 655, 1969].

L. Rosenfeld, Sur le Tenseur Impulsion-Energie [*Mém. Acad. Roy. Belg. Sci.* **18,** n° 6, 1940].

D.K. Ross, L.I. Schiff, Analysis of the Proposed Planetary Radar Reflection Experiment [*Phys. Rev.* **141,** 1215, 1966].

V.N. Rudenko, Relativistic Experiments in Gravitational Fields [*Soviet Phys. Usp.* **21,** 893, 1978].

M.P. Ryan, L.C. Shepley, *Homogeneous Relativistic Cosmologies* [Princeton University Press, Princeton, 1975].

J. Rzewuski, *Field Theory* [Polish Scientific Publishers, Warszawa, 1958].

R.K. Sachs, Gravitational Waves in General Relativity, VI The Outgoing Radiation Condition [*Proc. Roy. Soc.* **A264,** 309, 1961].

M.P. Savedoff, Physical Constants in Extra-Galactic Nebulae [Nature, **178,** 688, 1956].

L.I. Schiff, Sign of the Gravitational Mass of a Positron [*Phys. Rev. Letts.* **1,** 254, 1958].

L.I. Schiff, On Experimental Tests of the General Theory of Relativity [*Am. J. Phys.* **28,** 340, 1960].

L.I. Schiff, *Comparison of Theory and Observation in General Relativity,* in *Relativity Theory and Astrophysics* [Vol. 1]; J. Ehlers, ed. [Am. Math. Soc. Publ., Providence,1967].

L.I. Schiff, J.M. Weisberg, Sign of the Gravitational Mass of a Positron [*Phys. Rev. Letts.* **1,** 254, 1958].

A. Schild, *Lectures on General Relativity,* in *Relativity Theory and Astrophysics* [Vol. 1]; J. Ehlers, ed. [Am. Math. Soc. Publ., Providence, 1967].

P. Schneider, *Gravitational Lenses* [Springer, Berlin, 1992]

B. Schutz, Gravitational Waves on the Back of an Envelope [*Am. J.* **52,** 412, 1984].

B. Schutz, *Geometrical Methods of Mathematical Physics* [Cambridge University Press, Cambridge, 1985].

B. Schutz, *A First Course in General Relativity* [Cambridge University Press, Cambridge, 1988].

J. Schwartz, M. McGuinness, *Einstein for Beginners* [Unwin, 1979].

V.F. Sears, On the Verification of the Universality of Free Fall by Neutron Gravity Refractometry [*Phys. Rev.* **D25,** 2023, 1982].

H. Seeliger, Ueber das newton'sche Gravitationsgesetz [*Astron. Nachrich.* **137,** 9, 1894].

R. Sexl, H. Sexl, *White Dwarfs and Black Holes (An Introduction to Relativistic Astrophysics)* [Academic Press, New York, 1979].

I.I. Shapiro, Fourth Test of General Relativity [*Phys. Rev. Letts.* **13,** 789, 1964].

I.I. Shapiro, Testing General Relativity with Radar [*Phys. Rev.* **141,** 1219, 1966].

I.I. Shapiro, Fourth Test of General Relativity: Preliminary Results [*Phys. Rev. Letts.* **20,** 1265, 1968].

I.I. Shapiro, M.E. Ash, R.P. Ingalls, W.B. Smith, D.B. Campbell, R.B. Dyce, R.F. Jurgens, G.H. Pettengill, Fourth Test of General Relativity: New Radar Results [*Phys. Rev. Letts.* **26,** 1132, 1971].

I.I. Shapiro, W.B. Smith, M.B. Ash, Gravitation Constant: Experimental Bounds on its Time Variation [*Phys. Rev. Letts.* **26,** 27, 1971].

I.I. Shapiro, C.C. Counselman, R.W. King, Verification of the Principle of Equivalence for Massive Bodies [*Phys. Rev. Letts.* **36,** 555, 1976].

I.I. Shapiro,Ross–Schiff Analysis of a Proposed Test of General Relativity: a critique [*Phys. Rev.* **145,** 1005, 1966].

S.L. Shapiro, S.A. Teukolsky, *Black Holes, White Dwarfs and Neutron Stars* [Wiley, New York, 1983]

H.L. Shipman, Masses and Radii of White Dwarfs Stars. III Results for 110 Hydrogen-Rich and 28 Helium-Rich Stars [*Ap. J.* **228,** 240, 1979].

A.I. Shlyakhter, Direct Test of the Constancy of Fundamental Nuclear Constants [*Nature,* **264,** 340, 1976].

P. Sisterna, H. Vucetich, Time Variation of Fundamental Constants: Bounds from Geophysical and Astronomical Data [*Phys. Rev.* **D41,** 1034, 1990].

J.L. Snider, New Measurement of the Solar Gravitational Redshift [*Phys. Rev. Letts.* **28,** 853, 1972].

J.L. Snider, Comments on Two Recent Measurements of the Solar Gravitational Redshift [*Solar Phys.* **36,** 233, 1974].

R. Spero, J.K. Hoskins, R. Newman, J. Pellam, J. Schulz, Tests of the Gravitational Inverse-square Law at Laboratory Distances [*Phys. Rev. Letts.* **44,** 1645, 1980].

F.D. Stacey, G.J. Tuck, Geophysical Evidence for Non-Newtonian Gravity [*Nature* **292,** 230, 1981].

F.D. Stacey, G.J. Tuck, S.C. Holding, A.R. Maher, D. Morris, Constraint on the Planetary Scale Value of the Newtonian Gravitational Constant from the Gravity Profile within a Mine [*Phys. Rev.* **D23,** 1683, 1981].

N. Straumann, *General Relativity and Relativistic Astrophysics* [Springer-Verlag, Berlin, 1984].

J.L. Synge, *The Relativistic Gas* [North Holland, Amsterdam, 1957].

J.L. Synge, *Relativity: The Special Theory* [North Holland, Amsterdam, 1958].

E.F. Taylor, J.A. Wheeler, *Spacetime Physics* [W.H. Freeman, San Francisco, 1966].

J.H. Taylor, L.A. Fowler, P.M. McCulloch, Measurements of General Relativistic Effects in the Binary Pulsar PSR 1913 + 16 [*Nature,* **277,** 437, 1979].

J.H. Taylor, J.M. Weisberg, A New Test of General Relativity: Gravitational Radiation and the Binary Pulsar PSR 1913+16 [*Ap. J.* **253,** 908, 1982].

J.H. Taylor, J.M. Weisberg, Further Experimental Tests of Relativistic Gravity Using the Binary Pulsar PSR 1913+16 [*Ap. J.* **345,** 434, 1989].

W. Thirring, An Alternative Approach to the Theory of Gravitation [*Ann. Phys. (N.Y.)* **16,** 96, 1961].

K.S. Thorne, Gravitational-wave Research, Current Status and Future Prospects, [*Rev. Mod Phys.* **52,** 285, 1971].

K.S. Thorne, *Relativistic Stars, Black Holes and Gravitational Waves,* in *General Relativity and Cosmology,* R.K. Sachs, ed.; Varenna Course, 1969 [Academic Press, New York, 1971].

K.S. Thorne, *The Theory of Gravitational Radiation: an Introductory Review,* in *Rayonnement Gravitationnel,* NATO Advanced Institute; Les Houches, 1982; N. Deruelle, T. Piran, eds. [North Holland, Amsterdam, 1983].

K.S. Thorne, *Gravitational Radiation,* in *300 Years of Gravitation,* S.W. Hawking, W. Israël, eds. [Cambridge University Press, Cambridge, 1989].

M. Tinto, The Search for Gravitational Waves [*Am. J. Phys.* **56,** 1066, 1988].

M.A. Tonnelat, *Les Vérifications Expérimentales de la Relativité Générale* [Masson, Paris, 1964].

M.A. Tonnelat, *Histoire du Principe de Relativité* [Flammarion, Paris, 1971].

A. Trautman, Sur la Théorie Newtonienne de la Gravitation [*Comptes Rendus Acad. Sci.* **257,** 617, 1963].

A.D. Tubbs, A.M. Wolfe, Evidence for Large-Scale Uniformity of Physical Laws, [*Ap. J.* **236,** L105, 1980].

L. Valentin, *Physique sub-atomique: Noyaux et Particules* [Hermann, Paris, 1975].

C. Vanderriest, *Close-up on Gravitational Lensing: the Gravitational Mirages,* in *Gravitation, Geometry and Relativistic Physics,* Lab. "Gravitation et Cosmologie", ed. [Springer, Berlin, 1984].

C. Vanderriest, Les mirages gravitationnels: un outil de choix pour sonder l'Univers [*l'Astronomie,* January 1986].

R.S. Van Dyck, P.B. Schwinberg, H.G. Dehmelt, Precise Measurements of Axial, Magnetron, Cyclotron, and Spin-Cyclotron-Beat Frequencies on an Isolated 1MeV Electron [*Phys. Rev. Letts.* **38,** 310, 1977].

T.C. Van Flandern, Is the Gravitational Constant Changing? [*Astrophys. J.* **248,** 813, 1981].

G. de Vaucouleurs, Sur une Analogie de Structure Remarquable entre les Nébuleuses Elliptiques et les Amas de Nébuleuses Extragalactiques [*Comptes Rendus Acad. Sci., Paris,* **227,** 586, 1948]

R.F.C. Vessot, *Lectures on Frequency Stability and Clocks and on the Gravitational Redshift Experiment,* in *Experimental Gravitation,* B. Bertotti, ed.; Varenna Course, 1972 [Academic Press, New York, 1974].

R.F.C. Vessot, Tests of Gravitation and Relativity [*Contemp. Phys.* **25,** 355, 1984].

R.F.C. Vessot, M.W. Levine, A Test of the Equivalence Principle using a Space-Borne Clock [*Gen. Rel. Grav.* **10,** 181, 1979].

J.Y. Vinet, C. Man, Un Interféromètre Géant pour Détecter les Ondes Gravitationnelles [*Le Courrier du CNRS* n°76 Sup., p.29, 1991].

R.M. Wald, *General Relativity* [University of Chicago Press, Chicago, 1984].

A.G. Walker, On Milne's Theory of World Structure [*Proc. London Math. Soc.* **42,** 90, 1936].

D. Walsh, R.F Carswell, R.J. Weymann, 0957 + 61 A,B: Twin Quasistellar Objects or Gravitational Lens? [*Nature* **279,** 381, 1979].

A.H. Wapstra, G.J. Nijgh, The Ratio of Gravitational to Kinetic Mass for the Constituents of Matter [*Physica* **21,** 796, 1955].

J. Weber, Detection and Generation of Gravitational Waves [*Phys. Rev.* **117,** 306, 1960].

J. Weber, *General Relativity and Gravitational Waves* [Wiley, New York, 1961].

J. Weber, Evidence for Discovery of Gravitational Radiation [*Phys. Rev. Letts.* **22,** 1320, 1969].

J. Weber, *The Search for Gravitational Radiation* in *General Relativity and Gravitation One Hundred Years after the Birth of Albert Einstein,* A. Held, ed. [Plenum, New York, 1980].

G. Wegner, A New Gravitational Redshift for the White Dwarf σ^2 Eri B [*Astron. J.* **85,** 1255, 1980].

S. Weinberg, Photons and Gravitons in Perturbation Theory: Derivation of Maxwell's and Einstein's Equations [*Phys. Rev.* **B138,** 988, 1965].

S. Weinberg, *Gravitation and Cosmology* [J. Wiley, New York, 1972].

S. Weinberg, Les Trois Premières Minutes de l'Univers [*Le Seuil*, Paris, 1978].

S. Weinberg, The Cosmological Constant Problem [*Rev. Mod. Phys.* **61,** 1, 1989].

V.F. Weisskopf, The Visual Appearance of Rapidly Moving Objects [*Phys. Today*, **13,** n° 9, 24, 1960].

J.C. Wesley, A. Rich, High-Field g-2 Measurement [*Phys. Rev.* **A4,** 1341, 1971].

C.M. Will, Active Mass in Relativistic Gravity: Theoretical Interpretation of the Kreuzer Experiment [*Astrophys. J.* **204,** 224, 1976].

C.M. Will, *Theory and Experiment in Gravitational Physics* [Cambridge University Press, Cambridge, 1981].

C.M. Will, The Confrontation between General Relativity and Experiment: an Update [*Phys. Report* **113,** 345, 1984].

J.G. Williams, R.H. Dicke, P.L. Bender, C.O. Alley, W.E. Carter, D.G. Currie, D.H. Eckhardt, J.E. Faller, W.M. Kaula, J.D. Mulholland, H.H. Plotkin, S.K. Poultney, P.J. Shelus, E.C. Silverberg, W.S. Sinclair, M.A. Slade, and D.T. Wilkinson, New Test of the Equivalence Principle from Lunar Laser Ranging. [*Phys. Rev. Letts.* **36,** 551, 1976].

F.C. Witteborn, W.M. Fairbanks, Experiments to Determine the Force of Gravity on Single Electrons and Positrons [*Nature*, **220,** 436, 1968].

A.M. Wolfe, R.L. Brown, M.S. Roberts, Limits on the Variation of Fundamental Atomic Quantities over Cosmic Times Scales [*Phys. Rev. Letts.* **37,** 179, 1976].

P.W. Worden, F. Everitt, *Tests of the Equivalence of Gravitational and Inertial Mass Based on Cryogenic Techniques,* in *Experimental Gravitation,* B. Bertotti, ed.; Varenna Course, 1972 [Academic Press, New York, 1974].

I.M. Yaglom, *A Simple Non-Euclidean Geometry and its Physical Basis* [Springer, Berlin, 1979].

H.T. Yu, W.T. Ni, C.C. Hu, F.H. Liu, C.H. Yang, W.N. Liu, An Experimental Determination of the Gravitational Constant at Distances around Ten Meters [*Chinese J. Phys.* **16,** 201, 1978].

H.T. Yu, W.T. Ni, C.C. Hu, F.H. Liu, C.H. Yang, W.N. Liu, Experimental Determination of the Gravitational Forces at Separations around Ten Meters [*Phys. Rev.* **D20,** 1813, 1979].

E.C. Zeeman, Causality Implies the Lorentz Group [*J. Math. Phys.* **5,** 190, 1964].

Ya.B. Zeldovich, I.D. Novikov, *Relativistic Astrophysics (Vol. 1): Stars and Relativity* [The University of Chicago Press, Chicago, 1975].

Ya.B. Zeldovich, I.D. Novikov, *Relativistic Astrophysics (Vol. 2): The Structure and Evolution of the Universe* [The University of Chicago Press, Chicago, 1983].

F. Zwicky, Nebulae as Gravitational Lenses [*Phys. Rev.* **51**, 290, 1937a].

F. Zwicky, On the Probability of Detecting Nebulae Which Act as Gravitational Lenses [*Phys. Rev.* **51**, 679, 1937b].

Physical Constants[1]

In CGS units unless otherwise indicated

$c = 2.99792458 \ 10^{10}$

$\hbar = 1.05457266 \ 10^{-27} = 6.582122 \ 10^{-22}$ MeV s

$k_{\text{Boltz}} = 1.3806513 \ 10^{-16} = 8.617344 \ 10^{-11}$ MeV/K

$e = 4.8032068 \ 10^{-10}$

$\alpha = e^2/\hbar c = 137.0359895^{-1}$

$G = 6.67259 \ 10^{-8}$

$g = 980.665$

$m_{\text{electron}} = 0.91093897 \ 10^{-27} = 0.51099906$ MeV $= 5.92989 \ 10^9$ K

$m_{\text{proton}} = 1.6726231 \ 10^{-24} = 938.27231$ MeV $= 1.0888184 \ 10^{13}$ K

$m_{\text{neutron}} = 1.6749286 \ 10^{-24} = 939.56563$ MeV $= 1.0903193 \ 10^{13}$ K

$m_{\pi^\pm} = 2.488018 \ 10^{-25} = 139.5675$ MeV

$m_{\pi^o} = 2.406129 \ 10^{-25} = 134.9739$ MeV

$\lambda_{c_{\text{electron}}} = 3.86159323 \ 10^{-11}$

$\lambda_{c_{\text{electron}}}^{-3} = 1.73660252 \ 10^{31}$

$\lambda_{c_{\text{proton}}} = 2.10308937 \ 10^{-14}$

$\lambda_{c_{\text{proton}}}^{-3} = 1.07504542 \ 10^{41}$

$\lambda_{c_{\text{neutron}}} = 2.10019445 \ 10^{-14}$

$\lambda_{c_{\text{neutron}}}^{-3} = 1.0794971 \ 10^{41}$

$\lambda_{c_{\pi^\pm}} = 1.413847 \ 10^{-13}$

$\lambda_{c_{\pi^o}} = 1.461964 \ 10^{-13}$

MeV $= 1.60217733 \ 10^{-6} = 5.0677289 \ 10^{-3}$ fm^{-1} $= 1.16045 \ 10^{10}$ K

gram $= 5.6095862 \ 10^{26}$ MeV

fermi$^{-1} = 197.32705$ MeV

second$^{-1} = 1.5192669 \ 10^{21}$ MeV

Ry $= 1/2 \ \alpha^2 m_e c^2 = 2.1798741 \ 10^{-11} = 13.6056981$ eV

r_e **(classical electron radius)** $= e^2/m_e c^2 = 2.81794092 \ 10^{-13}$

σ_T **(Thomson cross-section)** $= 8\pi/3 r_e^2 = 0.66524616 \ 10^{-24}$

a_0 **(Bohr radius)** $= \hbar^2/m_e e^2 = 0.529177249 \ 10^{-8}$

---

[1] Compiled by H. Sivak, who is gratefully acknowledged.

μ_B **(Bohr magneton)** $= e\hbar/2m_ec = 9.2740154 \ 10^{-21}$
μ_N **(nuclear magneton)** $= e\hbar/2m_pc = 5.0507866 \ 10^{-24}$
N_A **(Avogadro's number)** $= 6.0221367 \ 10^{23}$
σ **(Stefan–Boltzmann constant)** $= \pi^2 k_B^4/60\hbar^3 c^2 \ = \ 5.670399 \ 10^{-5}$
$$= \ 6.418014 \ 10^{41} 1/\text{cm}^2\text{s MeV}^3$$

Absolute units

energy $= \sqrt{\hbar c^5/G} = 1.95633 \ 10^{16}$
time $= \sqrt{\hbar G/c^5} = 5.39056 \ 10^{-44}$
length $= \sqrt{\hbar G/c^3} = 1.61605 \ 10^{-33}$
mass $= \sqrt{\hbar c/G} = 2.17671 \ 10^{-5}$

Astrophysics

H_0 **(Hubble constant)** $= 100 h_0 \text{km/s Mpc} = (0.97781 \ 10^{10}\text{y})^{-1} h_0$
$$(0.4 < h_0 < 1)$$
t_0 **(age of the Universe)** $= 1.5 \ 10^{10}\text{y}$
ρ_0 **(critical density)** $= 1.87882 \ 10^{-29} h_0^2$
ρ_{local} **(local density)** $= 0.3 \text{MeV/cm}^3 \approx 3 \ 10^4 \rho_c$
$m_{\text{stellar}} = 3.676 \ 10^{33}$
$m_{\text{Sun}} = 1.989 \ 10^{33}$
$m_{\text{Earth}} = 5.977 \ 10^{27}$
$m_{\text{Moon}} = 7.35 \ 10^{25}$
$r_{\text{Sun}} = 6.9598 \ 10^5 \text{km}$
$r_{\text{Earth}} = 6.37817 \ 10^3 \text{km}$
$r_{\text{Moon}} = 1.738 \ 10^3 \text{km}$
parsec $= 3.0856775806 \ 10^{18} \ \text{s} = 3.2615 \ ly$
ly (light year) $= 0.946 \ 10^{18}\text{s}$